Growth Factors and Receptors

The Practical Approach Series

SERIES EDITOR

B. D. HAMES
Department of Biochemistry and Molecular Biology
University of Leeds, Leeds LS2 9JT, UK

See also the Practical Approach web site at **http://www.oup.co.uk/PAS**

★ indicates new and forthcoming titles

Affinity Chromatography
★ Affinity Separations
Anaerobic Microbiology
Animal Cell Culture (2nd edition)
Animal Virus Pathogenesis
Antibodies I and II
Antibody Engineering
★ Antisense Technology
★ Applied Microbial Physiology
Basic Cell Culture
Behavioural Neuroscience
Bioenergetics
Biological Data Analysis
Biomechanics - Materials
Biomechanics - Structures and Systems
Biosensors
Carbohydrate Analysis (2nd edition)
Cell-Cell Interactions
The Cell Cycle
Cell Growth and Apoptosis
★ Cell Separation
Cellular Calcium
Cellular Interactions in Development
Cellular Neurobiology
★ Chromatin
Clinical Immunology
★ Complement
Crystallization of Nucleic Acids and Proteins
Cytokines (2nd edition)
The Cytoskeleton
Diagnostic Molecular Pathology I and II
DNA and Protein Sequence Analysis
DNA Cloning 1: Core Techniques (2nd edition)
DNA Cloning 2: Expression Systems (2nd edition)
DNA Cloning 3: Complex Genomes (2nd edition)
DNA Cloning 4: Mammalian Systems (2nd edition)
★ Drosophila (2nd edition)
Electron Microscopy in Biology

Electron Microscopy in Molecular Biology
Electrophysiology
Enzyme Assays
Epithelial Cell Culture
Essential Developmental Biology
Essential Molecular Biology I and II
Experimental Neuroanatomy
Extracellular Matrix
Flow Cytometry (2nd edition)
Free Radicals
Gas Chromatography
Gel Electrophoresis of Nucleic Acids (2nd edition)
★ Gel Electrophoresis of Proteins (3rd edition)
Gene Probes 1 and 2
Gene Targeting
Gene Transcription
★ Genome Mapping
Glycobiology
★ Growth Factors and Receptors
Haemopoiesis
Histocompatibility Testing
HIV Volumes 1 and 2
★ HPLC of Macromolecules (2nd edition)
Human Cytogenetics I and II (2nd edition)
Human Genetic Disease Analysis
★ Immunochemistry 1
★ Immunochemistry 2
Immunocytochemistry
★ *In Situ* Hybridization (2nd edition)
Iodinated Density Gradient Media
Ion Channels
★ Light Microscopy
Lipid Modification of Proteins
Lipoprotein Analysis
Liposomes
Mammalian Cell Biotechnology
Medical Parasitology
Medical Virology
★ MHC Volumes 1 and 2
★ Molecular Genetic Analysis of Populations (2nd edition)
Molecular Genetics of Yeast
Molecular Imaging in Neuroscience
Molecular Neurobiology
Molecular Plant Pathology I and II
Molecular Virology
Monitoring Neuronal Activity
Mutagenicity Testing
★ Mutation Detection
Neural Cell Culture
Neural Transplantation
Neurochemistry (2nd edition)
Neuronal Cell Lines
NMR of Biological Macromolecules
Non-isotopic Methods in Molecular Biology
Nucleic Acid Hybridisation
Oligonucleotides and Analogues
Oligonucleotide Synthesis
PCR 1
PCR 2

★ PCR 3: PCR In Situ Hybridization
Peptide Antigens
Photosynthesis: Energy Transduction
Plant Cell Biology
Plant Cell Culture (2nd edition)
Plant Molecular Biology
Plasmids (2nd edition)
Platelets
Postimplantation Mammalian Embryos
Preparative Centrifugation
Protein Blotting
Protein Engineering
Protein Function (2nd edition)
Protein Phosphorylation
Protein Purification Applications
Protein Purification Methods
Protein Sequencing
Protein Structure (2nd edition)
Protein Structure Prediction
Protein Targeting
Proteolytic Enzymes
Pulsed Field Gel Electrophoresis
RNA Processing I and II
★ RNA-Protein Interactions
★ Signalling by Inositides
Subcellular Fractionation
Signal Transduction
Transcription Factors
Tumour Immunobiology

Growth Factors and Receptors

A Practical Approach

Edited by

IAN A. McKAY
Queen Mary and Westfield College, London

and

KENNETH D. BROWN
The Babraham Institute, Cambridge

Oxford New York Tokyo
OXFORD UNIVERSITY PRESS
1998

Oxford University Press, Great Clarendon Street, Oxford OX2 6DP

Oxford New York
Athens Auckland Bangkok Bogota Bombay Buenos Aires Calcutta
Cape Town Chennai Dar es Salaam Delhi Florence Hong Kong Istanbul
Karachi Kuala Lumpur Madrid Melbourne Mexico City Mumbai
Nairobi Paris São Paolo Singapore Taipei Tokyo Toronto Warsaw
and associated companies in
Berlin Ibadan

Oxford is a trade mark of Oxford University Press

Published in the United States
by Oxford University Press, Inc., New York

A catalogue record for this book is available from the British Library

Library of Congress Cataloging in Publication Data
Growth factors and receptors : a practical approach / edited by Ian A. McKay and K. D. Brown.
(Practical approach series : 194)
Includes bibliographical references and index.
1. Growth factors–Laboratory manuals. 2. Growth factors––Receptors–Laboratory manuals. 3. Genetic engineering–Laboratory manuals. 4. Gene therapy–Research–Laboratory manuals. I. McKay, Ian (Ian A.) II. Brown, K. D. (Kenneth D.) III. Series.
QP552.G76G7445 1998 571.8'4–dc21 98–9879

ISBN 0 19 963647 8 (Hbk)
0 19 963646 X (Pbk)

Typeset by Footnote Graphics, Warminster Wilts
Printed in Great Britain by Information Press, Eynsham, Oxon

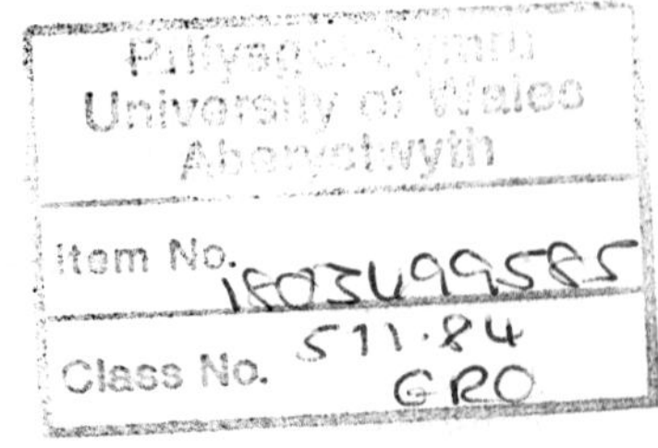

Preface

The rapid discovery of novel growth factors over the past 30 years led to a need for standard protocols describing methods for their isolation, identification and characterization. In 1993, one of us co-edited a book in the Practical Approach Series designed to fulfil that need. Since the publication of *Growth Factors: A Practical Approach*, the demands of many researchers working in this rapidly developing field have expanded and now include not only the analysis of physiological function, but also the engineering of factors with novel activities and applications. Our aim in producing this new book, *Growth Factors and Receptors: A Practical Approach*, was to build on the success of the previous volume and to meet those new demands.

In keeping with the aims of the Practical Approach Series, our international cast of authors has provided detailed experimental protocols that describe everything from basic analytical techniques to complex *in vivo* applications. While the protocols are exemplified by reference to the proteins of greatest interest to the individual authors, they should, in most cases, be applicable to studies of a range of other growth factors.

In the Contents the reader will find a list of chapters that deal with different approaches to growth factor studies, including Chapter 1 which serves as an overall introduction to growth factors and receptors. We have included three appendices with useful information, including references to all the families and individual growth factors known to us at the time of going to press. In addition to chapters with protocols applicable to their own studies, we hope that readers will also find other chapters which will help them understand and apply techniques that they might previously have thought too esoteric.

Finally, we extend our warm thanks to all the contributors for their positive response to our, sometimes nit-picking, editing, to Geraldine Garnett-Frizelle for her excellent secretarial assistance, and to you for buying this book.

Good factoring!

Contents

9. Growth factor–toxin chimeras and their applications 199

Pamela A. Davol, A. Raymond Frackelton, Jr., and Paul Calabresi

10. Gene therapy applications of growth factors 227

Tor Svensjö, Feng Yao, Bohdan Pomahac, and Elof Eriksson

Appendices

Index

Contributors

JENNIFER ANDERSON
Biomedical Research Center and Department of Biochemistry and Molecular Biology, University of British Columbia, Vancouver, BC V6T 1Z3, Canada.

KENNETH D. BROWN
The Babraham Institute, Babraham Hall, Cambridge CB2 4AT, UK.

STEPHEN A. BUSTIN
Academic Department of Surgery, St Bartholomew's and the Royal London School of Medicine and Dentistry, Queen Mary and Westfield College, Turner Street, London E1 2AD, UK.

PAUL CALABRESI
Department of Medicine, Rhode Island Hospital and Brown University School of Medicine, Providence, RI 02903, USA.

G. CILIBERTO
IRBM P. Angeletti, Department of Genetics, Via Pontina Km 30.6, 00040 Pomezia, Roma, Italy.

IAN CLARK-LEWIS
Biomedical Research Center and Department of Biochemistry and Molecular Biology, University of British Columbia, Vancouver, BC V6T 1Z3, Canada.

PAMELA A. DAVOL
Department of Medicine, Roger Williams Hospital, Providence, RI 02908, USA.

ELOF ERIKSSON
Division of Plastic Surgery, Brigham/Children's/Harvard Hospitals, 75 Francis Street, Boston, MA 02115, USA.

A. RAYMOND FRACKELTON JR
Department of Medicine, Roger Williams Hospital and Brown University, Providence, RI 02908, USA.

JIANG-HONG GONG
Biomedical Research Center and Department of Biochemistry and Molecular Biology, University of British Columbia, Vancouver, BC V6T 1Z3, Canada.

LOUIS S. GREEN
Nexstar Pharmaceuticals, 2860 Wilderness Place, Boulder, CO 80301, USA.

CARLOS F. IBÁÑEZ
Laboratory of Molecular Neurobiology, Department of Neuroscience, The Karolinska Institute, S-171 77 Stockholm, Sweden.

LEOPOLD L. ILAG
Laboratory of Molecular Neurobiology, Department of Neuroscience, The Karolinska Institute, S-171 77 Stockholm, Sweden.

NEBOJŠA JANJIĆ
Nexstar Pharmaceuticals, 2860 Wilderness Place, Boulder, CO 80301, USA.

A. LAHM
IRBM P. Angeletti, Department of Biocomputing, Via Pontina Km 30.6, 00040 Pomezia, Roma, Italy.

A. E. G. LENFERINK
Department of Cell Biology, University of Nijmegen, Toernooiveld, 6525 ED Nijmegen, The Netherlands.

IAN A. McKAY
Centre for Cutaneous Research, St Bartholomew's and the Royal London School of Medicine and Dentistry, Queen Mary and Westfield College, 2 Newark Street, London E1 2AT, UK.

SIMON R. MYERS
Centre for Cutaneous Research, St Bartholomew's and the Royal London School of Medicine and Dentistry, Queen Mary and Westfield College, 2 Newark Street, London E1 2AT, UK.

HARSHAD A. NAVSARIA
Centre for Cutaneous Research, St Bartholomew's and the Royal London School of Medicine and Dentistry, Queen Mary and Westfield College, 2 Newark Street, London E1 2AT, UK.

PHILIP OWEN
Biomedical Research Center and Department of Biochemistry and Molecular Biology, University of British Columbia, Vancouver, BC V6T 1Z3, Canada.

G. PAONESSA
IRBM P. Angeletti, Department of Genetics, Via Pontina Km 30.6, 00040 Pomezia, Roma, Italy.

BOHDAN POMAHAC
Division of Plastic Surgery, Brigham/Children's/Harvard Hospitals, 75 Francis Street, Boston, MA 02115, USA.

R. SAVINO
IRBM P. Angeletti, Department of Genetics, Via Pontina Km 30.6, 00040 Pomezia, Roma, Italy.

TOR SVENSJÖ
Division of Plastic Surgery, Brigham/Children's/Harvard Hospitals, 75 Francis Street, Boston, MA 02115, USA.

C. TONIATTI
IRBM P. Angeletti, Department of Genetics, Via Pontina Km 30.6, 00040 Pomezia, Roma, Italy.

M. L. M. VAN DE POLL
Department of Cell Biology, University of Nijmegen, Toernooiveld, 6525 ED Nijmegen, The Netherlands.

M. J. H. VAN VUGT
Department of Cell Biology, University of Nijmegen, Toernooiveld, 6525 ED Nijmegen, The Netherlands.

E. J. J. VAN ZOELEN
Department of Cell Biology, University of Nijmegen, Toernooiveld, 6525 ED Nijmegen, The Netherlands.

LUAN VO
Biomedical Research Center and Department of Biochemistry and Molecular Biology, University of British Columbia, Vancouver, BC V6T 1Z3, Canada.

SABINE WERNER
Max-Planck-Institut für Biochemie, 82152 Martinsried, Bei München, Germany.

FENG YAO
Division of Plastic Surgery, Brigham/Children's/Harvard Hospitals, 75 Francis Street, Boston, MA 02115, USA.

Abbreviations

AAV	adeno-associated virus
Ad	adenovirus
AEC	3-amino-9-ethylcarbazole
APS	ammonium persulfate
ATCC	American Type Culture Collection
BDNF	brain-derived neurotrophic factor
BrdU	5′-bromo-2′-deoxyuridine
CMV-IE	cytomegalovirus immediate early region gene
CNBr	cyanogen bromide
CRP	C-reactive protein
3D	three-dimensional
DAB	diaminobenzidine
DEPC	diethyl pyrocarbonate
DMEM	Dulbecco's modified Eagle's medium
DMSO	dimethyl sulfoxide
DSP	dithio*bis*(succinimidyl proprionate)
DSS	disuccinimidyl suberate
EDC	1-ethyl-3-(3-dimethylaminopropyl)-carbodiimide
EDTA	ethylenediaminetetraacetic acid
EGF	epidermal growth factor
EGFR	epidermal growth factor receptor
EIVT	enhanced *in vitro* translation
ELISA	enzyme-linked immunosorbent assay
FGF	fibroblast growth factor
FGFR	fibroblast growth factor receptor
FITC	fluorescein isothiocyanate
GF	growth factor
GH	growth hormone
GM-CSF	granulocyte/macrophage colony stimulating factor
GTh	gene therapy
GTr	gene transfer
HB-EGF	heparin binding EGF-like growth factor
HPLC	high-pressure liquid chromatography
HRG	heregulin
IGF	insulin-like growth factor
IL	interleukin
K_d	dissociation constant
KGF	keratinocyte growth factor
LD	lethal dose
MCP-1	monocyte chemoattractant protein-1

MEM	minimal essential medium
MTD	maximum tolerated dose
NGF	nerve growth factor
NT	neurotrophin
OD	optical density
PAGE	polyacrylamide gel electrophoresis
PBS	phosphate-buffered saline
PCR	polymerase chain reaction
PE	*Pseudomonas* exotoxin
Pipes	piperazine-*N,N′*-*bis*(2-ethanesulfonic acid)
RAC	Recombinant DNA Advisory Committee
RIA	radioimmunoassay
RNase	ribonuclease
RP-HPLC	reverse-phase high-pressure liquid chromatography
RT-PCR	reverse transcription polymerase chain reaction
SAP	saporin
SAR	structure–activity relationships
SB^3	*bis*(sulfosuccinimidyl)suberate
SEAP	secreted alkaline phosphatase
SELEX	systematic evolution of ligands by exponential enrichment
TAMIA	tag-mediated immunoprecipitation assays
TCA	trichloroacetic acid
TEMED	*N,N,N′,N′*-tetramethylethylenediamine
TFA	trifluoroacetic acid
TGF	transforming growth factor
VEGF	vascular endothelial growth factor
VPF	vascular permeability factor

1

Engineered growth factors and receptors, and their applications

IAN A. McKAY, STEPHEN A. BUSTIN, and KENNETH D. BROWN

1. Introduction

This chapter serves as an introduction to the rest of the book. In it you will find described the various types of growth factors and receptors, and their potential applications. This is followed by a discussion of systems for getting expression of recombinant growth factors and receptors, with protocols describing specific examples of *in vitro* and mammalian cell expression. Other examples will be found in Chapters 3, 4, 5, 7, and 9 in this volume.

1.1 Known growth factors and receptors

In the first edition of *Growth factors: a practical approach* (1), published in 1993, were listed approximately 130 different known growth factors. By the time of the first reprint in 1995 this number had grown to over 170. At the time of going to press about 200 are known (see Appendix 1). Therefore it is apparent that new growth factors are still being discovered. Moreover, as the various genome sequencing projects continue, it seems probable that yet more factors, or even additional factor families will be discovered. In the meantime many groups have turned to investigating applications for the known factors and it is that area of growth factor research which this present volume largely addresses. Simultaneous with the discovery of new factors has been the description of a wide range of growth factor receptors and a huge increase in our knowledge of how they interact with their cognate ligands.

1.2 Modes of factor–receptor interaction

Five major modes of growth factor function have been postulated (summarized in ref. 2), these being paracrine, autocrine, juxtacrine, intracrine, and endocrine. While these divisions are helpful from a functional viewpoint, they provide little insight into the mechanisms of factor–receptor interactions which are often complex. For example some factors bind to matrix molecules to allow interactions with their receptors. Other factors, such as members of the transforming growth factor beta (TGFβ) family, require cleavage by matrix proteases before they can interact with receptors (3). Yet other factors, such as the neuregulins, must recruit more than one receptor molecule to a complex before signal transduction can occur (4).

The receptors themselves may show distinct patterns of activity, the interleukin-6 (IL-6) receptor, for example, consists of a soluble subunit which complexes with IL-6 and a transmembrane subunit, gp130, which transduces the signal into the cell (see Chapter 3). The gp130 subunit is also shared with other cytokine receptors, presumably reflecting some cellular economy of signal transduction mechanisms. Other receptor molecules, such as the members of the ErbB family, can form homodimers or heterodimers in combinations that appear to be determined by the activating ligand (4).

An understanding of the mechanisms by which factors interact with receptors is essential to design of factors and receptors with novel specificities. Several chapters in this book and its sister volume contain protocols designed to help the researcher find out where a factor or its receptor is expressed (Chapter 4), which ligands bind to particular receptors (Chapter 8), and which factor or receptor residues are involved in the binding and activation processes (Chapters 2–5). Once these interactions have been characterized, novel applications of engineered factors and receptors can be contemplated.

2. Applications of engineered growth factors and receptors

There is a whole range of potential applications for engineered growth factors and receptors and some of these are covered in the following chapters including:

- cancer therapy with factor–toxin or factor–factor chimeras (see Chapter 9)
- enhanced healing of chronic leg ulcers (see Chapter 10)
- improved bone repair following fracture
- enhanced central and peripheral nerve regeneration (see Chapter 2)
- regeneration of hair follicles in alopecia
- therapy of inflammatory bowel disease

3. Principal methods for engineering growth factors and their receptors

3.1 Mutagenesis

The main method used for engineering growth factors or receptors is to alter the sequence of the DNA encoding them—a process known as mutagenesis which can be classified according to the types of mutation made (substitution, insertion, deletion) and also by the method used (random or sequence-specific, i.e. site-directed or site-specific). Numerous kits are commercially available for introducing mutations into DNA sequences using both methods.

3.1.1 Random mutagenesis

Although no method achieves entirely random mutagenesis, the aim is to isolate, sequence, and test mutants for their functional activity in a particular assay of factor–receptor interaction, thereby identifying specific amino acid residues which are essential for this interaction. If mutants are obtained for each residue of the protein, this is known as saturation mutagenesis.

3.1.2 Sequence-specific mutagenesis

The principles of sequence-specific mutagenesis are more than adequately covered in another volume from this series (*Directed mutagenesis: a practical approach*). This technique is normally used to alter single residues with a known or suspected function. For example the cysteines in a growth factor might be replaced to assess what role disulfide cystine bridges play in the secondary or tertiary structure of the molecule (see Chapter 4).

Examples of the application of different mutagenesis techniques to investigating growth factor–receptor interactions may be found in the chapters of this volume.

3.2 Formation of chimeric proteins

The Chimaera was a mythical beast with the head of a lion the body of a goat and a serpent's tail. Chimeric growth factors or receptors are therefore proteins which incorporate whole proteins or domains of proteins with other functions. Examples include the chimeric mitotoxins described in Chapter 9, the hybrid chemokines in Chapter 5, and the chimeras between EGF and TGFα described in Chapter 4.

3.3 Manufacture of alternative ligands and receptors

Another approach to engineering new ligand and receptor specificities is to create them *de novo*. Two techniques for this purpose are described in this volume. In Chapter 6 the SELEX method for exponential evolution of RNA aptamers which can specifically block the binding of VEGF to its receptor is described, and in Chapter 3 the use of phage display libraries to develop new ligands for soluble IL-6 receptors is described.

Finally a technique for activating signalling pathways by the use of agents which induce dimerization of receptors through an intracellular mechanism has been described and there is obviously a great deal of scope for engineering such agents to have affinities for individual combinations of receptors (5).

4. Design of engineered growth factors and receptors

Clearly, with the exception of the last three classes of agents described, it is essential to design and/or modify cDNAs for expression of growth factors, and this has been described in *Growth factors: a practical approach* (6). However, since the publication of that volume, there have been numerous advances in the methodology and some of these will be described here.

4.1 DNA sequence of factor or receptor to be engineered

The design of a DNA sequence to encode a protein is restricted by the source of the DNA, whether it is purely synthetic or wholly or partially derived from a cloned gene, and the nature of the organism which will be used to express it. Note: whenever cloning or expressing nucleic acid sequences, researchers must follow all local and national guidelines for the safe generation and handling of genetically modified organisms.

The very first step in the design process is to get hold of the DNA sequence for the wild-type factor or receptor. Apart from the original publication of the sequence, there are many databases on the Internet from which this information can now be downloaded. These are listed in special issues of the journal *Nucleic Acids Research*, where their addresses may be found. Many of these are also available at mirror sites created by various institutions across the globe.

4.2 Analysis of existing DNA sequence for useful restriction sites

As a second step it may be pertinent to take the known cDNA sequence encoding the factor or receptor and to analyse this for restriction enzyme cutting sites. These sites may then be used to subclone all or part of the cDNA encoding factor or receptor into expression vectors for production of protein. There are many useful software packages designed for the task of handling DNA sequence information. Again many of these are also listed in special issues of *Nucleic Acids Research* and some may be obtained as shareware. Commercially available software suites are frequently advertised in the scientific press. If useful restriction sites are not present in the cloned cDNA, new sites may be designed into the DNA sequence using a synthetic approach.

4.3 Design of wholly synthetic growth factors and receptors

Creation of an entirely synthetic gene can be an attractive alternative to the use of a cloned cDNA copy when attempting to maximize expression in one particular host. It allows maximum flexibility and fine control with respect to codon usage, incorporation of restriction enzyme recognition sites, and avoidance of repetitive sequences or stem-loop structures that might interfere with efficient translation. It also allows easy assembly of chimeric molecules made up from sequences derived from different growth factors or receptors. With the advent of synthesizers which can easily make oligonucleotides of 100 bases or more—oligonucleotides which generally need no further purification—it is now feasible to construct even large cDNAs cheaply.

In principle, there are two strategies for assembly of a synthetic gene from single-stranded oligonucleotides: the first (see *Figure 1a*) and the one most commonly used involves synthesis of complementary oligonucleotides specifying the complete gene. These are annealed and ligated directly into a vector digested with appropriate restriction enzymes. The second strategy (see *Figure 1b*) uses partially overlapping oligonucleotides and relies on the 5′–3′ polymerase activity of the Klenow fragment of *E. coli* DNA polymerase (or any other polymerase that lacks a 5′–3′ exonuclease activity) to generate the complete double-stranded DNA sequence. While method (b) requires fewer oligonucleotides and can therefore be less expensive, there are two advantages to method (a):

(a) Since no polymerization step is involved, there is less likelihood of an introduction of a mutation.
(b) The synthetic gene can be designed with appropriate overlaps at its 5′ and 3′ ends that permit direct cloning, without prior restriction enzyme digestion, into compatible ends of the chosen vector.

When designing oligonucleotides for synthetic gene assembly, it is important to consider the length of the oligonucleotides, the composition of the sequence, and the potential for possible oligonucleotide:oligonucleotide interactions, and a few simple rules applying to sequence-related characteristics will maximize the likelihood of successfully obtaining a synthetic gene. The stability of a given duplex is based upon the free energy of adjacent

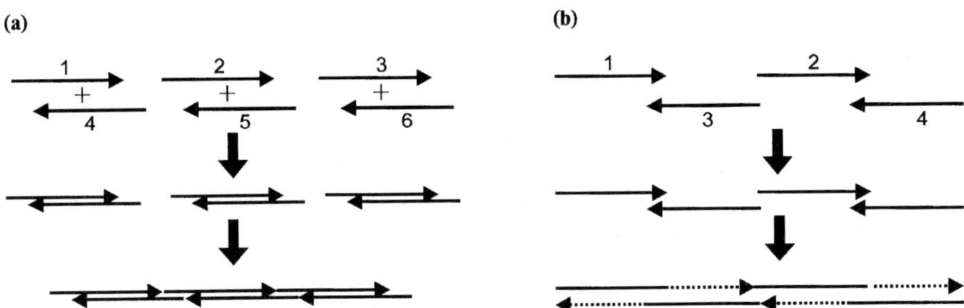

Figure 1. Alternative methods for synthesizing cDNAs encoding growth factors. (a) Complementary oligonucleotides specifying the complete sequence are annealed separately and ligated to a vector linearized with the appropriate restriction enzymes. (b) Oligonucleotides overlapping at their extreme 3′ ends only are annealed in a single reaction and extended using PolIK. The synthetic gene is digested with appropriate restriction enzymes for ligation to the vector of choice.

dinucleotides (7, 8), with the stability of a particular double-stranded sequence related to the number of hydrogen bonds formed. GC-rich sequences are more stable than AT-rich sequences although stability also depends on sequence order.

Design considerations:

(a) Palindromes, long segments of polypurines and polypyrimides, sequences complementary to those of the paired oligonucleotide, and repeated, contiguous use of the same codon should be avoided.
(b) Complementary oligonucleotides should pair with high stability and have little secondary structure to minimize the likelihood of introducing unwanted sequence alterations.
(c) Single-stranded overlapping regions used to assemble oligonucleotide blocks should be at least eight bases long and have no complementarity to other overlapping sequences.
(d) The ideal GC content generally lies in the range of 40–55%, although it can be higher in the overlap regions used to assemble the double-stranded oligonucleotides.

These conditions help prevent misaligning and generation of mutant sequences such as introduction of mutations, deletions, and insertions.

Apart from these general considerations, specific gene design also depends on the nature of the gene itself, where it is to be expressed, and what future experiments are likely to take place. A little bit of forethought at this stage will save a lot of time later:

(a) Should the gene contain restriction sites permitting easy domain swapping, insertions, deletions, duplications, and generation of protein fusions?
(b) Should codon usage be engineered for optimal expression in a particular organism?
(c) Is there a requirement for leader sequences allowing export of proteins to specific sites within cells or out of the cell entirely?
(d) Is an initiation codon (AUG) required or not? Several restriction enzyme recognition sites incorporate an initiation codon: *Sph*I (GCATG'C), *Nco*I (C'CATGG), *Bsp*H1 (T'CATGA), and *Nde*I (CA'TATG).
(e) Some restriction enzymes show preferential cleavage, with resistance to cleavage attributed to surrounding sequence context. This appears most pronounced when digesting with *Nar*I, *Nae*I, and *Sac*II, all of which recognize sites entirely composed of G and C, although others (*Sma*I, *Apa*I) do not show site preference. Unfortunately, such problems are difficult to predict but it always pays to incorporate restriction sites that are recogized by ubiquitous (and cheap) restriction enzymes.

An example of the design procedure is given in *Protocol 1*. Here the DNA sequence encoding human granulocyte/macrophage colony stimulating factor

(GM-CSF) is engineered for expression from *E. coli*. The genomic copy of this gene is transcribed with a 17 amino acid leader sequence. However, since the aim is to maximize expression in *E. coli* and since hydrophobic regions have been shown to have toxic effects on this host, the leader sequence will not be included in the synthetic version.

Protocol 1. Design of a synthetic gene specifying GM-CSF

Equipment

- DNA analysis software package
- Internet access software (optional)
- Computer

Method

1. Enter the DNA sequence encoding the factor into the DNA analysis program, either manually from the published sequence or by downloading from a DNA sequence database.
2. Use the software to translate the DNA sequence.
3. Reverse translate the amino acid sequence with strongly expressed non-degenerate *E. coli* backtranslation code to generate a DNA sequence that will specify a mRNA molecule allowing optimal translation in *E. coli* (*Figure 2*).
4. Design restriction enzyme recognition sites in the sequence that allow synthesis in the correct reading frame, domain swapping, and the generation of chimeric molecules.
5. Add restriction enzyme sites at the 5′ and 3′ ends to allow direct cloning into the expression vector pET-16b (*Nde*I and *Bam*HI in *Figure 2*).
6. Ensure there are no repetitive sequences or inverted repeats, and no stretches of polypurines or polypyrimidines.
7. Divide sequence into a number of overlapping oligonucleotides which can be annealed to give the entire coding sequence (*Figure 1a*).

5. Expression of DNAs encoding growth factors or receptors

5.1 Deciding on an expression system

The choice of an expression system is influenced by a variety of factors including:

(a) Quantity of product required.
  (i) Is the product for research or commercial use?

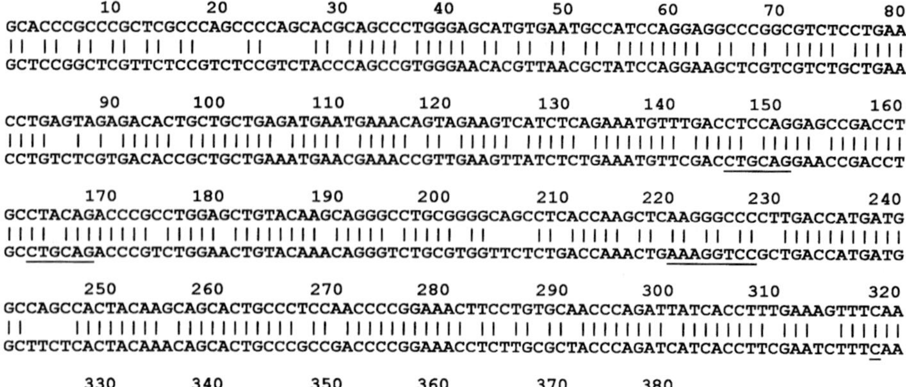

Figure 2. Manipulation of the DNA sequence of human GM-CSF for cloning and expression. The natural DNA sequence of the human GM-CSF cDNA is shown (*top*) alongside the sequence optimized for expression in *E. coli* (*bottom*). The sequence contains two *Pst*I sites (at positions 145 and 163), one of which (163) is removed by changing CTG to CTC. Changing two nucleotides in the sequence AA**A** GG**T** CCG (positions 220–228) to AA**G** GG**C** CCG generates an *Apa*I site at position 222. Finally, changing a C at position 321 to a T generates a *Dra*I site at position 316. In order to avoid restriction enzyme digestion of the synthetic genes (see *Figure 1*), oligonucleotides are annealed and ligated so that appropriate overhangs remain at the 5′ (5′-TA-3′) and 3′ (3′-CTAG-5′) ends. The complete gene, made up of 395 base pairs, is assembled from five blocks requiring the synthesis of seven 85-mers, one 84-mer, one 55-mer, and one 52-mer. Annealing of double-stranded oligonucleotide is made easy by allowing overlaps of 28 bases between blocks (a minimum of eight bases is advisable). Finally, a vector, in this case pET-16b, is double digested with *Nde*I and *Bam*HI restriction enzymes to generate sticky ends complementary to the overhangs on the assembled DNA.

   (ii) Most *in vitro* systems will not produce milligram quantities of protein.
   (iii) Availablity of fermenters or roller bottles etc. may influence choice between, for example, bacterial and mammalian expression systems.

(b) Post-translational modifications required. Bacteria and yeast will not necessarily glycosylate recombinant proteins correctly.

(c) The degree of purity required. Some systems, particularly bacterial systems, allow rapid isolation of proteins in relatively pure form, whereas recovery from the supernatants of mammalian cell cultures can be a long and difficult process.

Numerous methods for expressing recombinant growth factors and receptors have been described. These include baculovirus, Sindbis, microbial, *in vitro*, and eukaryotic systems. In this section we will describe one example of each of the last three.

5.2 Microbial expression

The most successful systems express recombinant proteins as fusion products that are more resistant to proteolytic degradation and are also more easily purified. Examples of such expression systems are the His-Tag, GST, and MalB systems available commercially for expression in *E. coli*.

5.2.1 Bacterial expression

Here the example using the synthetic GM-CSF DNA sequence is continued and a method is described in *Protocol 2* for assembly and subcloning of the DNA into a His-Tag expression vector (pET-16b) linearized with *Nde*I and *Bam*HI.

Protocol 2. Assembly and expression of a synthetic DNA encoding GM-CSF

Equipment and reagents

- 0.5 ml microcentrifuge tubes
- Thermal cycler for PCR amplification
- Microcentrifuge
- Ten single-stranded oligonucleotides at 1 pmol/μl (approx. 56 ng/μl)
- TE buffer: 0.01 M Tris pH 7.5, 0.001 M EDTA
- Appropriately double digested vector in TE buffer (250 ng/μl)
- T4 DNA ligase
- 10 × T4 DNA ligation buffer (supplied by manufacturer of T4 DNA ligase)
- Competent *E. coli* strain (e.g. XL1-Blue) (9)

Method

1. Take 10 μl (10 pmol) of each complementary pair of single-stranded oligonucleotides, and mix in five sterile, non-siliconized microcentrifuge tubes, one for each pair. In order to minimize misannealing it is best to anneal each complementary pair separately.
2. Place the tubes in a thermal cycler and heat to 95°C for 5 min. Leave on the heating block to slowly cool to room temperature.
3. Pulse-centrifuge tubes at 7000 *g* in a microcentrifuge to collect the annealed oligonucleotides at the bottom of each tube.
4. Mix 2 μl (2 pmol) of each annealed primer pair in a fresh microcentrifuge tube. Add 2 μl 10 × ligation buffer, 1 μl vector DNA, 1 μl DNA ligase, and water to bring the final volume to 20 μl.
5. Leave at room temperature for 1–12 h.
6. Heat to 70°C for 10 min to inactivate DNA ligase.
7. Use 1–10 μl to transform competent *E. coli* and select for appropriate antibiotic resistance.
8. Plate out on antibiotic-containing media and grow overnight according to standard procedures (9).

Protocol 2. *Continued*

9. Isolate single colonies, check for inserts of appropriate size, and prepare DNA from six individual clones (9).
10. Sequence the DNA inserts (9) and retain the clones which express DNA with the correct sequence.

5.3 *In vitro* expression

Another method of expression, commonly used for getting small quantities of protein for research purposes, uses *in vitro* translation (IVT). This essentially relies on translation of mRNA encoding the growth factor or receptor in a system supplemented with all the necessary agents to allow the translation process to proceed for as long as possible. Several commercial systems for IVT are available and some couple *in vitro* transcription to *in vitro* translation allowing continuous protein production. In its extreme form (enhanced *in vitro* translation—EIVT) the process can be carried on for days and milligrams of protein may be produced (10). *Protocol 3* describes the accumulation of recombinant IL-6 in an EIVT system (11). *Figure 3* presents a flow diagram of *Protocol 3*.

Protocol 3. Cumulative production of IL-6 by enhanced *in vitro* translation

Equipment and reagents[a]

- Micro-ultrafiltration translation chamber (8-MC, Amicon)
- Ultrafiltration membranes (XM-50, Amicon)
- Magnetic stirrer
- Peristaltic pump (capable of pumping solutions at 1 ml/h)
- DNA-dependent RNA polymerase (e.g. SP6, T7, T3 polymerases, the protocol below uses SP6)
- Linearized plasmid template (1–5 μg/μl)
- RNasin (Gibco BRL)
- Nucleoside triphosphates (NTPs) (100 mM stock solution, Boehringer Mannheim, 1 277 057)
- Wheat germ extract (Promega)[b]
- Transcription buffer: 80 mM Hepes–KOH pH 7.6, 16 mM $MgCl_2$, 2 mM spermidine, 20 mM dithiothreitol, 4 mM NTPs, 100 U RNasin
- [^{35}S]methionine (Amersham International)
- Translation mixture: 400 μl wheat germ extract, 1.5 mM ATP, 0.1 mM GTP, 8 mM creatine phosphate, 50 U creatine phosphate kinase, 20 μM amino acids (minus Met), 10 μM Met, 5.45 mM DTT, 100 μM spermidine, 0.7 mM EDTA, 90 mM KAc, 4.3 mM $MgAc_2$, 2.5 U RNasin, 30 mM Hepes pH 7.6
- Translation feeding buffer: as translation mixture except without the wheat germ extract and creatine phosphate kinase

Method

1. Add 4 μg of linearized plasmid template to 200 μl transcription buffer.
2. Add 200 U SP6 polymerase and leave at 37 °C for 2 h. This should result in the synthesis of approx. 200 μg RNA.
3. Phenol:chloroform extract (9), ethanol precipitate, and resuspend RNA in water at a concentration of 1 μg/μl.

4. Set-up ten 1.5 ml microcentrifuge tubes, each containing 7.5 pmol [^{35}S]Met.
5. Prepare translation mixture, remove a 25 μl aliquot, and add to the first of the microcentrifuge tubes. This constitutes the negative (minus mRNA) control.
6. Assemble ultrafiltration device according to the manufacturer's instructions and place 1.1 ml of translation mixture in the translation chamber (*Figure 4*).
7. Add 30 μg RNA to the translation buffer in the ultrafiltration chamber, mix well.
8. Immediately remove a 25 μl aliquot and add to the second of the microcentrifuge tubes from step 4. (This constitutes the positive control and is the equivalent of a conventional analytical cell-free translation.) Incubate the tube together with the negative control tube (from step 5) at 25 °C for 45 min and place in –20 °C freezer.
9. Pre-incubate the remainder of the translation mixture in the translation chamber at 25 °C for 30 min, before starting the flow of the translation feeding buffer.
10. Adjust the flow rate of the feeding buffer to 1 ml/h. As products are continuously removed through the ultrafiltration membrane, collect 1 ml fractions. Discard the fractions collected from the first 4 h of translation, as they will be contaminated with components of the translation machinery.
11. Withdraw 25 μl samples from the translation chamber every 12 h and add to the microcentrifuge tubes containing the 7.5 pmol [^{35}S]Met from step 4. Incubate the tubes at 25 °C for 45 min.[c]
12. Determine radioactive incorporation in control tubes as well as in sampling tubes by TCA precipitation with trichloroacetic acid (TCA) followed by polyacrylamide gel electrophoresis (9).
13. After completion, calculate the efficiency of protein synthesis from the radioactivity of the material precipitated by TCA.
14. Assay the biological activity of the products using bioassays (e.g. MTT assays) (11).

[a] All chemicals from Sigma unless otherwise stated.
[b] Extracts are stored according to manufacturer's instructions.
[c] This sampling of extract for activity avoids the need to use large amounts of [^{35}S]Met in the feeding buffer and allows the synthesis of non-radioactive product.

5.4 Eukaryotic gene expression

5.4.1 Yeast expression

As noted above the advantage of yeast expression systems is that there is the potential for glycosylation of the product after synthesis. However *S.*

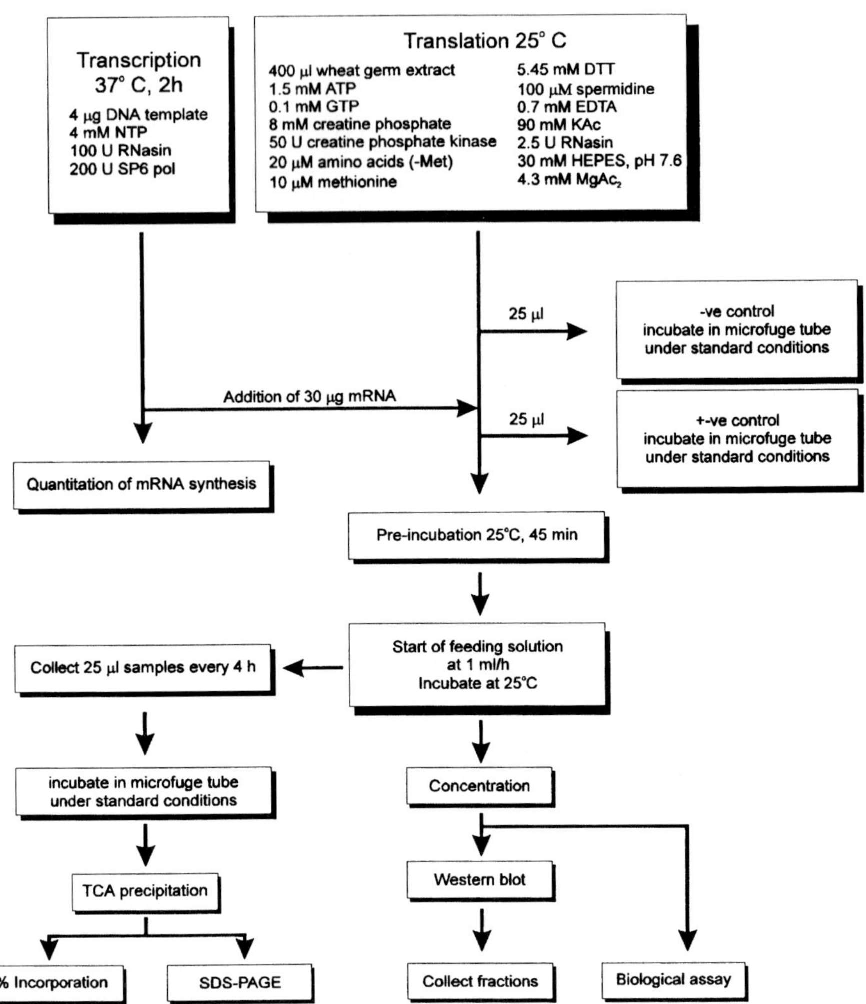

Figure 3. Flow diagram illustrating the essential steps in *Protocol 3*. (Reproduced with permission from ref. 11.)

cerevisiae may hyperglycosylate the protein product and *Pichia pastoris* may create inappropriate glycoforms. Numerous protocols have been described for getting proteins including growth factors and their receptors from yeast and these will not be described further here.

5.4.2 Expression of genes from mammalian cells *in vitro*

(i) DNA sequence design

Expression of growth factors or receptors in mammalian cells has the advantage that the proteins should show all the correct post-translational

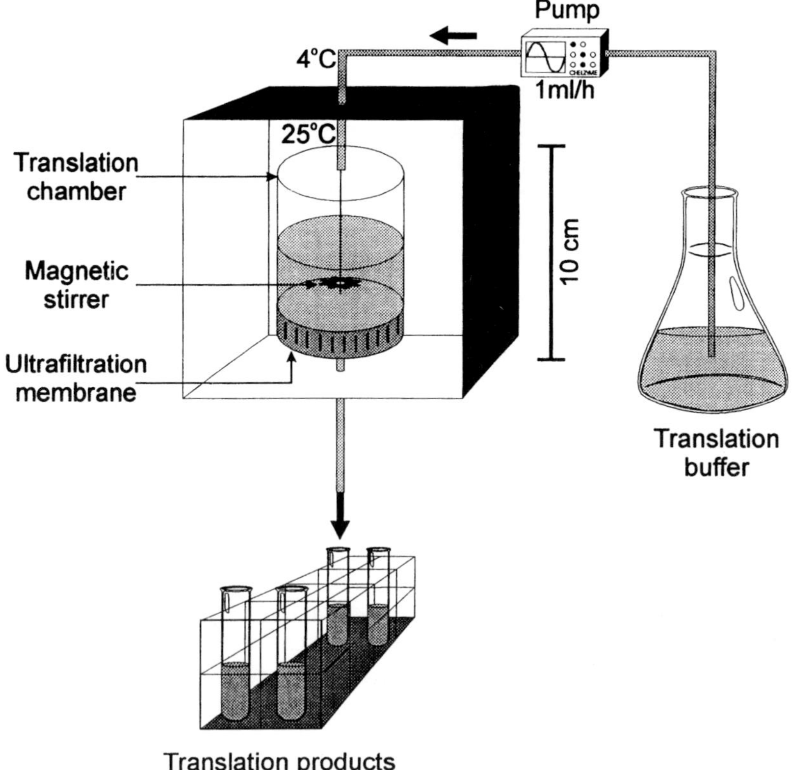

Figure 4. Amicon 8-MC-based continuous flow cell-free system. (Reproduced with permission from ref. 11.)

modifications. However, when expressing a protein from mammalian cells there are many additional considerations in terms of the design of the cDNA and choice of expression vector and expressing cell type. For example, to get secretion of a growth factor from mammalian cells it is likely that a protein will require an N terminal signal peptide which can be cleaved from the bulk of the protein by membrane-bound proteases. The recognition sequences for these proteases have not yet been fully characterized so that a limited number of them which have been shown to work empirically are commonly used. Moreover, once secreted, the factor may require further proteolytic cleavage so it is essential to include cleavage sites when designing the DNA sequence encoding such proteins. Likewise glycosylation and phosphorylation sites must be catered for.

The above points also apply when attempting to get expression of soluble receptors. However, when attempting to get expression of membrane-bound receptors a different set of signalling peptides may be required, but note that in the case of receptors which may also act as ligands there may be the need for two or more sets of signalling peptides for appropriate localization and subsequent activation.

(ii) Expression vectors for mammalian cells

A wide variety of expression vectors for mammalian cells is currently available. In each case the vector contains a cassette allowing expression of the factor or receptor cDNA. This cassette usually contains a promoter sequence, a multicloning site, and a polyadenylation signal which allows the cell to transcribe messenger RNA from the coding cDNA. Some of the promoters allow constitutive expression of the cDNA, an example would be the immediate early gene promoter of cytomegalovirus (CMV-IE promoter). Others can respond to addition of agents to the growth medium, allowing induction or repression of cDNA expression. An example of these would be the steroid-sensitive promoter from the mouse mammary tumour virus long terminal repeat sequence (MMTV-LTR promoter).

Vectors usually have a second cassette containing an antibiotic resistance marker allowing selection of expressing cells. This cassette often uses a promoter from the origin of simian virus 40 (SV40) coupled to the Tn5 transposon which encodes resistance to neomycin in bacteria and to geneticin (G418 sulfate) in mammalian cells. Other cassettes encode resistance to bleomycin, puromycin, and hygromycin. Some vectors are designed to repress transcription of the coding cDNA.

(iii) Delivery of expression vectors to mammalian cells

Again there are a whole host of methods for getting expression vectors into cells. These include physical methods such as electroporation and the use of the gene gun, use of lipid soluble DNA coats, and the old standby—calcium phosphate co-precipitation. Others include the use of viral vectors. In particular, adenovirus vectors are good for transient expression in a large number of infected cells, and retroviruses are useful for infecting only those cells which are dividing, and for longer-term expression studies.

(iv) Cell type used for expression

In the past fibroblasts, and in particular CHO fibroblasts, have been widely used for mammalian expression. They are particularly useful for expression of growth factor receptors as they do not carry many of their own (12). As an alternative COS cells, which express the SV40 large T protein and enhance expression from plasmids containing SV40 promoters, have also been widely used (13). It is worth bearing in mind that some cell types show preferences for particular promoters. For example, in a series of tests, the highest level of gene expression in keratinocytes was achieved using the cytomegalovirus immediate early (CMV-IE) gene promoter. This promoter also works well in other epithelial cell types. In *Protocol 4* an example is given of expression of heparin binding epidermal growth factor-like growth factor (HB-EGF) from rat intestinal epithelial cells using the CMV-IE promoter. The expression of multiple forms of HB-EGF by clones of the transfected cells is shown in *Figure 5*.

Protocol 4. Isolation of RIE-1 cell clones constitutively expressing proHB-EGF

Equipment and reagents

- Electroporation apparatus (e.g. BioRad Gene Pulser)
- Electroporation cuvettes (BioRad)
- Cloning rings for harvesting cell colonies
- Plasmid DNA: pcHBEGF(645),[a] linearized using a restriction enzyme that cuts outside the vector's expression cassettes
- Dulbecco's modified Eagle's medium (DMEM) containing 10% calf serum
- Geneticin (disulfate salt, Sigma)
- Phosphate-buffered saline (PBS): 140 mM NaCl, 2.7 mM KCl, 0.9 mM $CaCl_2$, 0.5 mM $MgCl_2$, 8 mM Na_2HPO_4, 1.5 mM KH_2PO_4 pH 7.4

Method

1. Harvest cells from culture dishes by trypsinization.[b] If necessary, pool cell suspensions from several dishes to obtain the required number of cells.
2. Pellet cells by centrifugation at 250 *g* for 5 min. Resuspend cells in 25 ml serum-free DMEM.
3. Repeat step 2.
4. Pellet cells by centrifugation at 250 *g* for 5 min.
5. Resuspend cells at 10×10^6 cells/ml[c] in serum-free DMEM, and place 450 µl of the cell suspension in a sterile, disposable electroporation cuvette.
6. Add 15 µg[d] of linearized plasmid DNA in PBS to the cell suspension, mix, cover with cuvette cap, and stand without further mixing for 5 min at room temperature.
7. Insert the cuvette in its holder and electroporate the cells. (We have found that settings of 300 V and 960 µF work satisfactorily for RIE-1 cells.)[e]
8. Without mixing, place the cuvette of cells on ice for 8 min.
9. Transfer the electroporated cells to 10 ml of DMEM containing 10% calf serum and mix thoroughly by pipetting. Add a further 90 ml of medium and plate the mixed cell suspension into ten 9 cm diameter culture dishes.
10. After 48 h, change the medium to fresh DMEM containing 10% calf serum and 800 µg/ml geneticin sulfate (G418).[f]
11. After 12–20 days of culture, isolate geneticin-resistant cell clones using sterile cloning rings.

Protocol 4. *Continued*

12. Subculture cell clones and screen for HB-EGF expression by Western blot analysis of whole cell lysates (see *Figure 5*).
13. Expand the required clones in culture and freeze stocks of cells according to standard procedures.

[a] An HB-EGF cDNA spanning the protein coding sequence was prepared by RT-PCR and cloned into the mammalian expression plasmid, pcDNA3 (Invitrogen) to generate plasmid pcHBEGF(645). Transcription of the cDNA insert is driven by a promoter sequence from the immediate early gene of the human cytomegalovirus. The vector includes a neomycin resistance gene allowing selection of G418-resistant clones. Note that DNA of PCR-derived clones should be sequenced to check that no mutations have been introduced during amplification.
[b] The cells used for transfection should be subconfluent and actively growing. The method described is for attached cells but could be adapted for use with suspension cultures.
[c] The cell concentration used can be varied, typically in the range 2–20 $\times$ 10^6 cells/ml.
[d] The amount of DNA used can be varied, typically in the range 5–30 μg.
[e] Transfection efficiency appears to increase with increasing applied voltage, but cell viability decreases. Thus, the optimal voltage for individual cell lines needs to be determined experimentally.
[f] The sensitivity of cells to G418 differs. The optimal concentration should be determined by testing the effects of the drug (200–1000 μg/ml) on survival of the untransfected cell line.

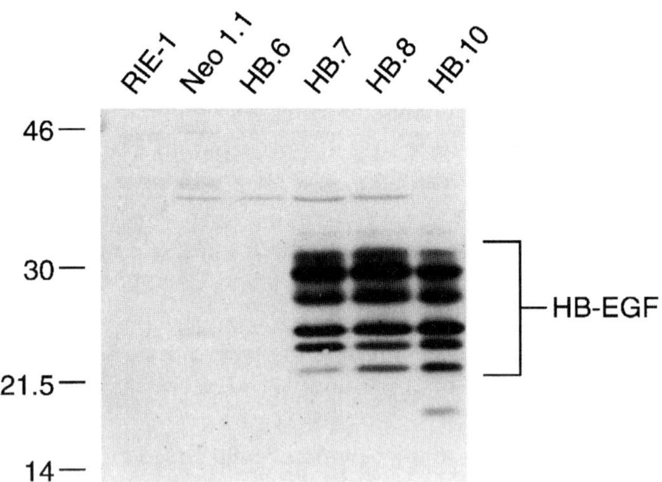

Figure 5. Expression of HB-EGF in an epithelial cell line by stable transfection. Parental cells (RIE-1), G418-resistant cells transfected with empty pcDNA3 vector (Neo 1.1), or with pcHB-EGF(645) (clones HB.6–HB.10) were grown to confluence and extracted into sample buffer for SDS–polyacrylamide gel electrophoresis (9). Proteins in the sample were separated by gel electrophoresis and then analysed by Western blotting using a rat anti-HB-EGF antiserum. Multiple forms of immunoreactive pro HB-EGF were detected. The migration positions of molecular weight markers (kDa) are indicated at left. Note that clone HB.6 had been shown to express no detectable HB-EGF mRNA.

It may be important, when getting expression from mammalian cells, that the cells themselves are not adversely affected by the expressed factor or receptor. This may require careful choice of expressing cell type.

5.4.3 Expression of genes from mammalian cells *in vivo*

Getting long-term expression of growth factors from cells *in vivo* has proven very difficult. Some groups have targeted skin for gene expression and both dermal fibroblasts and keratinocytes have proven useful for systemic delivery of proteins and for enhanced wound healing (see Chapter 10). Other groups have pursued the use of implants of factor-producing cells. In some cases these cells are enclosed in a semipermeable membrane and implanted into muscle where they slowly release the secreted factor. In other cases patients' tumour cells have been transduced with growth factor genes *ex vivo* and then re-implanted in the body in attempts to enhance the patients' immune responses to their tumours.

Acknowledgements

We are grateful to Pete Ellis for the production and analysis of HB-EGF expressing cell clones.

References

1. McKay, I. A. and Leigh, I. M. (ed.) (1993). *Growth factors: a practical approach.* IRL Press, Oxford.
2. McKay, I. A. (1993). In *Growth factors: a practical approach* (ed. I. A. McKay and I. M. Leigh), p. 1. IRL Press, Oxford.
3. Cai, J., Zheng, T., Lotz, M., Zhang, Y., Masood, R., and Gill, P. (1997). *Blood*, **89**, 1491.
4. Burden, S. and Yarden, Y. (1997). *Neuron*, **18**, 847.
5. Travis, J. (1993). *Science*, **262**, 989.
6. Dudgeon, T. J., Clements, J. M., and Hunter, M. G. (1993). In *Growth factors: a practical approach* (ed. I. A. McKay and I. M. Leigh), p. 157. IRL Press, Oxford.
7. Breslauer, K. J., Frank, R., Blocker, H., and Markey, L. A. (1986). *Proc. Natl. Acad. Sci. USA*, **83**, 3746.
8. Freier, S. M., Kierzek, R., Jaeger, J. A., Sugimoto, N., Caruthers, M. H., Neilson, T., *et al.* (1986). *Proc. Natl. Acad. Sci. USA*, **83**, 9373.
9. Sambrook, I., Fritsch, E. F., and Maniatis, T. (ed.) (1989). *Molecular cloning: a laboratory manual*, 2nd edn. Cold Spring Harbor Laboratory Press, NY.
10. Ryabova, L. A., Ortlepp, S. A., and Baranov, V. I. (1989). *Nucleic Acids Res.*, **17**, 4412.
11. Volyanik, E. V., Dalley, A., McKay, I. A., Leigh, I., Williams, N. S., and Bustin, S. A. (1993). *Anal. Biochem.*, **214**, 289.
12. Curtis, B. M., Gallis, B., Overell, R. W., McMahan, C. J., DeRoos, P., Ireland, R., *et al.* (1989). *Proc. Natl. Acad. Sci. USA*, **86**, 3045.
13. Laub, O. and Rutter, W. J. (1983). *J. Biol. Chem.*, **258**, 6043.

2

Engineering of novel neurotrophins

LEOPOLD L. ILAG and CARLOS F. IBÁÑEZ

1. Introduction

Neurotrophins constitute a class of growth factors that is essential for neuronal survival and differentiation in both the central and peripheral nervous systems. This class of proteins includes nerve growth factor (NGF), brain-derived neurotrophic factor (BDNF), neurotrophin-3 (NT-3), and neurotrophin-4 (NT-4). Neurotrophins influence the differentiation, survival, and regeneration of vertebrate neurones and although there are specific activities associated with each neurotrophin, overlapping functions also exist. Illustrating this point is the fact that all members of the family support, in varying degrees, the survival of embryonic neural crest-derived sensory neurones, NGF specifically supports the survival of embryonic sympathetic neurones but not placode-derived sensory neurones, which are supported by BDNF and NT-3 (1).

Modulation of specificity is believed to be achieved in part by the selective interaction between neurotrophins and the Trk tyrosine kinase receptors expressed on the cell surfaces of distinct neuronal populations. So far, there have been three Trk receptors characterized which interact with neurotrophins. TrkA, TrkB, and TrkC serve as receptors for NGF, BDNF, and NT-3 respectively. In addition, TrkB is shared by NT-4 and BDNF. Another receptor with which all the neurotrophins to similar extents interact, is the low affinity receptor, called p75. The role of p75 in neurotrophin signalling is not understood (2, 3). However, there have been recent studies suggesting a signalling mechanism involving NF-kB activation and production of ceramide, but the details of these pathways remain obscure (4, 5).

A milestone that greatly enlightened practitioners in the field came some six years ago when the crystal structure of NGF, the prototypic member of the family, was determined. It revealed a novel tertiary fold distinguished by a network of cystine bridges that form a knot in the lower half of the molecule. The elongated framework of the protein is dominated by two anti-parallel β-strands connected by flexible hairpin loops on which variable residues cluster (6, 7). The characteristic knot was later found to be a distinguishing feature of a class of structurally related factors like platelet-derived growth factor (PDGF), transforming growth factor-β (TGFβ), and chorionic gonadotropin which constitute what is now known as the cystine knot superfamily (8, 9). The recent determination of the crystal structure of a BDNF/NT-3 heterodimer molecule demonstrated that all members of the neurotrophin family share a common global fold (10).

The availability of the three-dimensional structure of NGF complemented efforts to understand structure–function relationships using site-directed mutagenesis. The latter is a technique that allows one to introduce specific alterations at the amino acid level in order to define critical residues important for a protein's particular function. However, there are a few requirements that have to be met before one can employ this technique:

(a) One must have cloned the gene of interest.
(b) The modified protein must be expressed and purified.
(c) Biochemical and/or biological assays are required in order to evaluate the protein's activity.

The potential use of neurotrophins in treating neurodegenerative disorders such as Alzheimer's and Parkinson's diseases makes it essential to understand more clearly their mechanism of action. The study of structure–function relationships in this family offers an approach to investigate neurotrophin function and may allow the design and engineering of analogues with desired pharmacological properties.

It is the aim of this chapter to take you step by step into systematically studying structure–function relationships and engineering chimeric proteins. The approaches we will focus on are of a biochemical and molecular/cell biological nature, given our laboratory's expertise. The protocols presented will be described as applied to our main molecules of interest, the neurotrophins.

2. Structure–function analyses of neurotrophins

A plethora of mutagenesis work has been put into the elucidation of structurally and functionally important residues in the neurotrophins. In the late 1980s, only primary structures were known and the sole guiding principle was that conserved residues may be important for maintaining the proper fold of the protein whereas variable residues may be determinants of specificity.

Two approaches are widely used in the systematic mutagenesis of proteins: alanine scanning and homologue scanning. The first one involves the replacement of individual residues or small domains with alanine. This small amino acid is chosen because it is the most commonly found amino acid in proteins, and it has no particular positional requirement, occurring in all manner of secondary structures (11). This allows for the elimination of certain functional interactions without disturbing local conformation. The second approach, homologue scanning, involves the replacement of parts or of whole variable regions with homologous sequences from other members of the same protein family. The rationale behind this is that, among proteins that have greater than 20% sequence identity, it is likely that they will have similar global folds despite functional diversity (11). Due to the need for some degree of identity, this approach is only applicable to families of proteins. Results from these two approaches may be classified as a loss- or gain-of-function.

The loss-of-function data are likely to identify only those residues that substantially contribute to receptor binding. Therefore, the contributions of other residues may be underestimated, especially if there is an extensive surface of binding. In neurotrophins for example, few residues in the variable regions were found to individually contribute enough binding energy to be identified from the loss-of-function data. Gain-of-function data, using the homologue scanning approach, led to the identification of functional domains in neurotrophins that determine specificity. Furthermore, homologue scanning has allowed the demonstration of swapping specificities among the neurotrophins (12, 13) which consequently widened the possibilities by which these factors may be engineered to enhance their therapeutic potential.

In our laboratory, we have embarked on a systematic study of neurotrophin structure–function and, although not all regions have been examined with equal intensity, a general picture emerges. Amino acid residues contributing to Trk receptor binding appear to be grouped on one side of the neurotrophin dimer, defining an extended surface that runs parallel to its twofold axis containing residues from both protomers. Identical binding surfaces are formed on both sides of the molecule because of its symmetry; this is crucial in facilitating receptor dimerization, essential for subsequent autophosphorylation and downstream signalling. Data accumulated over the past few years support a model in which conserved residues found on β-strands on one side of the dimer provide contacts with the highest binding energy, while variable residues in turns and loop regions along this surface determine biological specificity. The specificity of the interaction may be conferred either directly, by variable residues contributing contacts to cognate receptors, or indirectly by preventing binding to inappropriate receptors (14).

3. Site-directed mutagenesis

The ability to modify a protein's chemical and physical character through genetic engineering has revolutionized the field of molecular biology. Instrumental to this were the techniques of oligonucleotide-mediated mutagenesis and the polymerase chain reaction (extension of its use) which won for Dr Michael Smith and Dr Karry Mullis, respectively, the 1992 Nobel Prize in Chemistry (15, 16). Smith's and Kunkel's original concepts (16, 17) provided the basis for a popular method of performing site-directed mutagenesis: the Kunkel method. It is, however, important to note that this is only one of several ways by which site-directed mutagenesis may be performed. Variations of the PCR have also proven quite useful in this regard.

Combinatorial, random, and cassette mutageneses are becoming popular methods for creating a pool or 'library' of genetic variants. With these methods, several mutations are simultaneously introduced in the mutagenic oligonucleotide, by random incorporation of mononucleotides during oligonucleotide synthesis. Combining this strategy with a selection method for the desired variants would allow one to rapidly identify functional regions of a protein. Current methods of creating libraries of genetic variants using randomly introduced mononucleotides leads to incorporation of undesired amino acid codons (stop codons) and/or a biased distribution of certain codons because of the code's degeneracy. Advances in the development of trinucleotide- or codon-based mutagenesis should further improve the quality of these libraries (18). However, the availability of a good/efficient selection method is necessary to fully exploit these mutagenesis techniques. The emergence of powerful selection methods such as the yeast two-hybrid system and phage display (19, 20) has made these latter techniques more accessible. Unfortunately, no efficient means of selection is currently available for neurotrophins to employ these methods. In our laboratory we are currently developing a selection system using a novel phage display protocol (21) to take advantage of these more powerful mutagenesis methods.

If genetic manipulation is not possible because the recombinant form of the protein of interest is not available, chemical modifications may be used such as coupling with active esters, alkylation, reaction with tetranitromethane, and *N*-methylnicotinamide chloride (22, 23) in order to assess structurally and functionally important regions in the molecule. In fact, before the cloning of neurotrophins, important aspects of structure–function in NGF were addressed using related methods (23, 24).

3.1 Kunkel method

As mentioned above, we use the Kunkel method for generating mutant neurotrophins. This involves:

(a) The production of single-stranded DNA (template) containing uracils (substituting thymines) necessary for a subsequent selection step (i.e. degradation of the uracil-containing strand upon introduction of the plasmid into bacteria with functional uracil-*N*-glycosylase).

(b) An annealing reaction to form a heteroduplex between the template and the mutagenic oligonucleotide (primer).

(c) A polymerization reaction that allows for the synthesis of a second strand containing the modification.

3.1.1 Preparation of single-stranded template

The key component in the production of uracil-containing single-stranded DNA (ssDNA) is the *E. coli* strain CJ 236 which contains the following markers: dut-1, ung-1, thi-1, rel A-1; pCJ105 (Cmr). The dut^- and ung^- mutations are non-reverting and are responsible for the incorporation of some uracils in all newly synthesized DNA molecules inside the bacterium. In these cells, the lack of dUTPase (dut^-) results in an increase of intracellular dUTP since it can't be converted to dUMP, hence allowing dUTP to be incorporated into the newly made DNA molecule. Misincorporation of uracil residues for thymine is maintained because there is a complementary absence of the enzyme uracil-*N*-glycosylase (ung^-), which normally removes uracils from DNA (25).

Single-stranded template DNA is first prepared by growth of an appropriate recombinant M13 bacteriophage (or a phagemid such as pBluescript in *E. coli* CJ 236; Stratagene) containing the gene of interest. *Protocol 1* describes the preparation of this uracil-containing single-stranded DNA which is then used in a mutagenesis reaction (*Protocol 2*).

Protocol 1. Production of uracil-containing single-stranded DNA (ssDNA) template

Reagents

- *E. coli* CJ 236 (BioRad)
- R408 phages (Promega)
- TE: 10 mM Tris–HCl pH 7.5–8.0, 1 mM EDTA
- PEG (Sigma) solution: 20% polyethylene glycol 6000, 3.5 M NH_4Ac

Method

1. Inoculate 50 ml of mid-log phase (OD_{600} = 0.3) *E. coli* CJ 236 (transformed with appropriate vector, e.g. pBluescript carrying the gene of interest) with excess R408 phage in a 1:20 ratio and grow the culture for 8 h at 37 °C.[a]
2. Transfer the cells to a 50 ml plastic Falcon tube and spin down (wash) two times at 4000 r.p.m. for 15 min at 4 °C. Decant the supernatant into Corex tubes.
3. Add 12.5 ml PEG solution to the supernatant and let the mixture stand overnight at 4 °C.
4. Centrifuge the mixture at 10 000 r.p.m. for 30 min at 4 °C and keep the pellet.
5. Resuspend the pellet in 1 ml TE and transfer the suspension to a 1.5 ml Eppendorf tube. Purify ssDNA by extracting once with an equal volume of phenol, then twice with phenol:chloroform, and finally once with chloroform only.

Protocol 1. *Continued*

6. Precipitate the ssDNA by bringing the solution to 0.25 M NaCl and adding 2.5 vol. of ethanol. Let this stand for at least 30 min at –20 °C, and then centrifuge at 12 000 r.p.m. in a microcentrifuge for 30 min. Discard the supernatant.
7. Resuspend the pellet in 50 μl TE or double distilled water, and quantify the yield by spectrophotometry at 260 nm (1 OD_{260} = 30 μg DNA/ml).

[a] At OD_{600} = 0.3, there are approx. 2.5 × 10^8 cells/ml.

3.1.2 Designing mutagenic oligonucleotides (primers)

Primer oligonucleotides are usually obtained from a commercial source. The basic element in designing these primers is to conserve as much complementarity as possible with the template DNA. Because the mutated region will bulge out as a result of non-complementarity, the general design of the primer necessitates that there should be about 12 nucleotides flanking each side of the mutated region (two or three nucleotides) for a standard point mutation where one amino acid is changed. In homologue scanning, it is usual to work with mutations involving stretches as long as seven to ten amino acids which usually requires flanking regions of at least five amino acids (15 nucleotides) on each side. For longer mutations, it is advisable that the modification be carried out in at least two steps. The most challenging mutation we have dealt with so far using the Kunkel method is an insertion of 28 amino acids (78 nucleotides). Although the Kunkel method is very efficient (usually the majority of the clones obtained will carry the desired mutation), the degeneracy of the genetic code often allows the introduction of a suitable restriction site in the mutated sequence which can later be used to facilitate screening for mutant clones (see below).

It is very important that when you design the sequence for the mutation, that you alter the nucleotide sequence as little as possible. For example, in changing Glu (GAG) into a Gly (GGA, GGC, GGG, or GGU) the codon GGG is most favourable since it necessitates changing only one nucleotide. It is also important to avoid having more than three identical nucleotides at the very 3′ end of the mutagenic primer; loops may bulge out during annealing which, after polymerization, will result in undesired deletions or insertions. Finally, be sure you are designing an oligonucleotide that is *complementary* to your single strand!

3.1.3 Mutagenesis reaction

Mutagenic oligonucleotides to be used for the mutagenesis reaction must be phosphorylated at their 5′ end. This is important since a 5′ phosphate group is essential for the ligation of the newly synthesized strand (i.e. the one with the desired mutation). Although most suppliers of synthetic oligonucleotides offer this modification, it often represents a substantial portion of the cost of the primer. The phosphorylation of the oligonucleotide can be done *in vitro* with little effort. *Protocol 2* describes the procedure to carry out the basic steps as outlined in Section 3.1.1 after the preparation of the template.

Protocol 2. Kunkel method of site-directed mutagenesis

Reagents

- 10 × annealing buffer: 200 mM Tris–HCl pH 7.4, 20 mM $MgCl_2$, 500 mM NaCl
- 10 × synthesis buffer: 5 mM dNTP, 10 mM ATP, 100 mM Tris–HCl pH 7.4, 50 mM $MgCl_2$, 20 mM DTT
- T4 DNA ligase (Promega)
- T4 DNA polymerase (Amersham)
- T4 PNK (polynucleotide kinase) (Promega)
- 10 mM ATP (Promega)
- ssDNA template (see *Protocol 1*)
- MV1190 or DH5α strains of *E. coli* (BioRad)

A. *Phosphorylation of oligonucleotide primers*

1. Dissolve the oligonucleotides and quantify using a UV spectrophotometer (OD_{260} = 1.0 is approx. 30 μg/ml).
2. Set-up the phosphorylation reaction by mixing in a plastic microcentrifuge tube 2.5 μg oligonucleotide, 5 μl 10 × PNK buffer, 1 μl T4 PNK (5 U), 5 μl 10 mM ATP, and add water up to 50 μl.
3. Incubate the reaction mixture for 15 min at 30 °C and then let it stand on ice until subsequent use (part B).

B. *Annealing reaction*

1. Set-up the annealing reaction by mixing 50 ng (1 μl) phosphorylated oligonucleotide (see part A), 100 ng ssDNA template (see *Protocol 1*), 1 μl 10 × annealing buffer, and add water up to 10 μl.
2. Incubate the reaction in a beaker with water heated to 70 °C and allow the system to cool to room temperature. Alternatively, a thermal cycler could be used to bring down the temperature from 70 °C to 25 °C in 1 °C decrements every minute, and then cool to 0 °C.

C. *Synthesis reaction*

1. Set-up the synthesis reaction by mixing 10 μl annealing reaction (part B), 1 μl 10 × synthesis buffer, 1 μl (3 U) T4 DNA ligase, 1 μl (4 U)T4 DNA polymerase.
2. Incubate the reaction for 5 min on ice, followed by another 5 min at room temperature, and then finally for 90 min at 37 °C.
3. Take 7 μl of the reaction mix and transform into MV1190 or DH5α competent cells (see below).

3.1.4 Bacterial transformation

The plasmid obtained from the mutagenesis reaction (see *Protocol 2C*) should now be transformed into an appropriate host such as MV1190 or DH5α strains of *E. coli*. Minipreps (small scale preparation of plasmids) from various transformants (colonies) are then made and double-stranded sequencing is performed to screen for the incorporation of mutations. Screening may also be performed using restriction enzymes, provided a novel site was introduced during mutagenesis. Although this in some cases may speed up the screening process, it is always necessary to sequence at least two different clones to confirm that only the desired mutation was introduced in the template. The coding fragment is then cleaved from the pBluescript(+) vector and subcloned into the expression vector pXM or other appropriate vectors (26) for subsequent over-production of protein in eukaryotic systems (e.g. COS cells). Some expression vectors now available contain an M13 or f1 origin of replication, allowing the site-directed mutagenesis to be carried out on the ssDNA produced directly from the final vector. Details on how to perform the transformation of bacteria as well as miniprep analysis, plasmid cleavage, and subcloning are described elsewhere (25).

3.2 PCR-based mutagenesis

Oligonucleotide-mediated mutagenesis as described above is not the sole means by which site-directed mutagenesis may be performed. A powerful alternative to this technique is by polymerase chain reaction (PCR) which is a primer directed, enzymatic amplification of specific DNA sequences (15, 27).

Early work on introducing mutations by PCR involved designing primers that carry mutations such as new restriction sites. However, this method limits the positioning of the modifications because the mutations are always in the primers which are found in the termini of target sequences. A way around this is through mutagenesis by overlap extension. This involves the generation of two DNA fragments with overlapping ends. In a subsequent reaction, the fragments are joined together using PCR by allowing the 3′ overlap to serve as a primer for the 3′ extension of the complementary strand. This method makes it possible to perform the mutagenesis using the original vector with close to 100% efficiency. This method may be used also for creating fused gene products (28, 29). However, utmost care must be taken in optimizing the conditions for the overlap extension PCR, to reduce the level of non-specific reaction products (29). Due to the inherent imperfection of thermostable polymerases, PCR is prone to misincorporation of bases making it essential to sequence the entire open reading frame. With the Kunkel method we normally pick sister clones that carry the mutation and only read close to the site of the mutation. The probability that many clones carry undesirable mutations is very small. Before large scale expression, we carry out *in vitro* translation to verify that the protein is properly translated.

4. Expression of recombinant protein

At this point we are now ready to check if all the mutants we have constructed are in fact expressed and are biologically active. It is obvious that all the manipulations we have done so far would be useless if we don't get viable proteins suitable for subsequent assays. There are various ways of over-expressing properly folded and modified recombinant proteins. To assure as close to wild-type protein processing as possible, the system of choice are eukaryotic systems. Here we will describe two such systems namely expression in COS cells and baculovirus-infected insect cells (*Spodoptera frugiperda*). Although eukaryotic expression is ideal, it is usually fraught with rather tedious and time-consuming steps; hence many find it more convenient to use prokaryotic systems. Note however, that this system is not without its own drawbacks. Bacterial systems do not have the necessary machinery to perform post-translational modifications which may sometimes be crucial for a protein's activity. More importantly, it is not uncommon for inclusion bodies to form. These insoluble entities necessitate solubilization with rather harsh agents like guanidinium hydrochloric acid (GuHCl) which denatures proteins. Consequently, a refolding step needs to be implemented. Since the basic principles governing protein folding are still unclear, there is no guarantee that a given protein could be successfully refolded and be biologically active. Expression in prokaryotic systems such as *E. coli* will be briefly discussed below.

4.1 COS cell expression

COS-7 cells are immortalized monkey fibroblast cells that have proven very efficient for transient transfections as described below in detail (see *Protocol 3*). This expression system is convenient for preliminary production of small amounts (a few micrograms) of recombinant protein for biochemical and biological assays (Sections 6 and 7) but more importantly it is a rapid way of assessing the feasibility of expressing the gene of interest. It requires that plasmids used have the SV40 origin of replication. SV40 is a member of the papova group of small, non-enveloped DNA viruses which causes lytic infection of permissive monkey cells (25).

COS cells are derived from transformation of simian CV-1 cells with an origin-defective SV40 genome (30) which constitutively express the wild-type SV40 large T antigen, and contains all the necessary factors required to drive the replication of SV40 ori-containing plasmids like pXM, a vector often used in our laboratory for this purpose.

Over the course of the transfection, each cell accumulates greater than 10^5 copies of the recombinant plasmid. It is presumably due to this high level of extrachromosomally replicating DNA that expression in the system is transient as cells die approximately 70–90 hours post-transfection.

In using the DEAE–dextran method of transient transfection, there are two major things that could affect the efficiency of the transfection. First, the concentration of the DEAE–dextran solution, and secondly the length of time the cells are incubated in this solution. At a concentration of about 1 mg/ml it is customary to have incubation times ranging from 30–60 min whereas with a concentration of 250 μg/ml, the incubation time may be as long as 8 h. Although using the higher concentration is more efficient, it requires very careful monitoring of cells for signs of distress which requires a bit of experience. This makes the milder treatment, though less efficient, more reliable (25).

Protocol 3. Over-expression of recombinant protein in COS-7 cells

Equipment and reagents

- Cell culture hood (Kojair)
- CO_2 incubator (Forma Scientific)
- Centriprep-10 concentrators (Amicon)
- Tissue culture dishes (10 cm, Falcon)
- Flat-bottom tissue culture plate (6-well, Falcon)
- COS-7 cells (ATCC)
- TBS (solution A): 40 g NaCl, 1.9 g KCl, 1 g Na_2HPO_4, 15 g Tris, make up to 500 ml ddH_2O[a]
- TBS (solution B): 1.5 g $CaCl_2$, 1 g $MgCl_2$[a]
- 100 μM chloroquine (Sigma) in complete medium
- Complete medium: Dulbecco's modified Eagle's medium (DMEM, Gibco), 2 mM glutamine, 10% fetal calf serum, 60 μg/ml gentamycin (Sigma)
- DEAE–dextran (Sigma): 100 mg/ml aqueous solution, filtered through 0.2 μm filters
- 100 × labelling cocktail: [^{35}S]cysteine 100 μCi/ml (Amersham) and supplements for minimal essential medium minus cysteine (Gibco)
- Transferrin (Sigma)
- Insulin (Sigma)
- 'Amplify ' fluorographic reagent (Amersham)

A. *Transfection with DEAE–dextran*

1. Prepare DEAE–dextran dilution in 1 × TBS in a 1:30 ratio.
2. Dilute plasmid constructs in 1 × TBS (15 μg plasmid/750 μl TBS per 10 cm tissue culture dish).[b]
3. Add 1 vol. of the DEAE–dextran solution (step 1) to 3 vol. of DNA solution (step 2).
4. Rinse plates (four tissue culture plates per plasmid type) of COS-7 cells grown in complete medium (approx. 60% confluent; i.e. about 0.6 × 10^6 cells/plate) with TBS.
5. Evenly cover the cells with 1 ml of the DNA–dextran mixture (step 3). Let the plates stand for 45 min at room temperature, mixing every 10 min to ensure no area of the plate dries.
6. Aspirate the DNA–dextran mixture, add 6 ml/plate of complete medium containing 100 mM chloroquine, and incubate at 37 °C for 4 h, or until the cells are distinctly studded with vacuoles and are beginning to detach from the plate.

7. Replace the chloroquine-containing medium with 10 ml of complete medium and incubate at 37 °C overnight.
8. Replace the complete medium with 10 ml of serum-free medium (supplemented with 10 μg/ml transferrin and 5 μg/ml insulin), incubate for three days at 37 °C, after which the media is harvested (referred to as conditioned media).
9. Concentrate the conditioned media (approx. 40 ml) through centriprep-10 filters (cut-off: 10 kDa) to about 500 μl for subsequent assays.

B. *Metabolic labelling of proteins with [^{35}S]cysteine*

1. Perform the same procedure as outlined in part A, steps 1–7, using 6-well flat-bottom tissue culture plates and one-fifth of the volume of all solutions.
2. Two days after the transfection, measure transfection efficiency using a β-galactosidase assay.[b]
3. Rinse each well with serum-free medium and add 1 ml/well of radioactive labelling cocktail. Incubate overnight at 37 °C.
4. Harvest the media and run a 50 μl aliquot in a 13% denaturing SDS gel.[c]
5. Fix the gel in 30% acetic acid:10% methanol solution for 30 min at room temperature and incubate in 'Amplify' solution for 15 min. Vacuum dry the gel and subject it to fluorography at –70 °C for evaluation of expression levels by densitometric scanning.[d]

[a] The pH of solution A is adjusted to pH 7.5. Autoclave solutions A and B. Mix 890 ml sterile distilled water with 100 ml of solution A and add 10 ml of solution B by drops while stirring. Keep everything sterile!
[b] Don't forget to use a control plasmid (15 μg) like pCH110 that expresses β-galactosidase to evaluate the transfection efficiency (see ref. 25). If the assay shows an OD_{420} = 0.5 and above, the transfection is considered good.
[c] Use rainbow protein marker (Amersham).
[d] Neurotrophins run at about 14 kDa in a 13% SDS gel.

4.2 Baculovirus expression

Another means of expressing recombinant proteins in eukaryotic systems is by using the baculovirus system which can produce proteins that are of equivalent biological activity, but of higher yield and lower cost compared with mammalian expression systems. In general, yields of 1–100 mg/litre have been observed (31).

One of the most often used baculoviruses is the *Autographa californica* polyhedrosis virus (AcMNPV). In this species, nearly all transcription of the host cell is shut off at the very late phase of viral infection (around 20 hours after infection) except for some viral genes that are transcribed at very high rates. One of these is the polyhedrin gene which codes for a structural component of the polyhedra that accounts for 30% of the total cellular protein and is important only when the virus goes through a lytic cycle. Therefore, in the case of AcMNPV-infected cells in tissue culture where the virus is lysogenic, the polyhedrin protein becomes superfluous. This allows for the polyhedrin coding region to be replaced by the gene fragment of interest. Because the viral genome is too large to be amenable for manipulation, recombinant baculovirus generation requires a transfer vector and a recombination event. Detailed protocols on how to carry out expression in this system are thoroughly discussed elsewhere (31).

4.3 Expression of recombinant proteins in prokaryotic systems (*E. coli*)

Over-expression of proteins in prokaryotic systems is commonplace, hence, only a brief overview of this system will be presented here.

The initial step in expressing eukaryotic proteins in bacteria is to choose an appropriate expression vector that carries a strongly regulated promoter like the phage T5 promoter, hybrid trp–lac promoter, or the bacteriophage T7 promoter. The next step is to ensure that your expression vector has an efficient ribosome binding site (25). Commercially available vectors with synthetic ribosome binding sites designed for optimal mRNA recognition and binding are available (32). Note that vectors require specific types of host, so be sure to use an appropriate *E. coli* strain.

In some cases, biologically active proteins are not produced in prokaryotes for a number of reasons:

(a) In *E. coli*, the amino terminal methionine is removed to different extents hence affecting the stability (half-life) of the polypeptide of interest.

(b) Sometimes the eukaryotic proteins produced intracellularly in bacteria are inactive due to improper folding.

(c) Bacterial proteases may cleave some foreign proteins.

It may be possible to avoid some of these problems:

(a) By directing the secretion of the foreign protein into the periplasm through fusion with a signal sequence that eventually gets cleaved by signal peptidases.

(b) By expressing a fusion protein that can subsequently be cleaved by chemicals (e.g. cyanogen bromide CNBr) or by proteases (e.g. thrombin or factor VIII) (33, 34). The fusion partner can be used not only to confer added stability but also to aid protein purification (e.g. proteins fused to glutathione *S*-transferase which are purified using glutathione affinity columns, or by the addition of histidine tails that facilitate isolation using metal chelation (35).

Although neurotrophins have been successfully expressed in *E. coli*, a refolding step is required after the solubilization of inclusion bodies by strong acid like guanidinium hydrochloric acid or chaotropic salts like urea (36). Some groups have even tried modifying sequences around the translation initiation region to improve secretion of heterologous proteins (37).

5. Protein purification

Obtaining pure protein is not always necessary in order to characterize some activities of the protein, or if the aim is to compare the behaviour of a mutant with wild-type protein. It is generally advisable, particularly for a long series of mutants, to assay the activity of proteins transiently produced in COS cell conditioned medium. Mutant proteins of interest may subsequently be purified for further characterization. In view of the advances in molecular biophysics, it has become commonplace to analyse protein:protein interactions right down to the atomic level. For such physicochemical characterizations, there is often a demand for highly pure (about 95–98%) samples, especially when the objective is to determine three-dimensional structures by X-ray crystallography or nuclear magnetic resonance (NMR) spectroscopy. Some commonly used purification methods are described below.

Cation or anion exchange chromatography relies on differences in electrostatic properties of proteins for separation. The net charge of a protein depends on its amino acid composition, hence it could be effectively used in discriminating one protein from another. Depending on the protein's isoelectric point (pI) it will either carry a net positive or negative charge depending on the pH of the buffer system used. Proteins that differ in pI by about 0.1 have been known to be separable by this method. Neurotrophins are quite basic with a pI greater than 9.0 making it almost possible to do a one-step purification with cation exchange at pH 7.5.

Size exclusion chromatography is a method that separates biomolecules by their size and to some extent their shape. Also known as gel filtration chromatography, separation is effected by allowing a mixture of proteins to pass through a column packed with porous beads. The actual volume traversed by each protein depends on its size because the smaller the proteins are, the more likely they will pass through the pores of the beads. Bigger proteins take a shorter path and come out first during the separation. This method is also an effective means of quickly exchanging the buffer composition of your sample.

Reverse-phase high-pressure liquid chromatography (RP-HPLC) is a common means of purifying peptides. Although it employs relatively harsh conditions (e.g. low pH) it has proven to be useful for purifying neurotrophins and other growth factors smaller than 30 kDa. The neurotrophins can withstand the low pH (approx. pH 2) presumably due to the added stability afforded by their network of disulfide bonds. Protein separation by RP-HPLC is based on hydrophobic interactions and several surface chemistries are available that allow one to optimize separation. From our experience, a C8 RP-HPLC column works well for neurotrophin purification.

Metal chelating columns, e.g. nickel (Ni) are special affinity columns of somewhat general application. The metal is attached to an immobilized chelating ligand like imidoacetic acid (IDA) or nitrilo-tri-acetic acid (NTA). The latter leaves two vacant sites in the nickel's coordination sphere allowing the interaction with nitrogens in the indole rings of histidines. The effective coordination of the metal allows high affinity (approx. 10^{-13} at pH 8.0) interactions with, for example, histidine-tagged proteins resulting in a very efficient purification step (32). This method requires the protein of interest to have several histidines which is not often the case; hence, proteins are usually artificially His-tagged. Around six consecutive histidines afford stable interactions with the Ni column.

Protocol 4. Chromatographic purification of neurotrophins

Equipment and reagents

- HiLoad™ chromatography system (Pharmacia)
- Fraction collector (e.g. RediFrac, Pharmacia)
- Peristaltic pump P-50 (Pharmacia)
- High performance S-Sepharose (Pharmacia)
- IE buffer A: 50 mM phosphate buffer pH 6.5, 5 mM EDTA
- IE buffer B: buffer A with 1 M NaCl
- Superdex 75 column (Pharmacia)
- SE buffer: 50 mM phosphate buffer pH 7.4, 5 mM EDTA, 0.5 M NaCl, 10% acetonitrile
- C8 reverse-phase HPLC column (Vydac)
- RP buffer A: 0.1% TFA/water
- RP buffer B: 0.1% TFA/acetonitrile
- RP buffer C: 70% acetonitrile

A. *Ion exchange chromatography*

1. Adjust the pH of the conditioned medium to 6.5 (same as IE buffer A).
2. Equilibrate the S-Sepharose column with six column volumes of IE buffer A at 8 ml/min which is the flow rate used for the whole purification. Apply the conditioned medium to the column and wash with six column volumes of IE buffer A.
3. Elute bound proteins with a 0–1 M NaCl (IE buffer A/IE buffer B) linear gradient and collect 10 ml fractions.[a]
4. Identify fractions containing the expressed protein by Western blot using 40 μl of each fraction. Pool positive fractions and concentrate the solution using centriprep-10 ultrafilters for subsequent size exclusion chromatography (part B). Transfer the concentrate to a 1.5 ml Eppendorf tube.

B. *Size exclusion chromatography*

1. Equilibrate the Superdex 75 column with one column volume SE buffer.
2. Inject sample into the injection loop.

3. Elute with one column volume of SE buffer.
4. Proceed as in part A, step 4.
5. Adjust the pH of the concentrated sample (step 3) to approx. 3.5 with concentrated acetic acid for subsequent purification using reverse-phase HPLC (part C).

C. *Reverse-phase HPLC*

1. Spin down the sample (concentrate of pooled fractions from part B) in a microcentrifuge for 2 min at maximum speed.
2. Equilibrate the C8 reverse-phase HPLC column with three column volumes of RP buffer A.
3. Inject the sample and elute with 0–60% gradient of acetonitrile (RP buffer A/RP buffer B).[a,b]
4. Proceed as in part A, step 4.[b]
5. Vacuum dry the protein and reconstitute in water. Typical final yield is about 50–100 μg of pure protein per litre of conditioned medium.

[a] The gradient may be adjusted for optimal separation.
[b] Neurotrophins usually elute at around 40% acetonitrile (RP buffer B).

6. Biochemical assays

Constructing and expressing viable proteins would be futile if we have no means by which to measure the effects of mutations on the function of the protein. We may also want to evaluate whether introduced mutations confer new activities to the parent molecule. Consequently, biochemical assays suitable for assessing protein funtion are essential. Some common biochemical assays used in evaluating neurotrophin function are described below. These assays are used to make preliminary observations about gain- or loss-of-function resulting from the introduced modifications.

6.1 Binding assay

According to Wells (11), disruptions of binding affinities by a single alanine substitution for example, usually range from 2- to 100-fold. It is therefore necessary that an assay capable of discriminating a binding event with a factor of two is available. Conventional binding assays would usually require about 0.1–10 μg of a protein analogue to displace the receptor-bound labelled protein of about 20 kDa in order to obtain a good binding isotherm for affinities in the range of 10^{-8} to 10^{-10} M (11). In the case of neurotrophins, where the affinity is about 10^{-11} M, we normally use 0.003–3 μg/ml. The relative binding affinity of the analogue can be estimated from the ratio of the analogue's IC_{50}

(concentration necessary to inhibit the tracer binding by 50%) to that of the wild-type ligand. The protein preparations used for displacement in these assays need not be pure, provided appropriate controls are used to make sure that the observed effects are not due to impurities. Care should be taken also that proteolytic activity is absent from the crude samples. Finally, one should have an accurate titre of the analogue to confirm the stoichiometry of binding.

Further characterization of the binding properties of the mutants generated could be obtained from equilibrium saturation binding assays and Scatchard transformation (38). Association and dissociation kinetic experiments can also give information on the nature of those aspects of the ligand:receptor interaction that were altered by the mutation. For example, association (specificity) is believed to be governed by electrostatic interactions, so it may be of interest to study the association binding kinetics of a particular mutant with substituted charged residues. Because both Scatchard and kinetic analyses require radiolabelling of the mutant ligand, the latter should be purified to homogeneity.

Below is a detailed protocol for a competitive binding assay normally carried out in our laboratory.

Protocol 5. Radioligand displacement binding assay

Reagents

- Neurotrophins (see *Protocol 3*)
- Lactoperoxidase (Sigma)
- 30% hydrogen peroxide (Merck)
- Bovine serum albumin (BSA) (USB)
- [^{125}I]Na (Amersham, IMS 30)
- G25 Sepharose (Pharmacia)
- Trichloroacetic acid (TCA) (Merck)
- Binding buffer: phosphate buffer pH 6.5, 2% BSA, 0.7 mM $CaCl_2$[a]
- Gamma counter
- Trk-expressing cells (Regeneron Pharmaceuticals, Inc.)

A. *Preparation of gel filtration column for purifying radiolabelled ligand*

1. Plug the tip of a 10 ml plastic pipette with a glass bead.
2. Pack the pipette (column) with G25 Sepharose.
3. Equilibrate the packed column with one column volume of binding buffer (about 4 ml) making certain it does not run dry.

B. *Lactoperoxidase method of labelling neurotrophins*

1. Set-up the labelling reaction in an Eppendorf tube as follows:[b] 10 μl 1 mCi/ml [^{125}I]Na, 20 μl 0.1 M phosphate buffer pH 7.4, 10–20 μg NGF, 10 μl 50 μg/ml lactoperoxidase, and 10 μl 0.003% hydrogen peroxide.
2. Incubate the reaction for 30 min at room temperature and then add 10 μl of 0.003% hydrogen peroxide.
3. Take a 2 μl aliquot of the labelling reaction into a clean 1.5 ml microcentrifuge tube (on ice) and add 500 μl 10% TCA and 200 μl of

BSA (1 mg/ml). To the rest of the labelling reaction, add 200 μl of 0.4% acetic acid to stop the reaction (keep on ice until step 8).

4. Centrifuge the reaction mixture (step 3) for 5 min at maximum speed (approx. 13 000 r.p.m.) in a microcentrifuge.
5. Wash the pellet by resuspending it in 100 μl of 0.5 M NaOH and 1 ml 10% TCA, keeping the tube on ice.
6. Repeat steps 4 and 5 then finally dissolve the pellet in 100 μl of 0.5 M NaOH.
7. Take 10 μl, count radioactivity in the sample using a scintillation counter, then calculate the specific activity of the labelled protein.
8. Purify labelled neurotrophins from step 3 by gel filtration through a G25 Sepharose column (see part A).
9. Collect fractions in 1.5 ml microcentrifuge tubes. Identify the elution position of the labelled protein using a gamma counter to count the radioactivity in aliquots of each fraction.

C. *Binding assay*

1. Prepare triplicate samples of the required dilutions of wild-type and mutant neurotrophins (spanning three orders of magnitude around the dissociation constant of the molecule, e.g. 2.6 ng to 2.6 μg at twofold increments) to a total volume of 100 μl/well in a 96-well microtitre plate. Leave one set of triplicate samples without competitor to measure the maximum binding of the labelled factor (HOT). To another set of triplicate samples, add 100-fold excess of unlabelled neurotrophins to measure non-specific binding (COLD).[c]
2. Harvest Trk-expressing cells and resuspend in binding buffer at 1.5×10^6 cells/ml.
3. Add 20 μl of the cell suspension from (from step 2) and 20 μl of iodinated neurotrophin (26 ng/ml) from part B into each well of the 96-well plate (step 1). The final volume should be 140 μl.
4. Incubate the 96-well microtitre plate in a shaker (vigorous shaking) for 90–120 min at 4 °C.
5. Centrifuge the plate at 3000 r.p.m. for 15 min at 4 °C.
6. Aspirate the supernatant, replace with binding buffer to wash the cells, then repeat step 5.
7. Finally, aspirate the supernatant, dissociate (manually) the wells, and count in a scintillation counter.

[a] Dilute the $CaCl_2$ in half the volume of water to be used in the solution to avoid precipitation.
[b] When working with [^{125}I]Na, always handle the materials in an efficient fume-hood and use appropriate shielding.
[c] Neurotrophin binding: COLD/(HOT – COLD) × 100 = % binding.

6.2 Cross-linking assay

Knowing the binding affinities of various mutants gives vital information regarding residues required for the specificity and activity of the protein of interest. An alternative to the previously described approach in analysing binding events is chemical cross-linking. Several cross-linking reagents are available; the most commonly used are EDC (1-ethyl-3-[3-dimethylamino-propyl]-carbodiimide hydrochloride), DSP (dithio*bis*[succinimidylpropionate]) also known as Lomant's reagent, SB^3 (*bis*[sulfosuccinimidyl] suberate), and DSS (disuccinimidyl suberate). DSP and DSS are both homobifunctional, amine reactive agents differing only by the fact that the disulfide bond in DSP allows for it to be cleaved whereas DSS is non-cleavable. Consequently, reducing agents (see *Protocol* 7B, step 11) should not be used with DSP. BS^3 on the other hand, is a water soluble analogue of DSS that is membrane impermeable. Among the four, EDC seems to be the most versatile. It is water soluble and is reactive towards available -COOH and $-NH_2$ groups (39). Although all of these cross-linkers are effective in coupling neurotrophins to Trk receptors, we have observed that only EDC is effective in cross-linking neurotrophins to the low affinity nerve growth factor receptor (p75).

The advantage of this technique over conventional binding assays is that it demonstrates the physical association of the molecules with their receptors in near natural contexts. One could derive binding affinity constants with this assay although it is more tedious and less reliable given the fact that the cross-linkers may influence binding events. Furthermore, quantitative analysis of binding is limited by the inherent inefficiency of the cross-linking event (around 1%).

Protocol 6. Cross-linking assay

Reagents

- Chemical cross-linking reagents: EDAC, DSP, DSS, SB^3 (Pierce)
- PBS: 137 mM NaCl, 2.7 mM KCl, 10 mM Na_2HPO_4, 1.7 mM KH_2PO_4, pH adjusted to 7.4 with HCl
- Binding buffer (see *Protocol 5*)
- Pan-Trk antibody (Santa Cruz Biotechnology, Inc.)
- Protein A–Sepharose (Pharmacia)
- Lysis buffer: 1% NP-40 (ICN Biomedicals, Inc.), 20 mM Tris pH 8, 137 mM NaCl, 10% glycerol, 2 mM EDTA, 1 mM PMSF, 0.15 U/ml aprotinin (Sigma), 1 mM Na orthovanadate (Sigma), 20 mM leupeptin (Sigma)
- Amplify fluorography reagent (Amersham)
- 2 × SDS–PAGE sample buffer: 100 mM Tris–HCl pH 6.8, 4% SDS, 0.2% bromophenol blue, 20% glycerol

A. *Preparation of cortical tissue for competitive cross-linking*

1. Homogenize two whole cortices from adult Sprague-Dawley rats (in 15 ml Falcon tubes) in 10 ml cold (4°C) PBS/glucose (5 mg/ml) solution using a 20 ml syringe fitted with an 18G needle.
2. Allow the homogenate to sediment for 1–2 min on ice.

3. Collect the supernatant into a 50 ml Falcon tube and spin down at low speed (5000 r.p.m.).
4. Resuspend the pellet in 5 ml of cold (4°C) binding buffer.

B. *Competitive cross-linking assay*

1. Take 1 ml aliquots of the samples prepared in part A into 1.5 ml microcentrifuge tubes.
2. Add unlabelled neurotrophins at 50-fold excess to each tube, except to one of the tubes leaving it as a control.
3. Add 1 nM of iodinated neurotrophin (see *Protocol* 5B), cognate to the receptor under study, to each tube.
4. Incubate the sample on ice for 2 h, resuspending the cells every 10–15 min.
5. Add 2 mM EDAC and incubate for 30 min at room temperature resuspending the cells every 10 min.
6. Centrifuge the tubes at 2000 r.p.m. for 5 min, wash the cell pellets twice with TBS, then resuspend in 500 μl of lysis buffer, and leave the tubes on ice for 10 min.
7. Vortex the samples briefly and centrifuge to remove debris. Collect the supernatant and transfer into fresh microcentrifuge tubes.
8. Add 2 μl of pan-Trk antibody to each tube and incubate the samples for 3 h at 4°C.
9. Add about 70 μl of protein A–Sepharose beads and incubate for 15 min at 4°C with shaking.
10. Wash the beads two times with lysis buffer.
11. Resuspend the beads in 40 μl SDS–PAGE sample buffer and add 1 μl β-mercaptoethanol.
12. Boil the samples for 5 min and run a denaturing SDS gel. Proceed as in *Protocol* 3B, step 5 except treatment with 'Amplify'.

6.3 Phosphorylation assay

Upon receptor binding, neurotrophins induce dimerization of Trk receptors triggering their autophosphorylation on tyrosine residues. It is likely that specific patterns of receptor phosphorylation determine the biological consequences of binding. It should be noted that some NGF mutants are incapable of effecting autophosphorylation despite successful binding (12).

Protocol 7 describes how phosphorylation assays are carried out in our laboratory. Some key requirements for these experiments are the availability of appropriate antibodies (e.g. anti-Trk and anti-phosphotyrosine) as well as of cell lines that over-express (and ideally express exclusively) the receptors of interest.

Some important points to bear in mind are the following. It is necessary to incubate (approx. 30–60 min) and wash the cells with serum-free medium before treatment (see *Protocol 7*, step 3) to ensure suppression of residual phosphorylation due to serum. This should get rid of background signals. It is also critical that incubation with the neurotrophic factors does not deviate from the optimal 5 min (see *Protocol 7*, step 4) which is the peak of phosphorylation.

Protocol 7. Phosphorylation assay on Trk-expressing cell lines

Equipment and reagents

- Equipment (see *Protocol 3*)
- Trk over-expressing cell lines (Regeneron Pharmaceuticals, Inc.)
- 10 cm tissue culture dishes (Falcon)
- Complete DMEM (see *Protocol 3*)
- Serum-free DMEM (see *Protocol 3*)
- Lysis buffer (see *Protocol 6*)
- 10 × TBST buffer: 0.1 M Tris pH 8.0, 1.5 M NaCl, 2% Tween
- Pan-Trk antibody (see *Protocol 6*)
- Anti-mouse IgG-POD (peroxidase) (Boehringer Mannheim GmbH)
- Anti-phosphotyrosine monoclonal IgG2bk (UBI)
- Protein A–Sepharose beads (see *Protocol 6*)
- 2 × SDS sample buffer (see *Protocol 6*)
- ECL kit (chemiluminiscence detection system) (Amersham)

Method

1. Grow the appropriate cell line (TrkA, B, or C) of interest almost to confluence (approx. 1×10^6 cells) in 10 cm tissue culture dishes with complete medium.
2. Prepare the desired dilutions (usually 1:3 or 1:5 with serum-free medium) of the conditioned medium, from transiently transfected cells (see *Protocol 3*A, step 9) and equilibrate at 37 °C in a water-bath. Dilutions should be made to a volume of 5 ml, sufficient to cover a 10 cm tissue culture dish.
3. Replace the complete medium with serum-free medium and incubate the cells for 30 min at 37 °C.
4. Replace the serum-free medium with the different dilutions (step 2) and incubate the plates at 37 °C for 5 min.[a]
5. Immediately remove the medium, add 1 ml of lysis buffer to each plate, then let the plates stand for 10 min at 4 °C.
6. Take up the lysates into 1.5 ml microcentrifuge tubes and centrifuge (14 000 r.p.m. for 1 min) to remove debris.
7. Transfer the supernatant into new microcentrifuge tubes, add 1 μl of pan-Trk antibody, and incubate for 4 h at 4 °C.
8. Add protein A–Sepharose beads (approx. 70 μl to each tube) and agitate gently for 30–60 min at 4 °C.
9. Allow the beads to settle and aspirate the supernatant. Wash the beads six times with lysis buffer.

10. Remove the final wash solution, add 50 μl of SDS sample buffer/ 2 mM dithiothreitol (DTT), and boil the samples for 5 min.
11. Run a 10% denaturing SDS gel and blot proteins to a nitrocellulose filter.[b]
12. Incubate the filters with 5% BSA/TBST and leave shaking gently at 4°C for 30 min.
13. Replace the BSA solution with anti-phosphotyrosine (diluted 5000 ×) in TBST and incubate(agitate gently) for 2 h at room temperature.
14. Wash the filters with two quick rinses and then wash four times 10 min with TBST.
15. Incubate with anti-mouse IgG-POD (diluted 10 000 x) in TBST shaking gently at 4°C for 45–60 min.
16. Wash as in step 14.
17. Visualize immunoreactive proteins using the ECL kit.[c]

[a] It is critical to incubate for 5 min as this is the peak time for phosphorylation.
[b] Refer to ref. 25. Don't forget to use rainbow protein marker (Amersham).
[c] Trk receptors run at about 140–145 kDa.

7. Biological assays

A final measure of the effects of the introduced mutations should be obtained by evaluating the biological activities of the molecules generated. As these may or may not parallel the behaviour of the mutants in receptor binding and activation, important and unexpected insights into functional consequences of the changes introduced may be obtained. Many neurotrophic factors display cell growth promoting activities when assayed on heterologous systems, as for example, fibroblast or haematopoietic cell lines transfected with Trk receptors. This mitogenic activity can be used to quantify downstream signalling from an activated receptor, and is easily adapted to a 96-well plate format, allowing simultaneous screening of many molecules and conditions. When performed in the absence of serum, an activated Trk receptor, as many other receptor tyrosine kinases, can provide for a survival signal to serum-deprived fibroblasts. Below is a detailed protocol of survival/growth assays in 3T3 fibroblasts stably transfected with Trk receptors.

Protocol 8. Survival/growth assay in 3T3 fibroblasts

Equipment and reagents

- TrkA over-expressing fibroblasts (Regeneron Pharmaceuticals, Inc.)[a]
- Serum-free DMEM (see *Protocol 3*)
- 96-well tissue culture plates (Falcon)
- Kit of reagents for detection of cell proliferation ('Abacus'; Clontech)

Protocol 8. *Continued*

Method

1. Seed TrkA-expressing MG-3T3 cells in a 96-well plate at 5×10^4 cells/well in serum-free DMEM.
2. Prepare factor dilutions, usually 12 or 24 dilutions in twofold steps starting at 1 μg/ml in serum-free DMEM.
3. Add factor dilutions to cells in quadruplicates—this will use up one whole (24 dilutions) or half (12 dilutions) 96-well plate. Leave four wells in serum-free DMEM.
4. Incubate at 37 °C in CO_2 incubator for three to four days (up to one week is necessary if standard NIH3T3 TrkA-expressing cells are used instead of MG-3T3 cells).
5. Assay cell proliferation/survival following 'Abacus' kit instructions.

[a] Adapted from ref. 40.

Because the fibroblast assay described above is an 'artificial system', we want to also assess the biological activities of our mutants in primary cultures of responsive cells. Neurotrophic factors are particularly interesting for bioassaying given their variety of actions. Thus, for example, a particular mutant may still be competent in stimulating some aspect of neuronal differentiation (i.e. induction of a specific biochemical marker or morphological phenotype), but not in promoting neurone survival. Neurite outgrowth assays measure the ability of a neurotrophic factor to elicit extension of neurites, often from an explanted piece of nervous tissue or peripheral ganglion. A classical assay of nerve growth factor activity is the neurite outgrowth assay in explanted embryonic sympathetic ganglia. This is a semi-quantitative assay (i.e. activity is estimated from the size of the neurite halo elicited by the trophic molecule). It is advisable to set-up a dose–response standard curve in parallel using known amounts of a purified neurotrophic factor, typically NGF for sympathetic neurones. The neurite outgrowth assay has been adapted to other peripheral ganglia, such as nodose, dorsal root, cilliary, trigeminal, and superior cervical, which allows the activities of other neurotrophic factors to be assessed. Because chick embryos come conveniently packed in individual eggs (!), they are a convenient source of primary neurones, so many of these assays are typically performed with ganglia explanted from chicken embryos. Techniques for the dissection of peripheral ganglia from chick embryos have been discussed elsewhere (41). A detailed protocol for the neurite outgrowth assay performed in our laboratory is described below.

Protocol 9. Neurite outgrowth assay in explanted chick peripheral ganglia

Equipment and reagents

- Egg incubator
- Surgical instruments and stereomicroscope
- Inverted microscope
- Fertilized chicken eggs
- 24-well tissue culture plates (Falcon)
- DMEM complete medium (see *Protocol 3*)
- 7.5% $NaHCO_3$ solution
- Pen–Strep solution: final concentration 0.1 mg/ml streptomycin, 100 U/ml penicillin
- 0.15 M NaOH solution
- Concentrated HCl solution
- Rat tail collagen equilibrated by dialysis in 1:10 MEM–HCl pH 4.0 (see below)

A. *Preparation of rat tail collagen*[a]

1. Collect tails from adult rats and store (frozen) in plastic bags.
2. Thaw four tails for about 15 min in 95% ethanol.
3. Fracture each tail starting from the tip using sterile forceps.
4. Draw out the tendons attached to the distal piece of the tail, cut them free, and collect them in sterile distilled water.
5. Transfer the tendons to a flask containing 200 ml 0.5 M acetic acid. Let them dissolve for two to three days at 4°C with occasional shaking. The solution should be viscous, but not jelly-like. If required, the solution may be diluted with more 0.5 M acetic acid.
6. Clarify the solution by centrifugation at 4°C (1600 *g* for 1 h).
7. Transfer the supernatant into autoclaved dialysis bags.
8. Dialyse at 4°C overnight against 10 vol. of 0.1 × MEM (20 ml 10 × MEM, 4.5 ml 7.5% $NaHCO_3$, 2 ml 200 mM L-glutamine, and sterile water up to 2000 ml).
9. Repeat the overnight dialysis once as above.
10. Dialyse overnight two additional times against 0.1 × MEM adjusted to pH 4.0 with HCl (3–4 ml concentrated HCl per 2000 ml 0.1 × MEM).
11. Clarify the dialysate by centrifugation and aliquot into labelled bottles.
12. Titrate the amount of 0.15 M NaOH necessary to neutralize each batch of collagen stock by adding 5, 10, 20, and 50 µl of 0.15 M NaOH each to 1 ml aliquots of pH 4.0 collagen stock solution. Look for the minimal amount of 0.15 M NaOH that produces a shift in colour of the phenol red indicator in the MEM medium.

B. *Neurite outgrowth assay*

1. Dissect the required peripheral ganglia from chick embryos at the appropriate stages, and place them in 1 × MEM.[a]

Protocol 9. *Continued*

2. Prepare concentrated medium mix (sufficient for 4 ml collagen) by pipetting 455 μl 10 × MEM, 112 μl 7.5% $NaHCO_3$, 50 μl 200 mM L-glutamine, 55 μl serum, and four times the required volume of 0.15 M NaOH taken to neutralize 1 ml of the collagen stock solution. Adjust the final volume of this mixture to 1 ml with sterile water and keep on ice.
3. For each millilitre of collagen gel, add 0.2 ml of the ice-cold concentrated mixture to 0.8 ml ice-cold collagen stock solution, then leave it on ice until subsequent use.
4. With a Pasteur pipette, take one ganglion in a small drop of medium and carefully place it in the centre of an empty well of a 24-well tissue culture plate with as little medium as possible. Six ganglia at a time can conveniently be arranged in this way.
5. Carefully add 200 μl of the gel mix to each well, on top of each ganglion and incubate at 37 °C in a CO_2 incubator for 5 min until the gel polymerizes.
6. Add an additional 200 μl of the gel mix to each well and allow to polymerize.
7. Finally, add serial dilutions of the factor to be tested (as duplicates or triplicates) in 400 μl 1 × MEM (supplemented with 1% serum, 2 mM L-glutamine, and Pen–Strep). Note that this final step will dilute the concentration of factor twofold.
8. Incubate 24–48 h at 37 °C in a CO_2 incubator and score the halo of neurite outgrowth by visual inspection in an inverted microscope under phase-contrast or dark-field illumination. A good way to interpret the results is to start by setting a standard curve with known amounts of NGF (from 10 pg/ml to 100 ng/ml—this latter concentration will show saturation of the response, a normal phenomenon in this type of assay).

[a] See ref. 41.

PC12 is a pheochromocytoma cell line that, in response to NGF, stops dividing and differentiates into a sympathetic neurone-like cell. NGF treatment elicits neurite outgrowth after 48 to 72 hours in culture. This outgrowth can be quantified (for example, as percentage of cells bearing neurites of a certain length) and used as an assay of neurotrophic activity. PC12 cultures are simple, and sometimes a good substitute for primary cultures. It is however good to keep in mind that PC12 cells are after all transformed cells, and as such, could carry abnormalities in precisely some of the processes that we may want to study. A protocol for the neurite outgrowth assay using PC12 cells is described below.

Protocol 10. Neurite outgrowth assay in PC12 cells

Equipment and reagents

- Inverted microscope
- 6-well tissue culture plates (Falcon)
- PC12 cells (ATCC)[a]
- Complete DMEM (see *Protocol 3*), supplemented with 10% horse serum and 5% fetal calf serum (Gibco)

Method

1. Seed 10 000 PC12 cells/well in 1 ml of complete DMEM (use 6-well tissue culture plates).
2. Add serial dilutions of the factor to be tested (in triplicates or quadruplicates) in 1 ml complete DMEM. Note that this final step will dilute the concentration of factor twofold.
3. Incubate at 37 °C in CO_2 incubator for three to four days.
4. Score outgrowth in inverted microscope under phase-contrast illumination. Count the number of cells in a field (10 × magnification) and the number of cells bearing neurites longer than two cell diameters. Repeat the mesurement in five different fields of each well. Average measurements from each well, and express results as percentage of neurite-bearing cells.

[a] See ref. 42.

An important function of neurotrophic factors is to promote the survival of neurones. Neurone survival assays may be established using the same populations of peripheral neurones used for the neurite outgrowth assay, and may be adapted to virtually any type of neurone. Because neurones from older developmental stages are more difficult to maintain in culture, survival assays are usually performed with embryonic neurones, preferably at the peak of the naturally occurring cell death period. In addition to the peripheral populations mentioned above, several central populations are also used in survival assays. Because of their vulnerability in several neurodegenerative diseases, many laboratories have developed conditions to culture basal forebrain cholinergic neurones (affected in Alzheimer's disease), ventral midbrain dopaminergic neurones (which degenerate in Parkinson's disease), and spinal cord motorneurones (compromised in amyotrophic lateral sclerosis or ALS). Below, we describe a protocol to perform survival assays for neurotrophins using cultures of dissociated dorsal root ganglia, which contain different classes of sensory neurones responsive to different members of the neurotrophin family.

Protocol 11. Neuronal survival assays

Equipment and reagents

- Chicken egg incubator
- Surgical instruments and stereomicroscope
- Inverted microscope
- Fertilized chicken eggs
- 24-well tissue culture plates (Falcon)
- 5 mg/ml solution of poly-ornithine (Sigma) in 10 mM borate buffer pH 8.3 (sterilize by filtration)
- MEM medium, supplemented with 2 mM L-glutamine, 1% fetal calf serum, 1:100 dilution Pen–Strep mix (see *Protocol 9*; Gibco), and 0.1% bovine serum albumin
- Trypsin solution (Gibco)
- 1 mg/ml sterile solution of laminin (Sigma)
- 1% Trypan blue solution in PBS

Method

1. Coat 24-well plates with 0.5 ml/well of 0.5 mg/ml poly-ornithine (in 10 mM borate buffer) and leave overnight at room temperature.
2. Wash wells twice with sterile water and coat each well with 0.4 ml of 10 μg/ml laminin in PBS. Incubate for at least 30 min at 37 °C.
3. Dissect the required peripheral ganglia from chick embryos at the appropriate stages, and place them in medium.[a]
4. Wash the ganglia (typically E9) twice with PBS, add trypsin up to 0.1%, then incubate for 11 min at 37 °C.[b]
5. Stop trypsinization by adding an equal volume of fetal calf serum. Wash ganglia several times with serum, and then several times with serum-free MEM medium containing supplements indicated above.
6. Dissociate the explants by mechanical trituration with a fire-polished Pasteur pipette. Best results are obtained by starting with ten strokes, then decant the supernatant with cells into a fresh tube. Add additional medium to the explants and repeat for an additional round of trituration and decantation, pooling the fractions containing free cells. Assess cell viability and number by Trypan blue exclusion in a cell counting chamber.
7. Wash laminin-coated wells twice with PBS, once with serum-free MEM medium, and seed cells at 2000–10 000 cells/well in 0.4 ml serum-free medium.[c]
8. Add serial dilutions of the factor to be tested (in triplicates or quadruplicates) in 0.4 ml serum-free medium. Note that this final step will dilute the concentration twofold.
9. Incubate at 37 °C in CO_2 incubator for two or three days.
10. Score survival using an inverted microscope under phase-contrast

illumination by counting the number of phase-bright neurite-bearing neurones in a defined area of the well.[d]

[a] See ref. 41.
[b] Ganglia from older chick embryos may require longer incubation times (e.g. up to 30 min for E18 DRGs).
[c] DRG contain substantial numbers of non-neuronal cells which may be removed by pre-plating in uncoated tissue culture plastic dishes for 1 h at 37 °C. Neurones can be recovered by gently pipetting out the supernatant.
[d] Alternatively, the 'Abacus' kit (see above) from Clontech can be adapted to these neuronal survival assays using 96-well plates with 5000–10 000 neurones per well.

Recently, we have developed in our laboratory a novel assay to measure survival-promoting activities on subpopulations of peripheral neurones expressing specific types of Trk mRNAs. In this assay, the levels of different Trk mRNAs in a heterogeneous population of neurones, such as DRG sensory neurones growing in culture with different neurotrophins, are assessed by RNase protection analysis (RPA). The sensitivity of this assay allows small size explant or dissociated cultures to be used, often four to five DRG or 10^5 neurones are enough for subsequent quantitative and statistical analysis. Our protocol for RPA follows the descriptions provided by the manufacturer of the RPA kits (Ambion).

8. Therapeutic potential of engineered neurotrophins

The ability of neurotrophins to prevent neuronal degeneration, and to promote the survival of peripheral and central neurones, suggests that they may be valuable therapeutic agents for treatment of peripheral nerve injury and brain diseases. In fact, several brain nuclei have been shown to be affected in various neurodegenerative disorders. Such neuronal death can be prevented by neurotrophic factors (43). Taking into account the distribution and developmental regulation of neurotrophins in the context of various disease states, it is obvious that these molecules would play crucial roles in preventive as well as interventive therapies. A summary list of various diorders that may potentially benefit from neurotrophins or other growth factors is found elsewhere (43, 44).

In the central nervous sytem (CNS), NT-3 has been shown to support cortical and hippocampal neurones *in vitro*; and to promote the survival of nigral dopaminergic and cortical neurones. NT-3 is also able to rescue *locus coeruleus* noradrenergic neurones *in vivo*, which is an activity unique to NT-3 among neurotrophins (43). While the above mentioned neurotrophin is clearly of substantial pharmacological potential, it is NGF that has made headway into clinical trials for potential application in the treatment of Alzheimer's disease (AD) which is partially characterized by the loss of cholinergic neurones. NGF has for some time been known to be crucial for the normal development of cholinergic systems and thus possibly prevent degeneration, and its associated behavioural deficits in the adult animals (43). Furthermore, NGF could counter peripheral neuropathies resulting from chemotherapy (e.g. with cisplatin) and disorders like diabetes. NT-4 and BDNF seem most crucial for neurones projecting to the periphery where they have been shown to affect survival of motoneurones both *in vitro* and *in vivo*. In fact NT-4 has been shown to have the unique role of inducing sprouting of motoneurones as a consequence of increased muscle activity (45).

Because of the involvement of several subpopulations of neurones in several disorders, a combination of neurotrophic factors appears to be the most effective therapeutic strategy to implement. This is even more clearly seen in the peripheral nervous sytem wherein different neuronal populations may extend fibres through a single nerve. Administration of such a cocktail may be difficult given the different physical properties of the molecules such as diffusibility and stability. Such concerns have spurred us to explore the possibility of creating multifunctional neurotrophins and this has given rise to molecules like Pan-neurotrophin 1 (PNT-1).

By employing the methods described in this chapter, we have determined the structures which are critical for the biological activity of the different neurotrophins and consequently used these data for the construction of PNT-1. True to its design, it combines the functions of NT-3, BDNF, and NGF making it potentially useful in the treatment of peripheral nerve injuries (46). Other chimeric molecules that show therapeutic promise are the NGF/BDNF chimera that has been shown to possess the synergistic activities of its parental molecules and promote the survival of basal forebrain cholinergic neurones (47).

The creation of chimeric neurotrophins is, however, not limited to employing site-directed mutagenesis. As described in the beginning, these molecules exist as non-covalently linked homodimers. Through chemical manipulation of denaturation/renaturation, it has been shown that one can generate heterodimers with some combinations more stable than others. Due to the symmetric nature of neurotrophin dimers, the heterodimers have been shown to effectively mimic the combined activities of their constituents (48).

9. Conclusion

Neurotrophins are essential for the proper development and maintenance of the central and peripheral nervous system. To study the molecular details underlying their biological activity not only allows us to understand basic mechanisms of cellular physiology and signal transduction, but it also opens the possibility of preventing neuronal degeneration. The ability to manipulate the physicochemical properties of molecules has revolutionized not only our understanding of nature but also our possibilities of implementing long overdue treatments to some of the most debilitating diseases that plague humankind.

Acknowledgements

The authors would like to express their gratitude to colleagues at the Laboratory of Molecular Neurobiology, Karolinska Institute, Stockholm, Sweden. In particular to Ernest Arenas, Kaia Palm, Isabelle Neveu, Sofie Nilsson, Peter Lönnerberg, Mikael Ryden, and Leodevico L. Ilag (Morphosys GmbH) for critical reading of the text and for helpful comments.

References

1. Thoenen, H. and Barde, Y. A. (1980). *Physiol. Rev.*, **60**, 1284.
2. Ibáñez, C. F., Ebendal, T., Barbany, G., Murray-Rust, J., Blundell, T. L., and Persson, H. (1992). *Cell*, **69**, 329.
3. Jing, S. Q. and Tapley, P. (1992). *Neuron*, **9**, 1067.
4. Carter, B. D., Kaltschmidt, C., Kaltschmidt, B., Offenhauser, N., Bohm-Matthaei, R., Baeuerle, P. A., *et al.* (1996). *Science*, **272**, 542.
5. Dobrowsky, R. T., Werner, M. H., Castellino, A. M., Chao, M. V., and Hannun, Y. A. (1994). *Science*, **265**, 1516.
6. McDonald, N., Lapatto, R., Murray-Rust, J., Gunning, J., Wlodawer, A., and Blundell, T. L. (1991). *Nature*, **354**, 411.
7. McDonald, N. and Murray-Rust, J. (1995). In *Life and in the nervous system: role of neurotrophic factors and their receptors* (ed. C. F. Ibáñez, T. Hokfelt, L. Olson, K. Fuxe, H. Jornvall, and D. Ottoson), p. 3. Elsevier Science Ltd., Oxford.
8. McDonald, N. Q. and Hendrickson, W. A. (1993). *Cell*, **73**, 421.
9. Murray-Rust, J., McDonald, N. Q., Blundell, T. L., Hosang, M., Oefner, C., Winkler, F., *et al.* (1993). *Structure*, **1**, 153.
10. Robinson, R. C., Radziejewski, C., Stuart, D. I., and Jones, E. Y. (1995). *Biochemistry*, **34**, 4139.
11. Wells, J. A. (1991). In *Methods in enzymology* (ed. J. J. Langone), Vol. 202, p. 390. Academic Press, London.
12. Ibáñez, C. F., Ilag, L. L., Murray-Rust, J., and Persson, H. (1993). *EMBO J.*, **12**, 2281.
13. Ilag, L. L., Lonnerberg, P., Persson, H., and Ibanez, C. F. (1994). *J. Biol. Chem.*, **269**, 19941.
14. Ibáñez, C. F. (1995). *TIBTECH*, **13**, 217.
15. Mullis, K. B. (1993). In *Les Prix Nobel* (ed. T. Frangsmyr), p. 107. Almqvist and Wiksell International, Stockholm.
16. Smith, M. (1993). In *Les Prix Nobel* (ed. T. Frangsmyr), p. 123. Almqvist and Wiksell International, Stockholm.
17. Kunkel, T. A. (1985). *Proc. Natl. Acad. Sci. USA*, **82**, 488.
18. Virnekaes, B., Ge, L., Plueckthun, A., Schneider, K. C., Wellenhofer, G., and Moroney, S. E. (1994). *Nucleic Acids Res.*, **22**, 5600.

19. Allen, J. B., Walberg, M. W., Edwards, M. C., and Elledge, S. J. (1995). *Trends Biochem. Sci.*, **20**, 511.
20. Colas, P., Cohen, B., Jessen, T., Grishna, I., McCoy, J., and Brent, R. (1996). *Nature*, **380**, 548.
21. Krebber, C., Spada, S., Desplanq, D., and Plueckthun, A. (1995). *FEBS Lett.*, **377**, 227.
22. Offord, R. E. (1992). In *Protein engineering: a practical approach* (ed. A. A. Rees, M. J. E. Sternberg, and R. Wetzel), p. 231. Oxford University Press, Oxford.
23. Frazier, W. A., Hogue-Angeletti, R. A., Sherman, R., and Bradshaw, R. A. (1973). *Biochemistry*, **12**, 3281.
24. Bradshaw, R. A., Murray-Rust, J., Ibanez, C. F., McDonald, N. Q., Lapatto, R., and Blundell, T. L. (1994). *Protein Sci.*, **3**, 1901.
25. Sambrook, I., Fritsch, E. F., and Maniatis, T. (ed.) (1989). *Molecular cloning: a laboratory manual*, 2nd edn. Cold Spring Harbor Laboratory Press, NY.
26. Jones, S. (1995). *Stratagene Catalog* (ed. B. Johnson-Brown). Stratagene Cloning Systems, California, USA.
27. Higuchi, R., Krummel, B., and Saiki, R. K. (1988). *Nucleic Acids Res.*, **16**, 7351.
28. Ho, S. N., Hunt, H. D., Horton, R. M., Pullen, J. K., and Pease, L. R. (1989). *Gene*, **77**, 51.
29. Ali, S. A. and Steinkasserer, A. (1995). *Biotechniques*, **18**, 750.
30. Gluzman, Y. (1981). *Cell*, **23**, 175.
31. O'Reilli, D. R., Miller, L. K., and Leukow, V. A. (1994). *Baculovirus expression vectors and laboratory manual.* Clontech Lab. Inc., California, USA.
32. Gentz, R. (1984). PhD Thesis, University of Heidelberg, BRD.
33. Nilsson, B., Forsberg, G., Moks, T., Hartmanis, M., and Uhlen, M. (1992). *Curr. Opin. Struct. Biol.*, **2**, 569.
34. Fontana, A. and Gross, E. (1986). In *Practical protein chemistry—a handbook* (ed. A. Darbre), p. 67. John Wiley and Sons Ltd., NY.
35. Lindner, P., Guth, B., Wuelfing, C., Krebber, C., Steipe, B., Mueller, F., *et al.* (1992). *Methods: a companion to methods in enzymology*, **4**, 41.
36. Dicou, E. (1992). *Neurochem. J.*, **20**, 129.
37. Simmons, L. C. and Yansura, D. G. (1996). *Nat. Biotech.*, **14**, 629.
38. McGonigle, P. and Molinoff, P. B. (1989). In *Basic neurochemistry*, 4th edn (ed. G. Siegel, B. Agranoff, R. W. Albers, and P. Molinoff), p. 183. Raven Press, New York.
39. *Pierce Catalog and Handbook.* (1994). p. 90. Pierce Chemical Company, USA.
40. Ip, N. Y., Stitt, T. N., Tapley, P., Klein, R., Glass, D. J., Fandl, J., *et al.* (1993). *Neuron*, **10**, 137.
41. Ebendal, T. (1989). In *Nerve growth factors* (ed. R. A. Rush), p. 81. John Wiley and Sons, New York.
42. Green, L. A. and Rein, G. (1977). *Nature*, **268**, 349.
43. Rocamora, N. and Arenas, E. (1996). In *International encylopedia of pharmacology and therapeutics: chemical factors in neural growth, degeneration and repair* (ed. C. Bell), p. 219. Elsevier Science B.V., Oxford.
44. Hefti, F., Gao, W., Nikolics, K., Rosenthal, A., Shelton, D., Phillips, H. S., *et al.* (1995). In *Life and in the nervous system: role of neurotrophic factors and their receptors* (ed. C. F. Ibáñez, T. Hokfelt, L. Olson, K. Fuxe, H. Jornvall, and D. Ottoson), p. 379. Elsevier Science Ltd., Oxford.

45. Funakoshi, H., Belluardo, N., Arenas, E., Yamamoto, Y., Casabona, A., Persson, H., *et al.* (1995). *Science*, **268**, 1495.
46. Ilag, L. L., Curtis, R., Glass, D., Funakoshi, H., Tobkes, N. J., Ryan, T. E., *et al.* (1995). *Proc. Natl. Acad. Sci. USA*, **92**, 607.
47. Friedman, W., Black, I., Persson, H., and Ibáñez, C. F. (1995). *Eur. J. Neurosci.*, **7**, 656.
48. Treanor, J. J., Schmelzer, C., Knusel, B., Wislow, J. W., Shelton, D. L., Hefti, F., *et al.* (1995). *J. Biol. Chem.*, **270**, 23104.

3

Designing new agonists/antagonists of growth factor receptors—the rational design of a superantagonist of the IL-6 receptor

G. CILIBERTO, A. LAHM, G. PAONESSA, R. SAVINO, and C. TONIATTI

1. Introduction

Growth factors and cytokines assemble multisubunit receptor complexes stabilized by a variety of protein:protein interactions (1), each of which plays a distinctive role in signalling and activating biological responses. Identifying functional epitopes on the cytokine surface thus opens up possibilities for engineering variants with predefined properties for potential use as therapeutic agents. We have applied this concept to the rational design of receptor antagonists of human interleukin-6 (IL-6).

IL-6 is one member of a large group of secreted proteins, the helical cytokines, which have been predicted to share a common fold, a tightly packed bundle of four α-helices (2). The 3D structure of several helical cytokines has been determined, thus allowing prediction of the structure of those members, including IL-6, for which 3D data are not yet available (3). The transmembrane receptors for helical cytokines also have a common fold in the domain responsible for interaction with their ligand(s), the cytokine binding domain (CBD). This is a 200 amino acid region predicted to fold into two consecutive 'barrel-like' subdomains of seven β-strands, similar to immunoglobulin constant domains (4).

The first published X-ray structure of a helical cytokine complexed with its receptor CBDs was that of human growth hormone (GH) which binds a homodimer of the GH receptor (GHbp) to give a complex which we will call the GH-trimer. This information provided an important insight into the mode of interaction established by this class of molecules (5); GH acts as a bridging ligand between two GHbps, and two interacting surfaces have been identified on GH which are sites for binding two receptor chains (site1 and site 2), formed by non-overlapping epitopes on opposite sides of the four helix bundle.

The structural and functional information available from the study of GH and GHbp can be used for modelling receptor complexes for other helical cytokines (see also discussion of modelling Chapter 5, Section 1). These models predict which cytokine residues constitute receptor binding epitopes, and can be used in mutagenesis experiments designed to generate cytokine variants with predefined properties.

The active IL-6 receptor is assembled via sequential interaction of the cytokine with two transmembrane receptors, IL-6Rα and gp130 (6), each of which carries one CBD in its extracytoplasmic region (7, 8). IL-6 binding to IL-6Rα alone is not sufficient to activate intracellular responses, but requires further interaction with gp130 (6). The GH-trimer has been used as the basis for a molecular model to predict the interaction of IL-6 with the CBDs of IL-6Rα and gp130, and to identify also in IL-6 sites 1 and 2 specialized in binding IL-6Rα and gp130, respectively. This model postulates the formation of an IL-6-trimer topologically similar to that of GH, the only difference being the formation of a heterodimeric rather than a homodimeric receptor (9–11).

The signal transducer gp130 has been shown to undergo homodimerization on the cell surface and this event is crucial for the activation of intracytoplasmic signalling pathways (12). In order to study the molecular basis of IL-6/IL-6Rα-driven gp130 dimerization, we developed *in vitro* binding assays between IL-6, IL-6Rα, and gp130 to determine the stoichiometry of receptor complexes. From these studies we discovered that the functional IL-6 receptor complex is a hexamer composed of two IL-6, two IL-6Rα, and two gp130 molecules (13). The hypothetical model of the hexamer, supported by the results of several experiments, is that two GH-like IL-6-trimers each constituted by one IL-6, one IL-6Rα, and one gp130, flank each other in inverted orientations. *Figure 1* shows a schematic view of the IL-6 hexamer compared with the GH-trimer.

Besides the initially postulated site 2 we identified in IL-6 an additional gp130 binding site (site 3) specialized in the interaction of the cytokine of one IL-6-trimer with the gp130 chain of the opposing IL-6-trimer. Both gp130 binding sites are involved in gp130 dimerization and signalling activation (13, 14). The introduction of disruptive substitutions in either site 2 or 3 gives rise to IL-6 variants whose binding to IL-6Rα remains intact, but whose ability to dimerize gp130 is selectively altered. As expected, these variants behave as receptor antagonists (13, 14).

Finally, we applied phage display technology to select amino acid substitutions in site 1 which specifically increase affinity for IL-6Rα (15, 16). When these superbinder substitutions are combined with the antagonistic mutations, 'superantagonists' are obtained, which block IL-6 biological activity at low doses (17). In this chapter we describe the various techniques developed in our laboratory to study receptor assembly by helical cytokines,

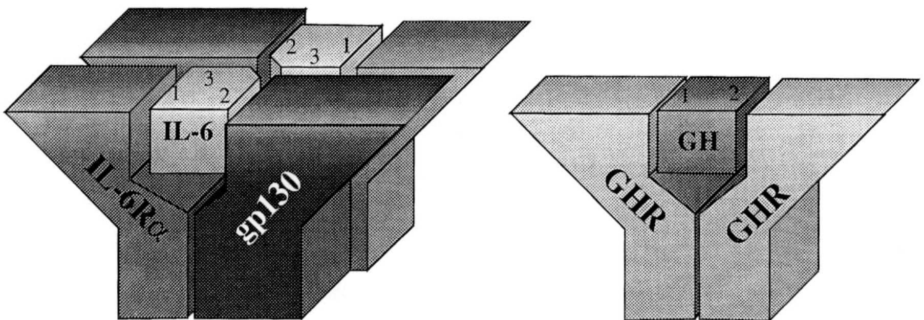

Figure 1. Comparison of the trimeric GH and the hexameric IL-6 receptor complexes. For receptor molecules only the cytokine binding domains (CBDs) are represented. The hexamer is stabilized by several protein:protein interactions. Besides those required for the formation of the GH-like IL-6-trimer (sites 1 and 2), the others arise as a result of the juxtaposition of the two trimers. Among these we identified a third receptor binding site (site 3) on IL-6, specialized in binding of the cytokine of one IL-6-trimer with the gp130 chain of the opposing IL-6-trimer.

and outline our strategy for the generation of potent cytokine receptor antagonists.

2. Molecular modelling of a trimeric IL-6/IL-6Rα/gp130 complex

2.1 Model construction

The GH-trimer X-ray structure provides a template from which to build the model for the IL-6 receptor complex. Whereas conserved sequence motifs within class I cytokine receptors clearly indicate homology between the CBDs of IL-6Rα, gp130, and GHbp, the GH ligand itself does not show sufficient sequence homology to IL-6. A model for the latter can however be derived from the X-ray structure of granulocyte colony stimulating factor (G-CSF) (18), a cytokine with more significant sequence homology to IL-6. Steps in the construction of the 3D models are schematically represented in *Figure 2*.

Generation of the molecular model for the IL-6 receptor complex is described in detail elsewhere (9–11) and makes use of standard sequence analysis and molecular modelling techniques, an overview of which can be found in recent reviews (19, 20).

2.2 Identification of the putative IL-6 site 1 and site 2 epitopes

The IL-6 residues predicted to form site 1 (interaction with IL-6Rα) or site 2 (gp130), and their location in the 3D IL-6 model are shown in *Figure 3*. Site 1 residues are: A56, L57, E59, N60, K66, A68, E69, K70, D71, F74, Q75, S76 (AB-loop), and K171, E172, Q175, S176, R178, R182, Q183 (helix D). Site 2

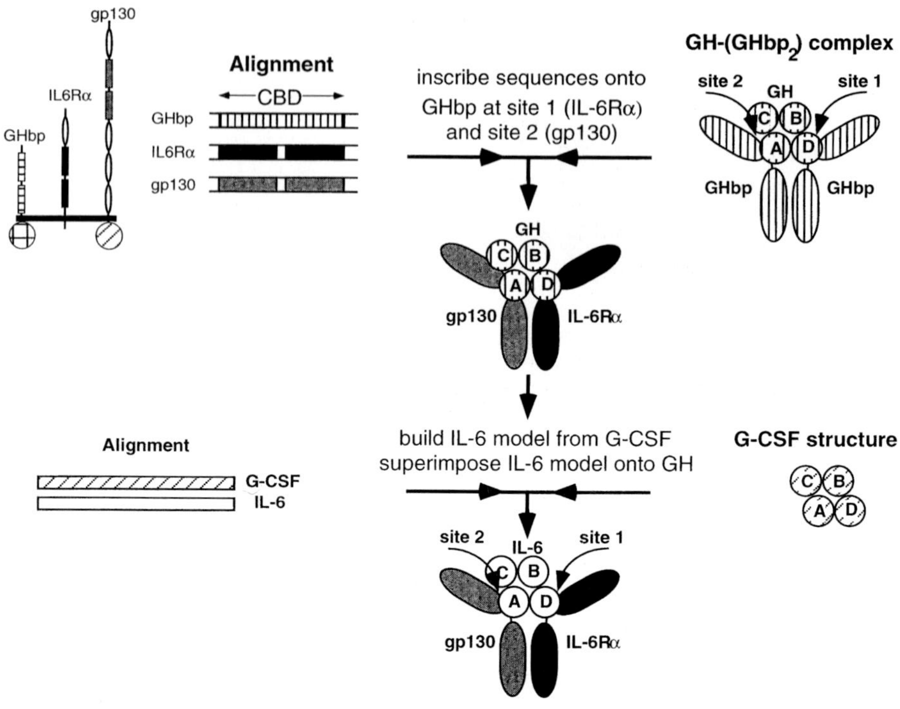

Figure 2. Construction of a computer model for the IL-6 receptor complex. Initially, the amino acid sequences of IL-6 and the CBDs of IL-6Rα and gp130 are aligned against the sequence of the corresponding template molecules, G-CSF and GHbp, respectively. The model for an IL-6Rα/gp130 heterodimer is then obtained by inscribing their sequences onto the GHbp structural template according to the alignment. The orientation and relative position between the CBDs of IL-6Rα and gp130 are maintained as in the GHbp dimer. The IL-6 model is then generated by inscribing its sequence onto the G-CSF structural template according to the alignment. In the last step the model for a trimeric IL-6/IL-6Rα/gp130 assembly is completed by superimposing the IL-6 model onto GH within the context of the GH-trimer. Since the position and orientation for the IL-6Rα and gp130 CBD models is kept identical to the GHbp dimer, the result is a model for the trimer. In a final step, loop and side chain conformations are optimized at the sites of interaction between all three molecules to reduce steric overlaps.

residues: K27, Q28, Y31, G35 (helix A), and S118, V121, F125 (helix C) (see Appendix 2 for single letter amino acid and code).

3. IL-6 biological assay

The rational design of IL-6 superantagonists involves generating a large collection of mutant proteins whose biological activity needs to be carefully

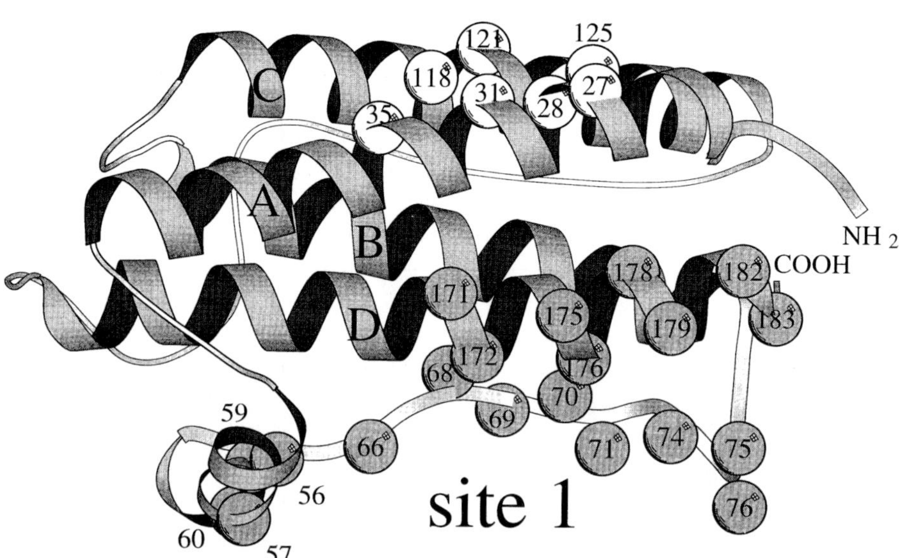

Figure 3. Schematic representation of the IL-6 model. Residues predicted to interact with (or be close to) either IL-6Rα (site 1, light grey spheres) or gp130 (site 2, white spheres) are indicated. Both sites are composite in that residues from more than one secondary structure element are involved, the AB-loop and helix D for site 1, and helices A and C for site 2, respectively. The figure was produced using the *MOLSCRIPT* program by E. Kraulis (21).

evaluated and compared with that of the wild-type (wt) cytokine. Hence, the need for a reliable and large scale bioassay on human cells. IL-6 is known to stimulate gene expression of acute phase response proteins in human hepatoma cells (22). IL-6 activity on hepatoma cells can be measured using:

(a) Immunoelectrophoresis to detect release in the culture medium of acute phase proteins (23).
(b) Northern blot analysis to detect increase of steady state hepatocyte-specific mRNA levels (24).
(c) Transient assays of gene expression to detect enhanced transcription of a reporter gene such as CAT under the control of an IL-6-inducible promoter (25).

However, none of these assays are optimal for our purpose because they are time-consuming, not very sensitive, or require extensive sample manipulation. We have therefore developed a new bioassay for IL-6 in human hepatoma cells which improves sensitivity and simplifies the manipulation and handling of samples (26). The assay is based on the transfection of a fusion gene formed from the IL-6-inducible promoter of the human C-reactive protein (CRP) gene and the coding region for a secreted form of alkaline phosphatase (SEAP). CRP–SEAP transfected human hepatoma Hep3B or HepG2 cells secrete SEAP after exposure to human but not to mouse IL-6 in a dose-dependent manner (mouse IL-6 does not bind human IL-6Rα, see ref. 27). SEAP activity can be directly measured in the culture medium of cells in multiwell plates with a fast and sensitive colorimetric assay. The major advantages of this SEAP assay are the ease of sample handling (up to 400 samples/day processed by a single operator) and the possibility of extending it to quantification of the activity of other cytokines (28).

Protocol 1. SEAP assay

Reagents

- DMEM: Dulbecco's modified Eagle medium supplemented with 2 mM L-glutamine (Gibco)
- Phosphate-buffered saline (PBS): dissolve 8 g NaCl, 0.2 g KCl, 1.44 g Na_2HPO_4, 0.44 g KH_2PO_4 in 800 ml of distilled H_2O, adjust the pH to 7.4 with HCl, add H_2O to 1 litre, sterilize by autoclaving
- Trypsin/EDTA solution: 0.5 g/litre trypsin (1:250), 0.2 g/litre EDTA, in 1 × PBS (Gibco)
- Hep3B cells and HepG2 cells (American Type Tissue Culture Collection)
- Hepes-buffered saline (HBS): 10 g/litre Hepes, 16 g/litre NaCl, adjust the pH to 7.1, filter sterilize through a 0.22 μm filter[a]
- Phosphate buffer: mix 1:1 (v/v) sterile solutions of 70 mM Na_2HPO_4 and 70 mM NaH_2PO_4
- 2 M $CaCl_2$ filter sterilized through a 0.22 μm filter
- 2 × SEAP buffer: 2 M diethanolamine pH 9.8, 1 mM $MgCl_2$, 20 mM L-homoarginine, stored at 4°C
- *p*-nitrophenylphosphate (Sigma)

A. *Cell plating (day 1)*

1. Plate 2 × 10^9 Hep3B or HepG2[b] cells in a 15 cm diameter culture dish (i.e. at a density of roughly 10^4 cells/cm^2) with 20 ml of DMEM containing 10% fetal calf serum.

B. *Transfection (day 2)*

1. Replace the DMEM medium with fresh DMEM containing 10% fetal calf serum and incubate cells at 37°C in 5% CO_2 for 4–8 h.
2. Prepare the calcium phosphate–DNA precipitate. For a 15 cm culture dish containing 20 ml of medium, set-up the following solutions: in tube A (15 ml, plastic, Falcon), dispense 500 μl of 2 × HBS together with 10 μl of 100 × phosphate buffer. In tube B (same as above), dispense 440 μl of H_2O containing 45 μg of DNA together with 60 μl of $CaCl_2$. Add dropwise the content of tube B to tube A while bubbling the solution in tube A to ensure uniform mixing.[c]
3. Let the mixture stand for 30 min at room temperature, then add dropwise the precipitate onto the cells.
4. Incubate the cells at 37°C in 5% CO_2 for 14–18 h.

C. *Cell splitting and incubation with cytokine (day 3)*

1. Wash the cells five times with PBS.
2. After removal of the last PBS wash, add 1 ml of trypsin/EDTA solution (for a 15 cm culture dish) and incubate at 37 °C (usually for 3–5 min) until the cells assume a round shape and detach from the dish (monitored using a microscope).
3. Recover cells in DMEM supplemented with 10% fetal calf serum, count cell numbers, dilute them in DMEM supplemented with 10% fetal calf serum to 5×10^4 cells/ml, and replate 1 ml aliquots in 24-well microtitre plates (corresponding to a density of 2.5×10^4 cells/cm^2).
4. Incubate at 37 °C in the presence of 5% CO_2 for 3–6 h, to let the cells attach to the culture dish.
5. Prepare serial dilutions of the test samples (IL-6 wild-type and/or mutant proteins) in DMEM supplemented with 10% fetal calf serum[d] and with 500 U/ml of IL-1β.[e]
6. Pre-warm at 37 °C and equilibrate with 5% CO_2 all the samples thus prepared.
7. Wash cells once with PBS and add 250 μl/well of each of the samples.
8. Incubate at 37 °C in 5% CO_2 for 60–64 h to let the SEAP enzyme accumulate in the cell culture medium.

D. *SEAP detection (day 6)*

1. Transfer the media samples to 1.5 ml microcentrifuge tubes.
2. Heat at 65 °C for 5 min, centrifuge for 3 min at top speed in a microcentrifuge, and recover the supernatant.
3. In a flat-bottom 96-well microtitre plate, dispense in each well 100 μl of 2 × SEAP buffer pre-warmed to 37 °C, plus 100 μl of each supernatant.
4. Start the reaction by adding 20 μl of 120 mM *p*-nitrophenylphosphate substrate dissolved in 1 × SEAP buffer.
5. Cover the microtitre plate with a plate sealer and incubate at 37 °C.
6. Monitor the reaction by reading the absorbance at 405 nm in an automatic plate reader[f] at various times and plot dose–response curves.

[a] Great care is needed in making up this buffer since pH is very critical.
[b] For HepG2, pass the cell suspension after trypsinization four or five times through an 18G needle to disperse cell clumps and obtain a uniform cell suspension.
[c] This order of addition is crucial.
[d] Previously heat inactivated by treatment for 1 h at 65 °C.
[e] IL-1β is required to obtain optimal translation of the chimeric CRP–SEAP mRNA (26).
[f] In positive control samples (cells incubated with 500 U/ml of IL-1β plus 10 ng/ml of IL-6), absorbance values of 1.0 A_{405} above background (which is around 0.25–0.3 A_{405}) are reached in 2–3 h depending on the efficiency of the transfection.

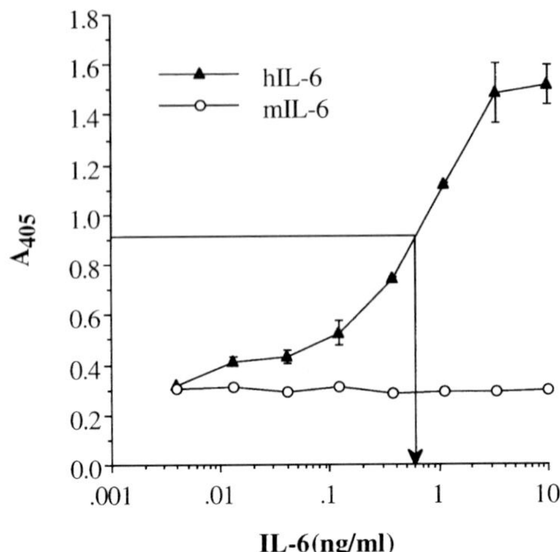

Figure 4. CRP–SEAP assay. Colorimetric quantification of SEAP enzymatic activity secreted by Hep3B cells transfected with p5′Δ-219 CRP–SEAP (26) and induced with various concentrations of human interleukin-6 (hIL-6) or murine IL-6 (mIL-6) in the presence of 500 U/ml of IL-1β. The EC_{50} is defined as the cytokine concentration able to elicit half-maximal response. Absorbance for half-maximal response is calculated according to the formula: (maximal abs. – abs. in IL-1β only control)/2 + abs. in IL-1β only control. In this particular experiment the absorbance for half-maximal response is: (1.52 – 0.3)/2 + 0.3 = 1.22/2 + 0.3 = 0.61 + 0.3 = 0.91, and is indicated as a horizontal line in the figure. Using the value of absorbance for half-maximal response, the EC_{50} can be derived from the dose–response curve indicated in the figure by the vertical arrow. In this particular experiment EC_{50} value is 0.6 ng/ml.

A typical CRP–SEAP assay result is shown in *Figure 4*. Exposure of transfected cells to human IL-6 (hIL-6) but not to mouse IL-6 (mIL-6) induces the secretion of SEAP enzyme (and colour development) in a dose-dependent manner up to 3 ng/ml. EC_{50} is defined as the cytokine concentration able to elicit a half-maximal response and can be derived from the dose–response curve as shown in *Figure 4*.

4. *In vitro* binding assay of IL-6 to IL-6Rα

Receptor:cytokine interactions are commonly studied by analysing the binding of radioiodinated proteins to the surface of receptor-expressing cells (29). The use of this technique is, however, not convenient when the binding activity of several protein mutants has to be measured and the major goal is that of determining their binding to the receptor compared to the wild-type ligand. In this case the best strategy is to produce recombinant forms of the extracellular domains of receptor chains (otherwise called soluble receptors).

If appropriately folded, these proteins maintain the same binding affinities as the native transmembrane receptors (30) and can be produced on a large scale, thus allowing the biochemical analysis of the binding reaction *in vitro*.

IL-6 binding to recombinant soluble IL-6Rα (sIL-6Rα) can be run in large scale 96-well format ELISA assays, using three reagents:

(a) *E. coli* produced bioactive IL-6 (31).
(b) sIL-6Rα from the culture supernatant of a CHO cell stable transformant (see ref. 9 and Chapter 1).
(c) A monoclonal antibody directed against a domain of sIL-6Rα that is not involved in receptor binding (32).

hIL-6 is first coated on 96-well ELISA plates. sIL-6Rα is allowed to react and binding is revealed by reaction with anti-IL-6Rα monoclonal I6R1/9.G11 (32) followed by a secondary antibody. *Figure 5A* represents the binding curve measured when the reaction is read after 3–5 min. Binding is inhibited in the presence of increasing concentrations of soluble human IL-6 which titrates out the added sIL-6Rα (*Figure 5B*). Competition experiments run in parallel with wt and mutant IL-6 allow relative binding potency (RBP) to be readily assessed (see Sections 5 and 6). In order to obtain a good signal for competition experiments (i.e. A_{405} values of 0.6–0.8) even in the presence of a low amount (2 ng/ml) of receptor, the reaction is left for a longer time (up to 30 min).

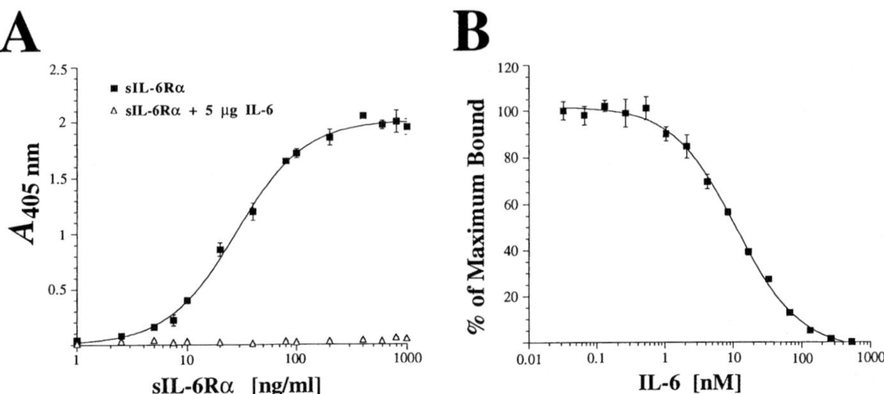

Figure 5. IL-6/sIL-6Rα ELISA binding assay. Each point represents the mean ± SD of six different experiments in triplicate. Data were fitted by non-linear least square fitting. (A) Direct binding of sIL-6Rα added in solution to IL-6 coated on plastic surface. (B) sIL-6Rα binding competition between coated IL-6 and IL-6 added in solution. The IC_{50} (nM) value, calculated by non-linear regression analysis using the four-parameter logistic model (33), was 11.2 ± 0.3.

Protocol 2. IL-6/sIL-6Rα ELISA binding assay

Equipment and reagents

- Automatic plate reader (Multiskan Bichromatic, Labsystems, Finland)
- 96-well microplates (Maxi-Sorp, Nunc)
- Tris-buffered saline (TBS): 50 mM Tris pH 7.5, 150 mM NaCl
- TBST: TBS with 0.5% Tween 20
- Alkaline phosphatase-conjugated goat anti-mouse IgG (Promega)
- Alkaline phosphatase substrate solution: 1 mg/ml *p*-nitrophenylphosphate dissolved in 10% diethanolamine pH 9.5

Method

1. Add 100 μl of a 10 μg/ml purified recombinant IL-6 in 100 mM Tris–HCl pH 8.0 to each well of a 96-well microplate and incubate for 4 h. Carry out all steps at room temperature with constant shaking.
2. Wash wells three times with 200 μl TBST and incubate for 3 h with 200 μl TBST.
3. Wash wells thoroughly with TBST.
4. Add sIL-6Rα, diluted in TBST to each well in a final volume of 100 μl, and incubate for 3 h at room temperature with shaking. In competition experiments, add increasing amount of cytokines (wt or mutant hIL-6) along with sIL-6Rα at a fixed concentration (2 ng/ml).
5. Wash wells five times with 200 μl TBST.
6. Add to each well 100 μl 300 ng/ml purified anti-sIL-6Rα mouse monoclonal antibody I6R1/9.G11 (see ref. 17) in TBST, and incubate for 1 h with shaking.
7. Wash wells five times with 200 μl TBST.
8. Add to each well 100 μl of alkaline phosphatase-conjugated goat anti-mouse IgG diluted 1:3000 in TBST, and incubate for 1 h at room temperature with shaking.
9. Wash wells five times with 200 μl TBST.
10. Add 100 μl of alkaline phosphatase substrate solution to each well. The solution turns yellow, and the intensity (as measured by a spectrophotometer at 405 nm) is directly proportional to the amount of shIL-6Rα bound to coated hIL-6.

5. IL-6 antagonists mutated in site 2: biochemical and biological evaluation

5.1 Generation of IL-6 antagonists

Residues predicted by the model in Section 2.2 to constitute site 2 of binding to gp130 (K27, Q28, Y31, G35—helix A, and S118, V121, F125—helix C, see

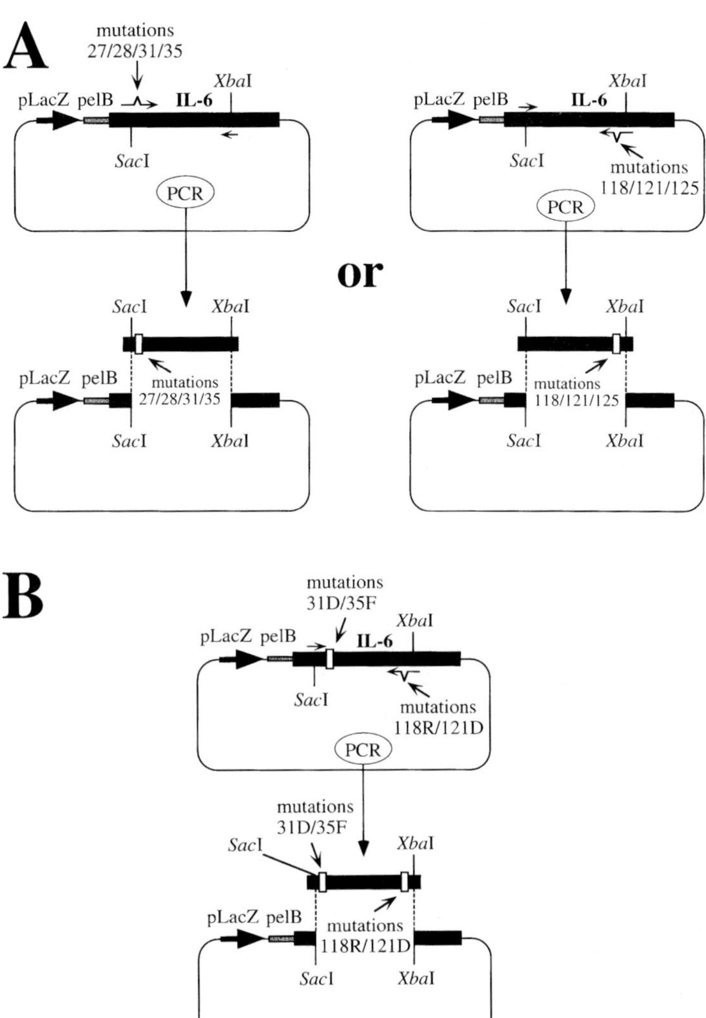

Figure 6. Mutagenesis of IL-6. In order to simplify replacement of wild-type IL-6 sequences by PCR amplified mutated fragments, a unique site for the restriction enzyme *Sac*I was introduced in the nucleotide sequence encoding hIL-6 amino acids 20–21–22 without changing their identity. (A) To generate IL-6 mutants by PCR, a pair of primers (one containing the desired mutations and the other wild-type) were used, one spanning the artificially introduced *Sac*I site described above, and the other spanning the natural *Xba*I site present in the hIL-6 cDNA at the level of codons 133–134. The amplified PCR fragment containing the desired mutations (derived from the mutagenetic oligonucleotide) was cut with *Sac*I and *Xba*I and subcloned in the vector cut with the same enzymes in order to replace the wild-type sequence with the mutated one. (B) In order to combine the mutations Y31D and G35F with the mutations S118R and V121D, the same PCR strategy described in (A) was used, amplifying this time not the wild-type IL-6 cDNA, but a cDNA coding for the mutant Y31D/G35F.

Figure 3) can be subjected to amino acid substitution either singly or combined. Mutagenesis is performed by PCR (34) according to the scheme described in *Figure 6*.

In order to simplify the expression of a large number of mutant proteins, mutagenesis is performed on an IL-6 cDNA cloned downstream of a secretion signal (*pelB*) which directs the protein to the periplasmic space of *E. coli*. The same vector can be used for the phage selection of IL-6 superbinder variants (described in detail in Section 7). Production and purification of IL-6 from bacterial periplasmic space is detailed in *Protocol 3*.

Protocol 3. Production of IL-6 (and variants) in the periplasmic space of *E. coli*

Reagents

- *E. coli* non-suppressor strain BL21 (DE3) *hsdS gal* (*LacIts*857 *ind*1 *Sam*7 *nin*5 *lac*UV5-T7 gene *1*)
- LB medium (per litre): 10 g bacto tryptone, 5 g bacto yeast extract, 10 g NaCl, adjust pH to 7.5 with 5 M NaOH, and autoclave to sterilize
- LB/amp/Glu: LB medium containing 100 μg/ml ampicillin and 1% glucose
- LB agar plates: prepare LB medium as described above; just before autoclaving, add 15 g/litre agar
- LB/amp/Glu plates: LB agar plates containing 100 μg/ml ampicillin and 1% glucose
- IPTG (isopropyl-β-D-thiogalactopyranoside): 400 mM stock solution in sterile, distilled H_2O (store at 4°C)
- TES buffer: 30 mM Tris–HCl pH 8.0, 1 mM EDTA, 20% sucrose
- Lysozyme solution: 10 mg/ml lysozyme in 50 mM Tris–HCl pH 8.0 (store at –20°C in small aliquots)[a]

Method

1. Transform competent BL21 cells with pHenΔhIL-6 or its derivatives (15) and plate on LB/amp/Glu plates,[b] before incubating the plates at 37°C overnight.
2. Inoculate a single colony in 10 ml of LB/amp/Glu and grow overnight at 25°C with shaking.
3. Dilute 800 μl of the overnight *E. coli* culture in 50 ml of LB/amp/Glu[c] medium in a 250 ml sterile glass flask, and grow at 25°C up to an OD at 600 nm of 0.6.
4. Centrifuge at 1200 *g* for 10 min at room temperature, wash once in sterile 50 mM NaCl, centrifuge again as above, and resuspend in 50 ml of LB/amp plus 0.4 mM IPTG.[c]
5. Grow the 50 ml culture for additional 3 h at 25°C with shaking.
6. Centrifuge at 1200 *g* for 15 min. Remove the supernatant carefully and resuspend the pellet in 1260 μl of TES buffer. Transfer the suspension in a 2 ml microcentrifuge tube and add 140 μl of ice-cold lysozyme solution. Incubate on ice for 10 min.
7. Centrifuge the tubes in a microcentrifuge at 14000 r.p.m. (~ 10000 *g*)

for 20 min at 4 °C. Collect the supernatant which contains partially pure IL-6. Store in aliquots at –80 °C.

8. To quantify the amount of IL-6 or its derivatives present in the periplasmic fraction, load 30 μl of each preparation on a 15% SDS–polyacrylamide gel along with increasing concentrations (from 0.1–3.2 μg, twofold serial dilutions) of purified rIL-6 (31). Following Coomassie staining, the gel is subjected to densitometric analysis, and the amount of IL-6 in the periplasmic space is interpolated from the standard curve obtained with the purified rIL-6.

[a] Do not refreeze thawed aliquots.
[b] Glucose is required to repress the LacZ promoter.
[c] IPTG induces the LacZ promoter thus allowing the transcription of the IL-6 cDNA.

Mutants expressed as described above are tested for biological activity using the SEAP assay (*Protocol 1*) and for IL-6Rα binding by ELISA (*Protocol 2*). The results are summarized in *Table 1*. Only mutations of Y31D and G35F on helix A or S118R and V121D on helix C generate variants with decreased bioactivity but with substantially intact binding to IL-6Rα.

In a second round of mutagenesis, performed by PCR according to the scheme described in *Figure 5B*, substitution of the four residues above are combined together. Mutant protein is expressed in *E. coli* periplasmic space as described in *Protocol 3*, and tested again for biological activity and for IL-6Rα binding by ELISA. The results are shown in *Table 1*. The IL-6 variant DFRD with the four substitutions (Y31D/G35F/S118R/V121D) has no residual biological activity, but binding to IL-6Rα is equal to wild-type. Thus the properties of DFRD are those expected for a potential receptor antagonist. Indeed, when Hep3B cells transfected with the IL-6-inducible CRP–SEAP vector are induced with 4 ng/ml of wt IL-6 in the presence of increasing amounts of DFRD (this time produced in large scale in *E. coli* and purified as described in ref. 31), IL-6 biological activity is inhibited in a dose-dependent manner (*Figure 7A*).

5.2 Specificity of IL-6 antagonists

The signalling chain gp130 participates in the formation of the receptor complexes in response to cytokines interleukin-11 (IL-11), leukaemia inhibitory factor (LIF), ciliary neurotrophic factor (CNTF), oncostatin M (OM), and cardiotrophin-1 (CT-1) (36). The presence of gp130 in the receptor complexes of this large group of ligands is responsible for the activation of a set of intracellular mediators common to all (37). This has best been shown in HepG2 cells which are responsive to IL-6, LIF, OM, and to the CNTF/s CNTFRα complex. In HepG2 cells all these cytokines cause a similar induction of acute phase protein genes (37). Nevertheless, the composition of their receptor complexes is specific to each cytokine, with receptor subunits, such as IL-6Rα or CNTFRα conferring cytokine specificity (36).

Table 1. Biological activity and receptor binding of IL-6 mutants[a]

Mutations							**Biological activity**		**RBP**[b]
Helix A				**Helix C**			**EC_{50} (ng/ml)**[c]	**Max. activity (% of wt)**	
K27	**Q28**	**Y31**	**G35**	**S118**	**V121**	**F125**	**0.8 ± 0.2**	**100 1.00**	
A							1.1 ± 0.36	100 ± 12	1.00 ± 0.01
	A						0.8 ± 0.14	100 ± 14	NT
		D					2.1 ± 0.2	105 ± 15	1.11 ± 0.19
			F				0.8 ± 0.25	100 ± 18	NT
			R				0.8 ± 0.1	100 ± 20	NT
			L				1.1 ± 0.1	100 ± 15	NT
			C				0.8 ± 0.01	100 ± 10	NT
A			R				0.7 ± 0.13	100 ± 13	NT
A			Q				1.0 ± 0.24	100 ± 17	1.00 ± 0.04
	A	A					1.1 ± 0.21	100 ± 19	NT
	A	S					0.8 ± 0.17	100 ± 18	NT
	A	D					10.0 ± 3.53	100 ± 10	0.19 ± 0.01
		D	H				2.1 ± 0.44	89 ± 5	0.85 ± 0.15
		D	L				15.4 ± 9.8	57 ± 9	1.88 ± 0.16
		D	Y				17.8 ± 3.95	53 ± 5	0.73 ± 0.17
		D	F				40.0 ± 9.9	51 ± 7	0.84 ± 0.09
		D	C				2.7 ± 0.09	81 ± 17	0.82 ± 0.18
				R			1.2 ± 0.2	100 ± 9	0.81 ± 0.18
					D		1.4 ± 0.5	100 ± 5	0.78 ± 0.02
					Y		0.6 ± 0.27	100 ± 20	NT
						V	0.5 ± 0.05	110 ± 8	1.02 ± 0.02
				L	D		2.2 ± 0.2	100 ± 4	0.92 ± 0.34
				R	D		22.8 ± 3.3	66 ± 6	0.66 ± 0.05
					F	V	0.5 ± 0.01	100 ± 5	NT
					Y	V	0.3 ± 0.08	100 ± 10	0.80 ± 0.04
					D	V	0.4 ± 0.05	110 ± 7	0.94 ± 0.05
					D	Y	0.8 ± 0.05	110 ± 4	NT
					Y	Y	0.8 ± 0.02	100 ± 10	NT
					Y	D	0.6 ± 0.09	100 ±20	NT
					D	D	0.8 ± 0.05	100 ± 9	NT
					A	A	0.8 ± 0.02	100 ± 5	NT
		D	F	R	D[d]		> 4000	0	0.97 ± 0.15

[a] Biological activity is measured as activation of the IL-6-inducible CRP–SEAP construct transfected in human Hep3B hepatoma cells, as decribed in Section 2. The first row in the table shows the wt IL-6 sequence in positions 27, 28, 31, 35, 118, 121, and 125. Where no change is indicated the wild-type residue is present.

[b] RBP (receptor binding potency): the mutants RBP (relative to wt IL-6) was calculated as the ratio between the amount of wt IL-6 and the amount of mutant IL-6 necessary to displace 50% of recombinant shIL-6Rα bound to coated rhIL-6 (see Section 3).

[c] The EC_{50} is the concentration of each mutant which gives 50% of the wt IL-6 maximal stimulation, and is determined by dose–response curves (see Section 2).

[d] This mutant will be referred in the text as DFRD.

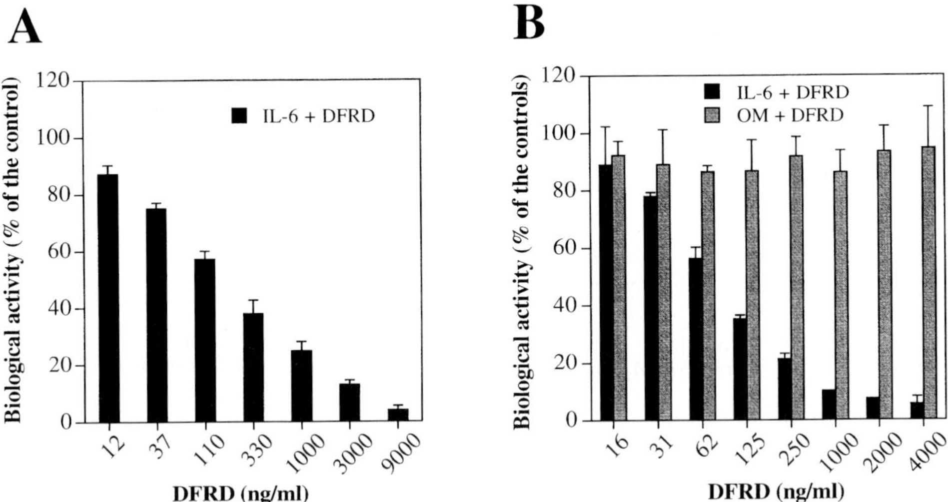

Figure 7. Efficacy and selectivity of an IL-6 antagonist. (A) The IL-6 variant with four substitutions (Y31D/G35F/S118R/V121D—known as DFRD) inhibits wt IL-6 activity on human Hep3B hepatoma cells. Human Hep3B cells were transfected with p5′Δ-219 CRP–SEAP (26) and induced with 4 ng/ml of IL-6 in the presence of increasing concentrations of DFRD, as indicated. SEAP activity was quantified (26) and expressed as per cent of the transcriptional efficiency in cells incubated with 4 ng/ml of IL-6 only. (B) DFRD inhibits IL-6-induced, but not OM-induced transcriptional activation. Human HepG2 cells were transfected with p5′Δ-219 CRP–SEAP (26) and induced with 4 ng/ml of IL-6 (filled columns) or of oncostatin M (OM shaded columns) in the presence of increasing concentration of the IL-6 mutant DFRD, as indicated. SEAP activity was quantified (26) and expressed as per cent of the transcriptional efficiency in cells incubated with 4 ng/ml of IL-6 only or OM only, respectively.

The site 2 IL-6 antagonist DFRD described in Section 5.1 is targeted to IL-6Rα, and is predicted by molecular modelling to impair binding to gp130 (see also Section 6), and thus expected to specifically interfere only with IL-6 biological activity and not with other gp130 signalling cytokines. The best way to assess the specificity of an antagonist is to measure its interference with the activity of other cytokines in a multi-cytokine responsive cell line utilizing the same bioassay. HepG2 are perfectly suited for this type of analysis because the same CRP–SEAP assay can be used to measure the activity of IL-6 and OM in parallel (28). An example is shown in *Figure 7B* in which DFRD effects on both cytokines are evaluated simultaneously.

6. Tag-mediated immunoprecipitation assays (TAMIA)

The assays we have developed to study the interaction of the IL-6/sIL-6Rα complex with gp130 and the IL-6-induced gp130 homodimerization *in vitro*,

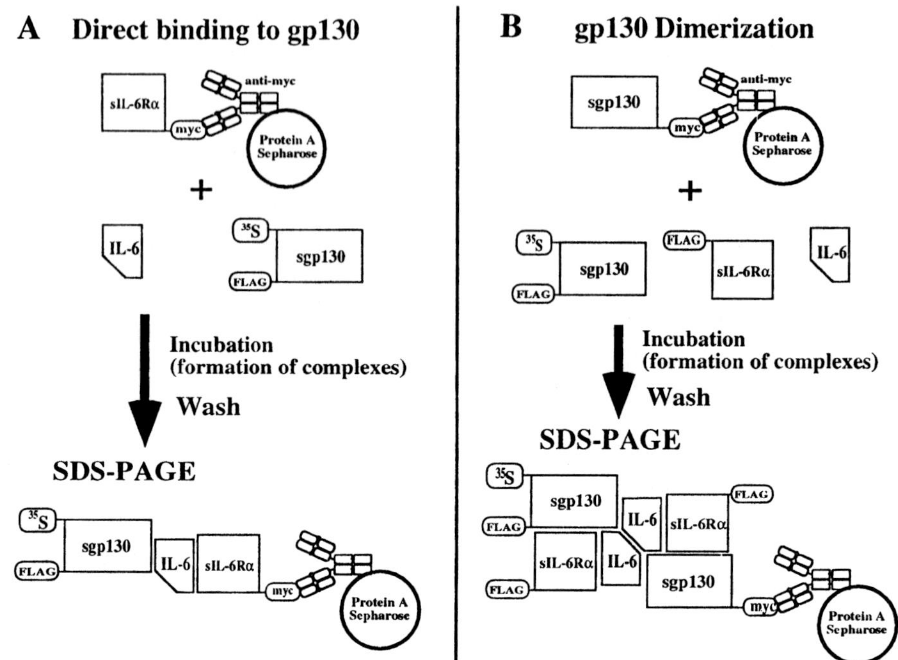

Figure 8. Schematic representation of tag-mediated immunoprecipitation assays (TAMIA). (A) Direct binding of gp130 to IL-6/IL-6Rα complex. Unlabelled sIL-6Rα–myc is immobilized on protein A–Sepharose beads via anti-myc 9E10 mAb and incubated with ^{35}S-labelled sgp130–FLAG and IL-6. After wash and SDS–PAGE the detection of labelled sgp130 indicates the formation of the gp130/IL-6/IL-6Rα complex shown at the bottom of the scheme. (B) Dimerization of gp130. Unlabelled sgp130–myc is immobilized on beads and incubated with ^{35}S-labelled sgp130–FLAG, sIL-6Rα–FLAG, and IL-6. After wash and SDS–PAGE the detection of labelled sgp130 indicates the formation of the gp130 dimer shown at the bottom as hexameric complex.

are known collectively tag-mediated immunoprecipitation assays (TAMIA) which are based on the production of soluble receptors carrying at their COOH terminus epitope tags which can be recognized by specific monoclonal antibodies (38, 39). Recombinant receptors are produced in the baculovirus expression system because of the abundant yield (from 0.5–2 mg of IL-6Rα or gp130 per litre of supernatant of infected cells) and the possibility to metabolically label the recombinant protein with ^{35}S with high specific activity and low background, both advantageous for binding studies. We generated soluble gp130 (sgp130–myc and sgp130–FLAG) and IL-6Rα (sIL-6Rα–myc and sIL-6Rα–FLAG) both tagged with myc or FLAG epitopes (13). *Figure 8* schematically illustrates the use of TAMIA to detect either the interaction of IL-6/sIL-6Rα complex with sgp130 or sgp130 dimerization.

Protocol 4. Direct binding of gp130 to the IL-6/IL-6Rα complex

Reagents

- Protein A–Sepharose CL-4B beads (Pharmacia)
- Anti-c-myc mouse mAb clone 9E10 (Boehringer)
- PBS (see *Protocol 1*)
- Brij 96 (polyoxyethylene 10 oleyl ether, Sigma)
- PBSTB: PBS, 0.05% Tween 20, 0.2% Brij 96
- Tween 20 (polyoxyethylene-sorbitan monolaurate, Sigma)
- SF-900 II: serum-free insect cell culture medium (Gibco)
- BSA (bovine serum albumin, Sigma)
- Amplify (Amersham)
- 3MM (Whatman)

Method

1. For each immunoprecipitation experiment in 1.5 ml Eppendorf tube, incubate 4 μg anti-c-myc 9E10 mAb with 50 μl protein A–Sepharose beads suspension (50% slurry in PBS) in a total volume of 300 μl PBS, for at least 2 h at 4 °C on a rotating wheel mixer.
2. Centrifuge the suspension at 5000 *g* for 30 sec; remove the supernatant and resuspend the beads in 1 ml of PBSTB. Repeat the centrifugation and remove the supernatant.
3. Incubate the beads with 500 μl of supernatant of infected insect cells expressing IL-6Rα–myc for at least 2 h at 4 °C on a rotating wheel mixer.
4. Wash the beads as in step 2.
5. Incubate the beads with 50 μl of ^{35}S-labelled gp130–FLAG plus increasing amounts of wt or mutant IL-6 (typically ranging from 10 ng to 10 μg). SF-900 II is added up to a total volume of 300 μl in the presence of 0.15% Brij 96 and 0.12 μg BSA final concentration. Incubate for at least 6 h at 4 °C on a rotating wheel mixer.
6. Wash the beads three times as in step 2.
7. Resuspend the beads in 50 μl SDS–PAGE loading buffer, heat 5 min at 95 °C, and separate proteins by SDS–PAGE on a 8% gel.
8. Incubate the gel for 30 min with Amplify, dry in 3MM, and expose with autoradiograph film.

Protocol 5. gp130 dimerization assay

1. Load protein A–Sepharose beads with anti-myc mAb as described in *Protocol 4*, steps 1 and 2.
2. Incubate the beads with 500 μl of supernatant from infected insect cells expressing gp130–myc for at least 2 h at 4 °C on a rotating wheel mixer.

Protocol 5. *Continued*

3. Wash the beads as in *Protocol 4*, step 2.
4. Incubate the beads with 50 μl ^{35}S-labelled gp130–FLAG, 500 μl of supernatant from infected insect cells expressing IL-6Rα–FLAG, plus increasing amounts of wt or mutant IL-6 (typically ranging from 10 ng to 10 μg). SF-900 II is added up to a total volume of 600 μl in the presence of 0.15% Brij 96 and 0.12 μg BSA final concentration. Incubate for at least 6 h at 4 °C on a rotating wheel mixer.
6. Wash the beads three times as in *Protocol 4*, step 2.
7. Resuspend the beads in 50 μl of SDS loading buffer, heat 5 min at 95 °C, and load on SDS–PAGE.

TAMIA can be used to compare wild-type and mutant IL-6 for their ability to bind and dimerize gp130. An example is shown in *Figure 9*, which illustrates the selective loss of gp130 dimerization by the IL-6 antagonist DFRD, this IL-6 variant is able to interact with gp130 at site 3, but is no longer capable of binding two gp130 molecules, thus explaining both receptor antagonism and loss of signalling, as described in Section 5.1.

TAMIA can be used to analyse any *in vitro* receptor assembly reaction,

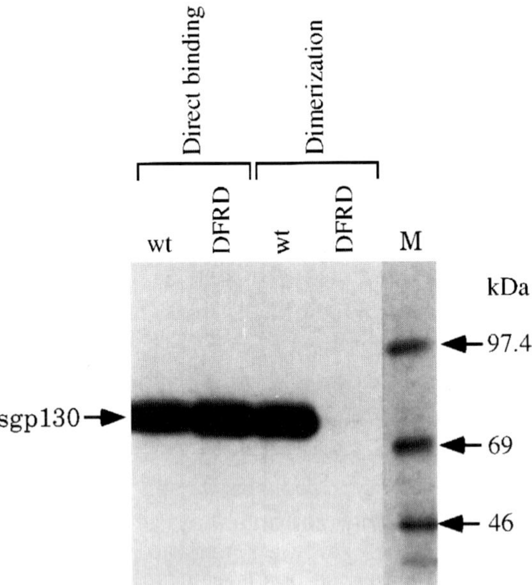

Figure 9. gp130 binding and dimerizing properties of wt IL-6 and IL-6 antagonist DFRD. Assays were performed as outlined in *Figure 8* and as described in *Protocols 4* and *5*. Immunoprecipitation of ^{35}S-labelled gp130 is revealed by SDS–PAGE followed by autoradiography.

particularly in the study of cytokines and cytokine receptor dimerization, stoichiometry, and topology of receptor complex formation. In fact this system has been recently extended to the study of CNTF (40) and IL-11 (41) receptor complexes.

7. Selection of amino acid substitutions which increase affinity of IL-6 for IL-6Rα

The identification of amino acid substitutions which increase the affinity of a protein ligand for its receptor (higher affinity binders) via conventional mutagenesis is a time-consuming and inefficient strategy. An alternative and highly efficient approach has been developed in the last years, the selection of variants from molecular repertoires expressed on the M13 phage surface (42). In this approach the ligand of interest is first expressed as a fusion protein on the phage surface, it is important to carefully check that in this form the ligand maintains the same receptor binding potential as the native molecule.

In a second step a library of variants is generated at the cDNA level in a region predicted to be part of the receptor binding site (see Section 1). By transfection into suitable bacterial hosts, the cDNA library is converted into a library of phage particles; in each particle there is a physical association between a variant protein and its encoding cDNA. The phage library is then subjected to multiple rounds of selection amplification, using the receptor itself as a selector. At the end of this selection, individual phage are isolated and characterized for binding properties and amino acid sequence in the mutagenized region. The most interesting mutants can then be produced as non-fusion proteins in the *E. coli* periplasmic space (Section 4) in order to better compare them with the wild-type ligand.

7.1 Generation of IL-6 phage particles

Selection of higher affinity ligands is facilitated by the use of 'monovalent display' technology, in which an average of less than one copy of a fusion between the protein of interest and the phage pIII gene product is displayed, thus avoiding avidity effects in the binding of the phage displayed ligand to its receptor (43). Monovalent display is achieved by constructing the pIII fusion gene in a packageable phagemid vector (44). Bacterial cells are first transfected with the phagemid encoding the fusion gene and then superinfected with helper phage (*Figure 10*).

For the construction of cytokine/pIII fusion proteins we have generated phagemid pHenΔ in which a multicloning site is inserted between the bacterial pelB leader and the C terminal part of M13 gene III (9). The whole gene is under the transcriptional control of the inducible LacZ promoter, and the pelB leader promotes localization of the gene product in the periplasm. Plasmid pHenΔhIL-6 is a derivative of pHenΔ (15) in which the hIL-6 cDNA

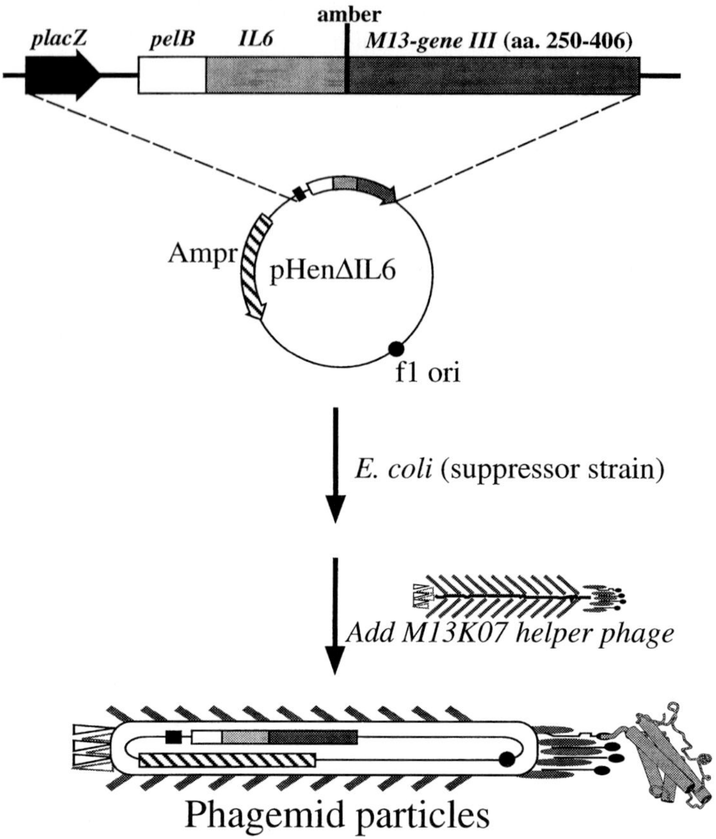

pHenΔIL6 and production of phagemid particles

Figure 10. Construction of IL-6 phage libraries. A mutagenic oligonucleotide, containing the desired substitutions, is annealed with a partially overlapping oligonucleotide which does not carry any mutations. The 3′ recessed ends are filled by treatment with Klenow enzyme: the resulting double-stranded (and blunt-ended) mutagenic cassette is then digested with the appropriate restriction enzymes and introduced into the IL-6 cDNA as a substitute for the corresponding wild-type sequence.

has been inserted in the polylinker upstream from the C terminus of gene III. The two coding sequences are cloned in-frame but separated by an amber (TAG) stop codon which can be suppressed in about 20% of translation processes as glutamine in *suII* ('suppressor') *E. coli* strains, thus producing an hIL-6/pIII fusion protein (*Figure 10*). hIL-6-phage obtained in this way show biological activity and receptor binding properties typical of wt hIL-6 (15). In non-suppressor strains, the TAG codon stops translation at the 3′ end of hIL-6 cDNA, thus producing 'free' hIL-6 (see Section 4).

Protocol 6. Purification of hIL-6/phage particles[a]

Reagents

- Suppressor *E. coli* strain, such as XL-1 blue (*recA*1 *endA*1 *gyrA*96 *thi hsdR*17 ($r_K^- m_K^+$) *supE*44, *relA*1 *lac*⁻ F′ [*proA*⁺*B*⁺ *lacI*^q, *lacZ*ΔM15, Tn*10*])
- 2 × TY medium (per litre): 16 g bacto tryptone, 10 g bacto yeast extract, 5 g NaCl, adjust pH to 7.0 with 5 M NaOH, and autoclave to sterilize
- 2 × TY/amp: 2 × TY medium containing 100 μg/ml of ampicillin
- Helper phage M13K07 (Pharmacia; New England Biolabs) (large quantities can be prepared as described in ref. 45)
- PEG/NaCl: 20% polyethylene glycol 6000, 2.5 M NaCl
- TBS (see *Protocol 2*)

Method

1. Transform competent XL-1 blue cells, an *E. coli* 'suppressor' strain, with pHenΔhIL-6 or its derivatives, and plate on LB/amp plates (see *Protocol 3*). Incubate overnight at 37 °C.
2. Inoculate a single colony in 10 ml 2 × TY/amp and grow at 37 °C with shaking to an OD at 600 nm of 0.5–0.7 (about 2–2.5 h).
3. Superinfect the 10 ml bacterial culture with helper phage M13K07 at a phage:bacteria ratio of 20–50:1 (1.0 OD unit of bacteria is approx. 8 × 10^8 bacteria/ml) for 30 min at 37 °C in a water-bath. Add 2 × TY/amp medium up to 1 litre and grow overnight at 37 °C with shaking.
4. Centrifuge the overnight culture at 3000 *g* for 30 min at 4 °C.
5. Collect the supernatant and add 0.2 vol. PEG/NaCl (i.e. add 250 ml PEG/NaCl solution to 1000 ml supernatant), mix well, and incubate for 4 h (or overnight) in ice.
6. Collect phage by centrifuging the PEG-treated supernatant at 10 800 *g* for 40 min at 4 °C. Carefully remove the supernatant, resuspend the pellet in 100 ml TBS.
7. Transfer the phage suspension to a water-bath at 70 °C and leave for 30 min.[b]
8. Centrifuge for 30 min at 12 000 *g* at 4 °C. Collect the supernatant, add 0.2 vol. PEG/NaCl, and incubate in ice for 4 h.
9. Centrifuge the PEG-treated supernatant at 10 800 *g* for 40 min at 4 °C, discard the supernatant, and resuspend the pellet in 8 ml of TBS.
10. Add 0.45 g CsCl/ml of phage supension and centrifuge in Beckman rotor SW40 at 37 000 r.p.m. (~ 180 000 *g*) for 48 h at 20 °C.
11. Place the tubes with a light coming from the top: phage will appear as

Protocol 6. *Continued*

a bright sky-blue band. Collect phage with a 1 ml syringe through a 20G × 1½ inch needle and place in a 2 ml microcentrifuge tube.

12. Add 1 vol. TBS to the phage suspension: spin in a bench-top centrifuge at 14 000 r.p.m. (~ 10 000 *g*) for 5 min at 4 °C. Collect the supernatant, add TBS up to 25 ml final volume, and centrifuge at 50 000 r.p.m. in Beckman rotor 70Ti (~ 200 000 *g*) at 20 °C for 4 h.
13. Remove the supernatant and resuspend the pellet in 0.5–1 ml TBS: this constitutes the caesium-purified phage preparation. Phage concentration can be determined as ampicillin-transducing units.[a]

[a] For detailed information on the basic technique in handling phage see ref. 45.
[b] This step is required to denature residual bacterial protein.

7.2 Generation of IL-6 phage libraries

Five different IL-6 phage libraries were generated (15, 16). In each library the sequence of four residues was fully randomized for a total of 20 residues, which covers most of the IL-6Rα binding surface (site 1) predicted by molecular modelling (Section 1.2). The complete randomization of four amino acid residues should result in the generation of 1.6×10^5 different protein variants. We constructed mutagenic cassettes by primer extension of two partially overlapping oligonucleotides according to the scheme illustrated in *Figure 11*. The cassette can be used to replace wt sequence in the receiving plasmid vector pHenΔhIL-6 or its derivatives (15, 16).

Protocol 7. Phage library construction

Equipment and reagents

- 'Gene Pulser' apparatus (BioRad)
- Micro-collodion bags (Sartorius Corporation, Cat. No. SM132 02)
- Large square plates (243 × 243 mm Nunc Bio-Assay dishes)
- Annealing buffer: 10 mM Tris pH 7.5, 50 mM NaCl, 10 mM $MgCl_2$, 1 mM DTT
- dGTP, dCTP, dATP, and dTTP (100 mM stock solutions)
- [α-^{32}P]dATP, 3000 Ci/mmol (Amersham)
- Sephadex G50 (Pharmacia)
- TE buffer: 10 mM Tris–HCl pH 7.5, 1 mM EDTA
- Ligase buffer: 200 mM Tris–HCl pH 7.6, 50 mM $MgCl_2$, 50 mM DTT
- SOC medium (per litre): 20 g bacto tryptone, 5 g bacto yeast extract, 0.5 g NaCl. For 1 litre of solution add 10 ml 250 mM KCl. Adjust the pH to 7.0 with 5 M NaOH and sterilize by autoclaving. Allow it to cool to 50 °C, add 20 ml of a sterile 1 M solution of glucose to a final concentration of 20 mM.
- LB/agar plates (see *Protocol 3*)
- LB/amp/gly: LB medium containing 100 μg/ml ampicillin and 10% glycerol

Method

1. Mix 200 pmol of each oligonucleotide in a final volume of 100 μl of annealing buffer in microcentrifuge tubes. Boil for 5 min and slowly cool the solution down to room temperature.

2. Mix 7 μl of [α-^{32}P]dATP (3000 Ci/mmole),[a] 50 U of *E. coli* polymerase Klenow subunit, and cold dGTP, dCTP, dATP, and dTTP (final concentration = 2.5 mM), in a final volume of 150 μl of annealing buffer at 37 °C for 1 h.
3. Prepare a 5 ml column of Sephadex G50 (equilibrated in TE buffer) in a sterile pipette. Load the primer extension reaction products on the column and elute with TE buffer in fractions of 150 μl. Collect fractions in microcentrifuge tubes, measure the radioactivity in a beta counter, pool fractions containing the first peak of radioactivity, and transfer these to a 2 ml microcentrifuge tube.
4. Extract DNA with phenol followed by phenol:chloroform 1:1 (v/v). Add sodium acetate pH 5.2 (0.3 M final) and 2 vol. 100 % ethanol. Mix well and leave at –20 °C for at least 1 h.
5. Centrifuge tubes in a microcentrifuge at 14 000 r.p.m. (~ 10 000 *g*) for 30 min at 4 °C.
6. Discard the supernatant and wash the DNA pellet with 70% ethanol. Resuspend the DNA in 100 μl H_2O, digest with restriction enzymes, and purify digested DNA following standard procedures.[b]
7. Set-up the ligation reaction by mixing 200 ng of the purified mutagenic cassette with 1 μg of digested vector plasmid, add 5 U T4 DNA ligase, and incubate for 16 h at 15 °C in a final volume of 100 μl ligase buffer.
8. Transfer the ligation mix into micro-collodion bags and dialyse for 8 h against 4 litres of high quality deionized H_2O with four changes (1 litre each). Electroporate 20 fractions, 5 μl fractions of the ligation mix, separately in 100 μl of XL-1 blue electrocompetent cells (42). Carry out electroporation using Gene Pulser apparatus set at 2.5 kV, 200 Ohm, 25 μF (average time constant 4.8 msec).
9. Pool all fractions in a 50 ml polypropylene tube, add 20 ml of SOC medium, grow for 1 h at 37 °C with shaking.
10. Spread the 20 ml suspension onto large square LB/amp agar plates. A small amount of the cell suspension must be reserved to make a titration on round dishes (10 cm diameter) in order to evaluate the number of independent clones present in the library.[c]
11. Following overnight incubation at 37 °C, scrape colonies from plates and resuspend in 30 ml LB/amp/gly. Add an aliquot of the suspension (about 1 ml) to 1 litre of LB/amp in a 2 litre flask up to an OD (600 nm) of 0.025–0.05. Store the remaining scraped cells at –80 °C.
12. Grow the 1 litre suspension at 37 °C with shaking at 250 r.p.m. up to an OD (600 nm) of 0.25.

Protocol 7. *Continued*

13. Superinfect the 0.25 OD_{600} culture with helper phage M13K07 at a ratio of 50:1 (see *Protocol 6*) for 30 min at 37 °C in a water-bath.
14. Prepare a purified library of phage-exposed IL-6 mutants according to *Protocol 6*.

[a] [α-^{32}P]dATP is included in order to follow the reaction product.
[b] For more detailed information on basic molecular biology techniques, refer to the laboratory manual by Sambrook *et al.* (35).
[c] In theory, the 1.6×10^5 different mutants potentially encoded by the library should be well represented by 1×10^6 independent clones. However, due to the mutagenesis strategy, a fraction of the clones will actually contain improper sequences in the mutated region (i.e. stop codons, undesired substitutions, and deletions) and it is thus preferable to obtain at least 3–4 $\times 10^6$ independent clones.

7.3 Selection of IL-6 superbinders from a phage library

For enrichment experiments with the IL-6 receptor complex, a soluble form of IL-6Rα (sIL-6Rα, residues 1–323) present in the supernatant of an established CHO cell clone (CsRh14) was used (9).

Protocol 8. Library selection

Reagents

- Magnetic beads (Dynabeads® M-450 Tosyl-activated, Dynal)
- PBS (see *Protocol 1*)
- PBSB: PBS containing 1 mg/litre bovine serum albumin (Fraction V)
- Conjugation buffer: 50 mM sodium borate pH 9.5
- TBSB: TBS (see *Protocol 2*) containing 1 mg/litre bovine serum albumin (Fraction V)
- TBST (see *Protocol 2*)
- Elution buffer: 100 mM sodium citrate pH 3.2
- LB/amp large square plates (see *Protocol 7*)
- 3 M Tris–HCl pH 8.9
- XL-1 blue bacteria (see *Protocol 6*)

Method

1. Place 100 μl Dynabeads® M-450 into a 1 ml microcentrifuge tube and concentrate them using a magnetic particle concentrator according to the manufacturer's instructions.
2. Wash the beads with 1 ml conjugation buffer, concentrate them again, add 400 μl of the same buffer containing 15 μg anti-hIL-6Rα monoclonal antibody (see *Protocol 2*), and incubate at room temperature for 24 h on a rotating wheel mixer.
3. Concentrate the beads, now covalently conjugated with the monoclonal antibody, resuspend them in 10 ml PBSB, and incubate for 12 h at 4 °C on a rotating wheel mixer.
4. Concentrate the beads and wash them with 1 ml PBSB.

5. Resuspend the beads in 400 μl serum-free conditioned medium from the CsRh14 cell clone containing 250 ng shIL-6Rα. Following a 4 h incubation at room temperature on a rotating wheel mixer, concentrate beads.
6. Mix shIL-6Rα coated beads with 2 × 10^{12} phage particles (input), estimated as colony-forming units (c.f.u.) on LB/amp plates, and incubate for 24 h at 4 °C on a rotating wheel mixer. (Keep the volume constant (1 ml) by adding the appropriate amount of TBSB.)
7. Wash the beads three times with 1 ml TBST and incubate for 2 min at 23 °C in 400 μl elution buffer.
8. Collect the supernatant, containing eluted phage, and neutralize with a few drops of 3 M Tris–HCl pH 8.9.
9. Infect fresh XL-1 blue bacteria with collected phage (30 min at 37 °C in a water-bath) and plate on LB/amp large square plates. A small amount of the cell suspension is also used to titrate the eluted phage by estimating the number of c.f.u. on LB/amp plates.[a]
10. Following overnight incubation at 37 °C, colonies are scraped from plates and phage particles are purified on caesium chloride as described in *Protocol 6*. At the same time, 20–30 individual colonies are picked, plasmid DNA is extracted, and the mutated region in the cDNA is sequenced according to standard procedures.

[a] This value is the output of the selection procedure: the input/output ratio gives an estimate of the enrichment obtained following the described selection scheme.

Purifed phage particles can be subjected to multiple rounds of selection by repeating *Protocol 8*, steps 1–10. In order to increase the ratio of high affinity sIL-6Rα binders, the concentration of the receptor is progressively reduced from 250 ng/ml of the first round to 5 ng/ml and 500 pg/ml in the following rounds. Selection is stopped when either a consensus sequence emerges among the clones, or the input/ouput ratio remains constant, thus indicating that no further enrichment has taken place. Individual isolates can be expressed as free protein (*Protocol 1*) and tested in the ELISA sIL-6Rα binding assay (*Protocol 3*).

7.4 IL-6 variants with higher affinity for IL-6Rα

Selection from phage libraries was successful in identifying substitutions which increased IL-6 affinity for IL-6Rα (15, 16). Representative substitutions are listed in *Table 2*, together with their effect on binding. IRA, the best binding variant obtained by substituting residues in helix D (15), has improved receptor binding approximately fivefold. Affinity selection from a library of residues D71/F74/Q75/S76 in the AB-loop (constructed in

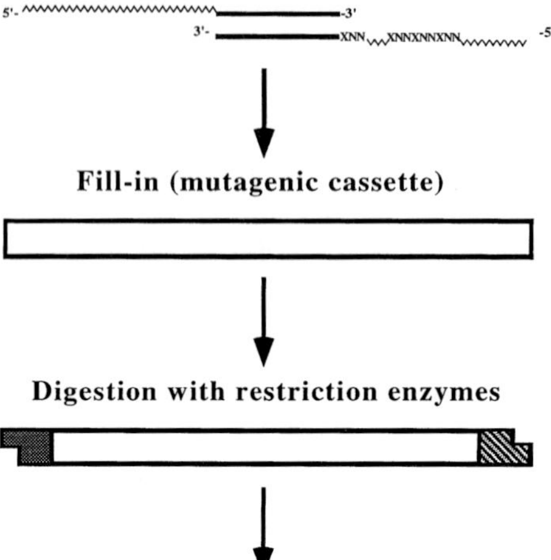

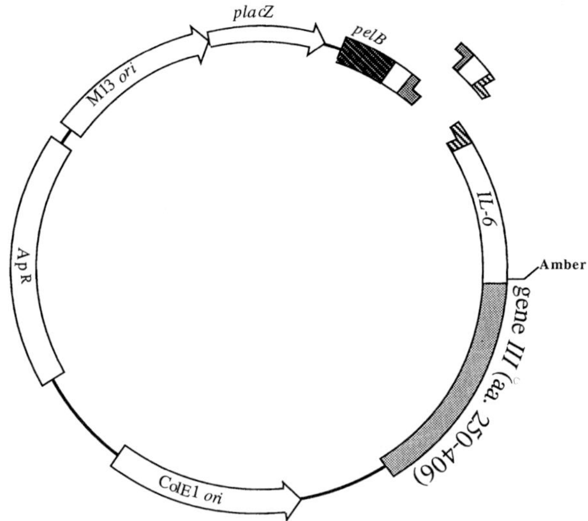

Figure 11. Strategy for the monovalent display of IL-6 on phage surface. Plasmid pHenΔIL-6, in which the coding sequences of IL-6 cDNA and pIII gene are in-frame but separated by an amber (TAG) stop codon, is introduced in XL-1 blue, an *E. coli* 'suppressor' strain. In this bacterial cells, the TAG codon is translated as a Gln in about 20% of translational events, thus allowing the production of the IL-6–pIII fusion protein. Transformed bacteria are then infected with helper phage M13K07: plasmid is thus packaged into phagemid particles which are recovered from the supernatant of the bacterial culture. About 1% of the viral particles expose on their surface one copy of the IL-6–pIII fusion protein.

Table 2. Sequences and sIL-6Rα binding activity of the three IL-6 mutants selected from the $D_{71}F_{74}Q_{75}S_{76}$ library

	Amino acid residues						
	AB-loop			D helix			RBP[a]
	74	75	76	175	176	183	
wt IL-6	F	Q	S	Q	S	Q	1
IRA	–	–	–	I	R	A	4.8 ± 0.3
D-2	Y	F	I	I	R	A	23.5 ± 3.0
D-4	–	Y	I	I	R	A	27.7 ± 1.6
D-6	–	Y	K	I	R	A	42.0 ± 3.6

[a] RBP = relative binding potency (see *Table 1*). Values represent the mean ± SD of five different experiments, performed in duplicate, with three different protein preparations.

the background of an hIL-6 cDNA already containing the mutations D175I/S176R/Q183A) led to the identification of more potent superbinders (D-2, D-4, and D-6) which improved receptor binding from 23- to 40-fold (see *Table 2*).

8. Combination of antagonistic and superbinder mutations to generate IL-6 superantagonists

The model of IL-6 interaction with its receptors (Section 2) suggests that sites 1 and 2 function independently from each other. This idea is strengthened by the observation that the site 2 IL-6 antagonist DFRD described in Section 5.1 binds IL-6Rα as well as wt IL-6, but loses the ability to dimerize gp130. It can also be predicted that further increasing the antagonist's affinity for IL-6Rα would increase its potency (i.e. decrease the amount of antagonist required to obtain IL-6 blockage) without any rise in bioactivity. To generate IL-6 superantagonists, the substitutions present in the antagonist DFRD (Section 5.1) have been combined with superbinder mutations isolated by phage selection (Section 7.4) using a PCR strategy similar to that illustrated above.

The combination of DFRD with IRA produces superantagonist Sant1, and combining DFRD with D-4, D-2, and D-6 generates superantagonists Sant3, 4, and 5, respectively. Superantagonists maintain the same receptor binding properties as their parent superbinders (compare *Table 2* and *Table 3*) and show a parallel increase of their ability to inhibit IL-6 biological activity, as measured by SEAP assay. Worthy of note is that the superantagonist with the highest affinity for IL-6Rα, Sant5, is approximately 80-fold more potent than the original DFRD.

Table 3. Amino acid sequence, binding, and antagonistic properties of IL-6 superantagonists[a]

Mutant	**Mutations**										**RBP**[b]	**IC_{50} (ng/ml)**[c]
	Antagonistic				**Superbinder**							
	Y 31	G 35	S 118	V 121	F 74	Q 75	S 76	Q 175	S 176	Q 183		
DFRD	D	F	R	D							1.0 ± 0.1	140.4 ± 38.4
Sant 1	D	F	R	D				I	R	A	4.5 ± 0.6	18.4 ± 4.8
Sant 3	D	F	R	D		Y	I	I	R	A	28.0 ± 2.5	2.2 ± 0.2
Sant 4	D	F	R	D	Y	F	I	I	R	A	23.5 ± 1.9	2.3 ± 0.5
Sant 5	D	F	R	D		Y	K	I	R	A	40.0 ± 3.4	1.8 ± 0.4

[a] The first row shows the amino acid present in wt IL-6 at the position indicated. For each mutant, all the amino acid sustitutions in the respective positions are indicated; where no change is indicated, the wild-type residue is present.
[b] RBP = relative binding potency (see *Table 1*).
[c] The IC_{50} is the concentration giving 50% IL-6 inhibition on Hep3B cells.

9. Potential use of IL-6 superantagonists in therapy

The general strategy used for the generation of IL-6 receptor superantagonists described in this chapter can be applied to all cytokines and growth factors which interact with at least two receptor subunits through distinct binding epitopes (1). On one epitope disruptive substitutions are introduced which impair receptor interaction, while on the other epitope, gain-of-function substitutions are selected which increase receptor binding potency. Combining the two sets of amino acid replacements engenders bifacial variants of the hormone which lock the receptor complex in a non-productive configuration, and do so at low concentrations. This latter feature is of practical importance in obtaining efficient cytokine blockage *in vivo* in pathological conditions where production of a cytokine may be greatly increased above normal physiological levels.

IL-6 is a central growth factor for B cell growth and differentiation (36). Although the physiological production of IL-6 is part of the development of a normal immune response (46), over-production of this cytokine is considered to play a pathogenic role during the development of autoimmune disorders, post-menopausal osteoporosis, and lymphoid malignancies (47). In particular, in multiple myeloma, IL-6 is a growth factor for tumour cells with a considerable increase in the daily production of the cytokine, often estimated to be up to 100 μg/day (48). IL-6 superantagonists fully inhibit growth and induce cell death of human myeloma cells *in vitro* (17, 49). These properties make them likely candidates for the therapy of myeloma and the potency displayed *in vitro* implies that effective dosage could also be achieved in those patients with the highest production of the cytokine.

Note added to the proofs

While this work was in press the X-ray structure of IL-6 at 1.9 Å resolution was published (50). This nicely confirms our previous modelling predictions (see Section 2) and the topological localizations of receptor binding sites 1, 2, and 3 indentified by site-directed mutagenesis.

Acknowledgements

We are grateful to Ilaria Bagni for invaluable help in editing this manuscript and to Janet Clench for text revision.

References

1. Heldin, C. H. (1995). *Cell*, **80**, 213.
2. Bazan, F. J. (1990). *Immunol. Today*, **11**, 350.
3. Sprang, S. R. and Bazan, J. F. (1993). *Curr. Opin. Struct. Biol.*, **3**, 815.
4. Bazan, J. F. (1990). *Proc. Natl. Acad. Sci. USA*, **87**, 6934.
5. De Vos, A. M., Ultsch, M., and Kossiakoff, A. A. (1992). *Science*, **255**, 306.
6. Taga, T., Hibi, M., Hirata, Y., Yamasaki, K., Yasukawa, K., Matsuda, T., *et al.* (1989). *Cell*, **58**, 573.
7. Yamasaki, K., Yaga, T., Hirata, Y., Yawata, H., Kawanishi, Y., Seed, B., *et al.* (1988). *Science*, **241**, 825.
8. Hibi, M., Murakami, M., Saito, M., Hirano, T., Taga, T., and Kishimoto, T. (1990). *Cell*, **63**, 1149.
9. Savino, R., Lahm, A., Salvati, A. L., Ciapponi, L., Sporeno, E., Altamura, S., *et al.* (1994). *EMBO J.*, **13**, 1367.
10. Salvati, A. L., Lahm, A., Paonessa, G., Ciliberto, G., and Toniatti, C. (1995). *J. Biol. Chem.*, **270**, 12242.
11. Lahm, A., Savino, R., Salvati, A. L., Cabibbo, A., Ciapponi, L., Demartis, A., *et al.* (1995). *Ann. N. Y. Acad. Sci.*, **762**, 136.
12. Murakami, M., Hibi, M., Nakagawa, N., Nakagawa, T., Yasukawa, K., Yamanishi, K., *et al.* (1993). *Science*, **260**, 1808.
13. Paonessa, G., Graziani, R., De Serio, A., Savino, R., Ciapponi, L., Lahm, A., *et al.* (1995). *EMBO J.*, **14**, 1942.
14. Ciapponi, L., Graziani, R., Paonessa, G., Lahm, A., Ciliberto, G., and Savino, R. (1995). *J. Biol. Chem.*, **270**, 31249.
15. Cabibbo, A., Sporeno, E., Toniatti, C., Altamura, S., Savino, R., Paonessa, G., *et al.* (1995). *Gene*, **167**, 41.
16. Toniatti, C., Cabibbo, A., Sporeno, E., Salvati, A. L., Cerretani, M., Serafini, S., *et al.* (1996). *EMBO J.*, **15**, 2726.
17. Sporeno, E., Savino, R., Ciapponi, L., Paonessa, G., Cabibbo, A., Lahm, A., *et al.* (1996). *Blood*, **87**, 4510.
18. Hill, C. P., Osslund, T. D., and Eisenberg, D. (1993). *Proc. Natl. Acad. Sci. USA*, **90**, 5167.
19. Computer Methods for Macromolecular Sequence Analysis. (1996). *Methods in enzymology* (ed. R. F. Doolittle), Vol. 266. Academic Press, San Diego, USA.

20. Sali, A. (1995). *Curr. Opin. Biotechnol.*, **6**, 437.
21. Kraulis, P. J. (1991). *J. Appl. Crystallogr.*, **24**, 946.
22. Ciliberto, G. (1989). In *Acute phase proteins in the acute phase response* (ed. M. B. Pepsys), p. 29. Springer–Verlag, London.
23. Mackiewicz, A., Ganapathi, M. K., Schultz, D., Brabenec, A., Weinstein, J., Kelley, M. F., *et al.* (1990). *Proc. Natl. Acad. Sci. USA*, **87**, 1491.
24. Gauldie, J., Richards, C., Harnish, D., Lansdorf, P., and Baumann, H. (1987). *Proc. Natl. Acad. Sci. USA*, **84**, 7251.
25. Morrone, G., Ciliberto, G., Oliviero, S., Arcone, R., Dente, L., Content, J., *et al.* (1988). *J. Biol. Chem.*, **263**, 12554.
26. Gregory, B., Savino, R., and Ciliberto, G. (1994). *J. Immunol. Methods*, **170**, 47.
27. Fiorillo, M. T., Cabibbo, A., Iacopetti, P., Fattori, E., and Ciliberto, G. (1992). *Eur. J. Immunol.*, **22**, 2609.
28. Savino, R., Ciapponi, L., Lahm, A., Demartis, A., Cabibbo, A., Toniatti, C., *et al.* (1994). *EMBO J.*, **13**, 5863.
29. Rickwood, D. and Hames, B. D. (1992). In *Receptor-ligand interactions: a practical approach* (ed. E. C. Hulme), p. 63. IRL Press, Oxford.
30. Fuh, G., Mulkerrin, M. G., Bass, S., McFarland, F., Brochier, M., Bourell, J. H., *et al.* (1990). *J. Biol. Chem.*, **265**, 3111.
31. Arcone, R., Pucci, P., Zappacosta, F., Fontaine, V., Macolni, A., Marino, G., *et al.* (1991). *Eur. J. Biochem.*, **198**, 541.
32. Sporeno, E., Paonessa, G., Salvati, A. L., Graziani, R., Delmastro, P., Ciliberto, G., *et al.* (1994). *J. Biol. Chem.*, **15**, 10991.
33. Rodbard, D. and Frazier, G. R. (1975). In *Methods in enzymology* (ed. Bert W. O'Malley and Joel G. Hardman), Vol. 37, p. 3. Academic Press, NY.
34. Kidd, K. K. and Guano, G. (1995). In *PCR 2: a practical approach* (ed. M. J. McPherson, B. D. Hames, and G. R. Taylor), p. 1. IRL Press, Oxford.
35. Sambrook, J., Fritsch, E. F., and Maniatis, T. (ed.) (1989). *Molecular cloning: a laboratory manual*, p. 6.24–6.27, 2nd edn. Cold Spring Harbor Laboratory, Cold Spring Harbor, NY.
36. Kishimoto, T., Akira, S., Narazaki, M., and Taga, T. (1995). *Blood*, **86**, 43.
37. Stahl, N., Boulton, T. G., Farruggella, T., Ip, N. Y., Davis, S., Witthuhn, B. A., *et al.* (1994). *Science*, **263**, 92.
38. Hopp, T. P., Prickett, K. S., Price, V. L., Libby, R. T., March, C. J., Cerretti, D. P., *et al.* (1988). *Bio/Technology*, **6**, 1204.
39. Evan, G. I., Lewis, G. K., Ramsay, G., and Bishop, J. M. (1985). *Mol. Cell. Biol.*, **5**, 3610.
40. De Serio, A., Graziani, R., Laufer, R., Ciliberto, G., and Paonessa, G. (1995). *J. Mol. Biol.*, **254**, 795.
41. Neddermann, P., Graziani, R., Ciliberto, G., and Paonessa, G. (1996). *J. Biol. Chem.*, **271**, 30986.
42. Clackson, T. and Wells, J. A. (1994). *Trends Biotechnol.*, **12**, 173.
43. Bass, S., Greene, R., and Wells, J. A. (1990). *Proteins*, **8**, 309.
44. Smith, G. P. (1985). *Science*, **228**, 1315.
45. Harrison, J. L., Williams, S. C., Winter, G., and Nissim, A. (1996). In *Methods in enzymology* (ed. J. N. Abelson) Vol. 267, p. 83. Academic Press, San Diego, CA.
46. Kopf, M., Lamers, M., Bluethmann, H., and Koehler, G. (1994). *Nature*, **376**, 339.
47. Akira, S., Taga, T., and Kishimoto, T. (1993). *Adv. Immunol.*, **54**, 1.

48. Bataille, R., Barlogie, B., Lu, Z. Y., Rossi, J. F., Lavabre-Bertrand, T., Beck, T., *et al.* (1995). *Blood*, **86**, 685.
49. Demartis, A., Barnassola, F., Savino, R., Melino, G., and Ciliberto, G. (1996). *Cancer Res.*, **56**, 4213.
50. Samers, W., Stahl, M., and Seeha, J. S. (1997). *EMBO J.*, **16**, 989.

4

A chimeric approach for studying receptor binding domains in EGF-like molecules

E. J. J. VAN ZOELEN, A. E. G. LENFERINK, M. J. H. VAN VUGT, and M. L. M. VAN DE POLL

1. Introduction

The use of chimeric proteins has gained increasing popularity in studies on the structure–function relationship of biologically active peptides. Using recombinant DNA technology, proteins can readily be made in which specific domains have been exchanged for corresponding domains from other proteins. The use of chimeras obtained by such a domain-exchange approach has been shown to be particularly fruitful when comparing the biological properties of two structurally related proteins. It is essential that the two proteins of interest have different biological activities, and that a direct assay is available to distinguish these activities. That is the reason why this approach has been used particularly for proteins with specific binding or signalling activity, including polypeptide growth factors, growth factor receptors, and nuclear transcription factors.

A chimeric approach for studying the structure–function relationship of biologically active proteins is often used in parallel with an approach of site-directed mutagenesis, in which the effects of individual amino acid mutations are studied. In general a chimeric approach has four major advantages over an approach using site-directed mutagenesis:

(a) By exchanging domains between two structurally very related molecules, it can be expected that the chimeras obtained will have a very similar three-dimensional structure. In the case of a single amino acid substitution, however, activity may be lost if, as a result of a mutation, the proteins are unable to fold into an active conformation.

(b) Studies using site-directed mutagenesis tend to focus on amino acids which are conserved between structurally related proteins, including the amino acids required for proper protein folding. In contrast, a domain-exchange strategy will leave these conserved amino acids invariant, and focuses on the amino acids which are different between the proteins investigated.

(c) A major difference between the two approaches results from the observation that most single amino acid substitutions have either no effect on biological activity or result in a loss or reduction of such activity. Since a reduction in biological activity can have many different reasons, including problems with protein expression, processing, and folding, it is difficult to draw final conclusions based only on a reduction in activity. In a domain-exchange strategy in which one protein has a specific biological activity that is not shared by a structurally related second protein, the second protein will gain biological activity by introduction of the domains relevant for the activity of the first protein. Such an enhanced activity can be interpreted in a more straightforward manner than a general reduction in activity.

(d) A final advantage of a domain-exchange approach is that in general a relatively limited number of mutants is required to identify a region important for a specific biological activity. In the case of site-directed mutagenesis each individual amino acid can be mutated in many ways, resulting in either an unlimited number of mutants, or a bias for mutations of specific amino acids. In general it can be stated, that using a domain exchange strategy a domain essential for a specific biological activity can readily be narrowed down, and that the specific amino acids involved in this domain can subsequently be identified by an exchange of individual amino acids. Essential for such an approach remains that an assay is available in which the biological activities of the two proteins compared differ in a significant manner (reviewed in ref. 1).

2. Chimeric growth factors

A chimeric approach for studying protein domains involved in a specific signalling function has been widely used for such molecules as receptors and transcription factors, which require only the preparation of chimeric DNA constructs for expression in cells. In the case of growth factors, the recombinant chimeric proteins have to be extensively purified in order to assess the effect of the domain-exchange on its biological activity. As a result the number of studies on chimeric growth factors is still limited. So far this approach has been used particularly for studying the structure–function relationship of cytokines acting as haematopoietic cells. This is due to the fact that interspecies differences strongly affect the ability of cytokines to bind to their receptors. Moreover, different cytokines often share the same receptors for their activity to which they may bind with different affinity (1). Domain-exchange strategies have been used in this respect to identify receptor binding domains in interleukin (IL)-3 (1, 2), IL-4 (3), IL-5 (4), IL-6 (5, 6), leukaemia inhibitory factor (LIF; 7, 8), IL-8 (9), IL-12 (10), and granulocyte/macrophage colony stimulating factor (GM-CSF) (1). Such and other studies on the structure–function relationship of cytokines have resulted in the development of a number of haematopoietic growth factor receptor antagonists (11–15), in addition to the identification of a naturally occurring IL-1 receptor antagonist (16, 17).

In the case of ligands which bind to receptors with intrinsic kinase activity, the number of studies in which chimeric growth factors have been used is limited. Chimeric growth factors have been made of transforming growth factor (TGF) β1 and TGFβ2 to identify the region in TGFβ1 responsible for binding to α_2-macroglobulin (18) and betaglycan (19), both properties that are not shared by TGFβ2. Chimeras have been made between insulin-like growth factor (IGF)-1 and insulin, to identify the sites responsible for selectivity in binding to the IGF-1 and insulin receptor, respectively (20, 21). Amino acid exchanges between nerve growth factor (NGF) and neurotrophin-3 have identified residues responsible for high affinity binding to the NGF receptor (22), while a domain-exchange strategy between acidic and basic fibroblast growth factor (FGF) has identified the region responsible for heparin-dependence of activity (23). Platelet-derived growth factor (PDGF) can occur as homodimeric products of the PDGF A-chain or PDGF B-chain genes, or as the heterodimeric product of both. The PDGF A-chain can only bind to the PDGF α-receptor, while the PDGF B-chain can bind both the PDGF α-receptor and the PDGF β-receptor. Using an exchange strategy between PDGF-A and PDGF-B, the regions in both the growth factor molecules and in their receptors responsible for this selectivity in binding have been identified (24–26). In addition chimeric molecules between growth factors and toxins have been employed to introduce toxins into cells by virtue of the growth factor receptor internalization process (27). Although the above studies have given clear information on the structure–function relationship of these growth factor molecules, they have not resulted in the development of antagonists for those receptors which have intrinsic kinase activity.

3. EGF-related factors and their receptors

3.1 EGF-related factors

Ever since its discovery by Stanley Cohen (28), epidermal growth factor (EGF) has been one of the most intensively studied polypeptide hormones (29). In its mature form the human EGF molecule consists of 53 amino acids, characterized by three disulfide bridges which are essential for biological activity. EGF binds with high affinity to the EGF receptor, a 170 kDa glycoprotein that is abundantly expressed in cells from most solid tissues. The extracellular part of the receptor consists of four distinct domains, of which domains II and IV are rich in cysteines, whereas domain III contains the ligand binding site (30). The intracellular part of the receptor molecule contains a domain with intrinsic tyrosine kinase activity, which is activated upon ligand binding. Various pieces of evidence have been presented to show that receptor domerization upon ligand binding is essential for activation of the receptor, although in the case of the EGF receptor (EGFR) it is unclear how a small, single chain polypeptide like EGF can induce this dimerization (31, 32).

Studies over the last 15 years have shown that EGF belongs to a family of growth factors which all exert their action by binding to the EGFR. All of these growth factors share conserved cysteine residues with EGF, in addition to a number of other amino acids essential for receptor binding (see below). The best characterized other member of this family is transforming growth factor-α (TGFα), which in its mature form consists of 50 amino acids. In contrast to EGF, which is mainly expressed in mature tissue including salivary glands and kidney, TGFα is particularly expressed in embryonic tissue and tumour cells (33). Expression of TGFα in tumour cells generally correlates with tumorigenic behaviour of these cells, most likely because secreted TGFα is involved in autocrine and paracrine growth stimulation the corresponding tumour tissue. Other members of the EGF family include amphiregulin (AR), heparin binding EGF (HB-EGF), betacellullin (BTC), and epiregulin (EPR) (34). Of these molecules AR and HB-EGF contain an N terminal extension with a high content of positively charged amino acids, which facilitates binding of these growth factors to heparin, but may also serve as a nuclear target signal once internalized by the cell. In one report it has been claimed that Schwannoma-derived growth factor, the rat form of AR, is able to induce mitogenesis in cells carrying kinase-deficient EGF receptors, by its ability to act as a nuclear protein (35). Moreover, the transmembrane precursor form of HB-EGF has been shown to serve as the cellular target site for the diphtheria toxin (36). Besides the above six ligands for the EGFR encoded by the mammalian genome, three additional ligands for the EGFR are known which are encoded by pox viruses. These vaccinia virus, Shope fibroma, and myxoma growth factors all share the overall structure of EGF and TGFα, including the spacing of the conserved cysteine residues (29, 37). In addition to the above ligands for the EGF receptor, a family of proteins with a so-called 'EGF-like domain' has been characterized. These proteins share the cysteine spacing with EGF receptor ligands, but lack amino acids essential for receptor binding, and as a consequence they do not interact with the EGFR. Proteins which contain such an EGF-like domain include laminin, proteolytic enzymes and lipoprotein receptors, as well as the precursor molecule for EGF itself and the *Drosophila* receptor-type protein Notch (29,37).

3.2 Receptors for the EGF-related factors

The EGF (EGFR) receptor itself, also known as the erbB1 protein, belongs to a family of receptors, which includes the erbB2, also known as neu or HER2, the erbB3, and the erbB4 proteins (38). These three other receptor molecules share the major characteristics of the EGFR, but in general do not interact with the above ligands for the EGFR. Moreover, an additional family of growth factors, known as the heregulins (HRGs), has been characterized which are able to bind to erbB3 and erbB4, but not to erbB1 (38, 39). It is currently unclear if a specific ligand for erbB2 exists, although putative ligands have been identified (40). Recent data also indicate the betacellulin (BTC) may not only be a ligand for erbB1, but also for erbB4, demonstrating that cross-reactivity of ligands is possible (41, 42). In addition it has been shown that EGF may not only induce homodimeric erbB1-erbB1 receptors upon receptor binding, but also heterodimeric receptors of erbB1 with the other receptor members, of which the erbB1–erbB2 dimer appears to be most potent in transducing mitogenic signals. Similarly HRGs may induce receptor dimers of erbB3 and erbB4 (39, 43).

3.3 Structure–function relationship of EGF and TGFα

Figure 1 shows the primary structures of human EGF (*top*) and human TGFα (*bottom*). Besides their similarity in size, both proteins appear to have the same overall structure due to the presence of three conserved disulfide bridges. Both molecules consist of a linear N terminal region, a so-called A-loop consisting of amino acids 6–20 in EGF, a B-loop consisting of amino acids 20–31, a hinge region involving asparagine 32, a C-loop consisting of amino acids 33–42, and a C terminal linear region. In spite of the fact that EGF and TGFα bind with similar affinity to the EGFR, the overall amino acid identity is only 40%. Moreover, there seems no conservation of either charged or hydrophobic residues at the same position of the two molecules. It can be suggested that these growth factors only need a certain conformation to bind to their receptor, and that the nature of the individual amino acids may be less relevant. In support of this hypothesis it has been shown by NMR techniques that the overall three-dimensional structures of EGF and TGFα are very similar (37, 44). However, mutation of the conserved leucine at position 47 in EGF has been shown to result in mutants with a three-dimensional structure identical to native EGF, but fully lacking biological activity (34, 45). This strongly indicates that a specific conformation of these growth factors is essential, but not sufficient for high affinity receptor binding.

Many studies have been performed on the structure-function relationship of EGF and TGFα by mutating individual amino acids (34, 46). These studies have shown that four types of mutations can be discriminated:

(a) Mutation of the conserved cysteines and the glycines 22 and 39 of EGF (grey residues in *Figure 1*) always results in a complete loss of activity, most likely because they directly affect the ability of the growth factor to fold correctly.
(b) Residues can be discriminated (shaded amino acids in EGF) which can

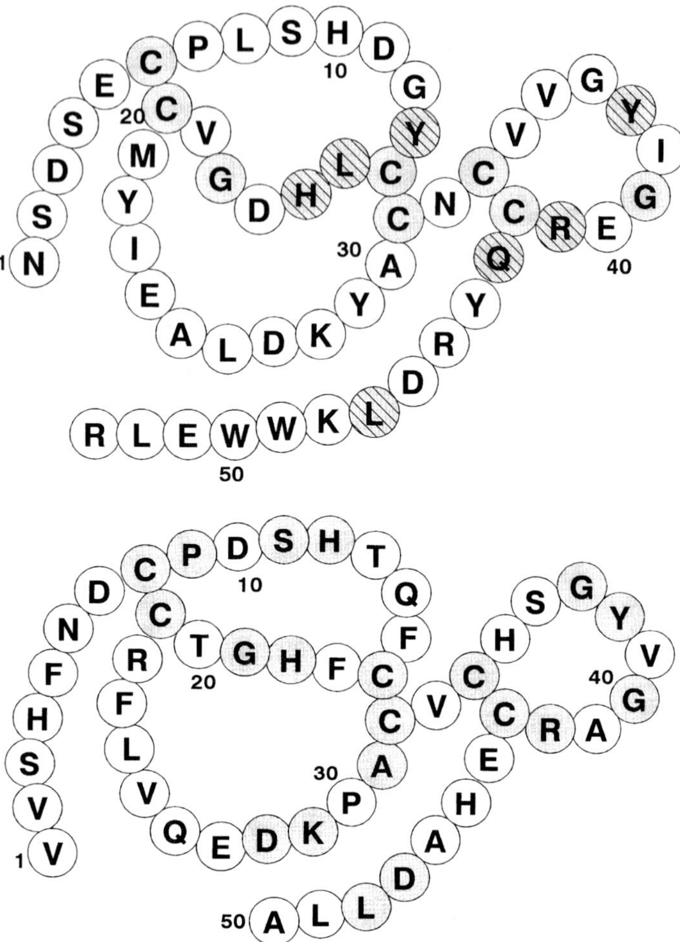

Figure 1. Primary structure of human EGF (*top*) and human TGFα (*bottom*). In the EGF sequence those cysteines and glycines that are conserved in all EGF like growth factors, and are essential for maintaining the proper three-dimensional structure, are shaded grey. The cross-hatched amino acids have been shown to be very sensitive for mutation, and are most likely involved in receptor binding. In the TGFα sequence the amino acids conserved with EGF are shaded in grey.

only be mutated to a very limited extent, and which are thought to be directly involved in receptor binding. Generally these amino acids are highly conserved in the various members of the EGF growth factor family. Based on these mutational studies in combination with NMR studies, it has been suggested that the initial part of the C terminal tail (amino acids 39–47) folds back to amino acids 13–16, to form together a non-linear receptor binding domain (34, 44).

(c) Amino acids can be discriminated which can readily be mutated, but only if an additional amino acid is altered as well. Both the B-loop and C-loop of EGF form a beta-sheet, in which, e.g. methionine 21 interacts with alanine 30, and tyrosine 22 with tyrosine 29. In TGFα different amino acids are present in these positions, which can interact two by two in a similar manner. As a consequence, only combined alterations of amino acids in this region permit the protein to fold in a correct way with the required beta-sheet and beta-turn structure.

(d) Amino acids which can be mutated with little or no effect. These are mainly located in the N terminal region of the molecule, and in the B-loop. This latter observation is remarkable, since early studies based on the use of synthetic peptides have indicated that a cyclic peptide corresponding to amino acids 20–31 of EGF could mimic the activity of EGF, although only when added at very high concentrations (47). From our own data, we also observed that the B-loop of EGF outside the region required for beta-sheet formation is relatively insensitive to mutations.

3.4 Chimeric proteins of the EGF family

Based on the overall structure of members of the EGF family, consisting of a linear N terminal region, three looped regions and a linear C terminal region, it is tempting to speculate that each of these domains may serve a specific function. In most domain-exchange studies on EGF-like molecules, regions have therefore been exchanged between the conserved cysteine residues. In an initial study by Purchio *et al.* (48), it has been shown that a chimera of vaccinia virus growth factor and TGFα is biologically active. We have made chimeras between human EGF and human TGFα to identify domains in these molecules involved in receptor binding (see below), while Richter *et al.* (49) have made chimeric molecules of murine EGF and human TGFα. Finally, Barbacci *et al.* (50) have made chimeras between heregulin and EGF, to show that the linear N terminal region of heregulin is essential for binding to erbB3 and erbB4, thereby creating ligands which bind both to erbB1 and erbB3 and erbB4. In addition many toxins have been linked to EGF, TGFα, and heregulin, to use the corresponding receptors as shuttles for entry of the toxins into the cells (27).

The use of EGF/TGFα chimeras is dictated by the observation that EGF and TGFα may bind with similar affinity to the human EGF receptor, but that, in contrast to TGFα, EGF binds only with low affinity to the chicken EGF receptor (30). By exchanging domains between the human and chicken EGF receptor, it has been established that the so-called domain III of the EGF receptor contains the ligands binding domain (30). This is an agreement with data of Wu *et al.* (51), who showed that an anti-EGFR antibody which competes with EGF for receptor binding, binds the peptide region corresponding to amino acids 352–366 of the human EGF receptor, which is fully located inside this domain III of the receptor.

We have used the above observation to analyse which domains of TGFα have to be introduced into EGF to make EGF a high affinity ligand for the chicken EGF receptor (52, 53). The rationale for this approach is that the region in TGFα thus identified should be part of the high affinity binding domain for the chicken EGF receptor. Since the human and chicken EGF are structurally very similar, however, it is anticipated that the amino acids thus identified by this chimeric gain-of-function approach may be generally important for binding of EGF-like molecules to their receptors.

4. Preparation and isolation of recombinant EGF-like growth factors

4.1 Introduction

EGF and TGFα have been made as recombinant proteins in both prokaryotes such as *Escherichia coli* (54), and eukaryotes including *Pichia pastoris* (55) and other yeast strains (56). Both EGF and TGF-α are derived from large transmembrane precursor molecules, which are cleaved at the cell surface by specific proteolytic enzymes to generate the mature growth factors (29). In all studies described so far, only the mature sequences of EGF and TGFα have been used to generate the recombinant growth factors. The formation of the proper disulfide bridges is essential for the biological activity of these growth factors, but it seems that a precursor form is not essential for obtaining properly folded EGF-like growth factors. Eukaryotic expression systems such as *Pichia pastoris* generally result in high production levels of properly folded EGF and TGFα, and are the systems of choice when only well-characterized growth factors have to be produced in milligram amounts, e.g. for structural analyses. When studying the structure–function relationship of growth factors, also mutants with low biological activity will have to be purified and characterized. Under those conditions it is of great importance to have a cleavable linker attached to the recombinant protein which can be used for subsequent isolation. Because of the small size of EGF and TGFα, non-reversible tags should be avoided, since they may strongly interfere with the biological properties of the obtained growth factors. The use of such reversible tags has been worked out particularly in *E. coli*, e.g. by making fusion proteins with bacterial protein A (see below).

In initial studies in which EGF and TGFα were made as recombinant proteins in *E. coli*, large quantities were obtained from inclusion bodies, which were however biologically inactive since no disulfide bridges are formed intracellularly in these prokaryotes. The obtained growth factors were subsequently refolded in the presence of reduced and oxidized glutathione, to yield a product in which approximately 5% of the growth factor molecules were in the active conformation (54). From our own experience, many commercial preparations of EGF and TGFα have similarly low specific activity, because active and inactive conformations have not been separated. It has been established that only EGF-like molecules with the proper disulfide bridges are biologically active, and studies by Winkler *et al.* (54) have shown that active TGFα can be separated from disulfide bridge mismatches using reverse-phase HPLC.

The use of protein A–fusion proteins of EGF and TGFα has made it possible to obtain growth factors with a high specific activity, since protein A is secreted by *E. coli* into the periplasmic space, a condition under which disulfide bridges will be formed (57). The presence of the protein A part facilitates purification of the growth factors, making use of the high affinity binding of protein A to immunoglobulins (IgGs). In our approach the bacterial protein A sequence has been linked to the sequence encoding mature EGF or TGFα through a spacer encoding a four amino acid sequence corresponding to the recognition sequence for the proteolytic enzyme factor X_a (Ile–Glu–Gly–Arg; see ref. 58). As a result, the protein A–fusion protein can first be isolated from the periplasmic space using IgG–Sepharose chromatography. After cleavage with factor X_a, the protein sample can be rerun on the same column, leaving protein A bound to the column, while eluting the growth factor molecules in a highly purified form. In our initial studies we have made use of the pHEMA 153 expression vector, containing the heat-inducible λP_R promoter and the coding sequence of *Staphylococcus aureus* protein A, including its signal sequence (59). In addition use was made of the degP protease-deficient *E. coli* strain KS474 to prevent proteolysis of the obtained growth factor molecules (60). Most likely due to the heat shock treatment, however, significant proteolysis still occurred, while in addition many disulfide bridge mismatches were obtained, resulting in a low specific activity of the derived growth factors (52). In our subsequent studies, we made use of the pEZZ18 vector (Pharmacia, Uppsala, Sweden), containing a constitutively active promoter, again expressed in the KS474 *E. coli* strain. Although the production of protein A was quite similar under both conditions, the major product formed was correctly folded growth factor, which could easily be separated from low amounts of disulfide bridge mismatches or dimeric growth factors by reverse-phase HPLC, eventually resulting in 10–100 μg amounts of fully active, highly purified growth factor (53). Details of the various procedures involved will be given in the following sections, when describing the EGF/TGFα chimeras used for analysis of receptor binding domains in these growth factors. The isolation of the protein A–fusion proteins from the periplasmic space of the KS474 *E. coli* bacteria is described in *Protocol 1*.

Protocol 1. Isolation of protein A–EGF from *E. coli*

Equipment and reagents

- Flasks (50 and 500 ml)
- Incubator–shaker GFL 3020 (New Brunswick Scientific)
- Spectophotometer (280 nm and 600 nm)
- Centrifuge (Hettich Universal)
- AP25 borosilicate microfibre glass filter (Millipore)
- P-3 peristaltic pump (Pharmacia)
- 2YTE: add 16g bacto tryptone, 10 g yeast extract, and 8 g NaCl to distilled water, 1 litre final volume—sterilize by autoclaving
- Sucrose buffer: 0.5 M sucrose, 0.1 M Tris–HCl pH 8.2, 1 mM EDTA
- 250 mg/ml ampicillin (Boehringer)
- TST buffer: 150 mM NaCl, 50 mM Tris–HCl pH 7.6, 0.05% Tween 20
- *E. coli* strain KS 474, containing pEZZ 18 vector encoding protein A–EGF (see Section 4.2)
- 20 mg/ml kanamycin (Gibco)
- 40% glucose solution
- 10 mg/ml lysozyme (Boehringer)
- 1 M $MgSO_4$
- IgG–Sepharose (20 ml bed volume; Pharmacia) in 2.5 × 50 cm column (Pharmacia)
- 5 mM NH_4Ac, adjust to pH 5.0 with HAc
- 0.5 M HAc, buffered with NH_4Ac to pH 3.4
- TST containing 20% ethanol

Method

1. To a sterile 50 ml flask add 10 ml 2YTE, 10 μl 250 mg/ml ampicillin, and 50 μl 20 mg/ml kanamycin. Add 200 μl of *E. coli* KS 474 pEZZ 18 stock, and incubate with shaking (200 r.p.m.) overnight at 30 °C.
2. The next day, prepare a 500 ml sterile flask containing 100 ml 2YTE, 1 ml 40% glucose, and 100 μl 250 mg/ml ampicillin. Add 5 ml of the pre-culture prepared in step 1, and incubate with shaking for approx. 3 h at 30 °C, until the optical density (600 nm) has reached a value of 1.5. Place the flask on ice.
3. Centrifuge the cooled *E. coli* suspension at 5000 r.p.m., for 7 min at 4 °C. Resuspend the pellet in 10 ml ice-cold sucrose buffer, and leave on ice for 10 min. Add 180 μl lysozyme (10 mg/ml), immediately followed by 10 ml ice-cold water. Mix, and leave on ice for 5 min. Add 400 μl 1 M $MgSO_4$, and centrifuge at 5000 r.p.m. for 20 min at 4 °C. Use the supernatant for subsequent purification steps.
4. Run at least 200 ml TST over the IgG–Sepharose column, and continue until the pH of the eluted buffer is 7.6. Also adjust the pH of the periplasmic solution at 7.6, if necessary. Filter the periplasmic solution through an AP25 filter to remove any insoluble material.
5. Apply the periplasmic material to the column at 4 °C, and pump at a speed of 20 ml/h.
6. Wash the IgG column with 200 ml TST buffer, followed by 40 ml 5 mM NH_4Ac pH 5.0, using the same pump speed.
7. Elute the IgG binding fraction by applying 100 ml 0.5 M HAc pH 3.4 to the column at a speed of 20 ml/h and collect 10 ml fractions. Measure

the absorbance of the column fractions at 280 nm (absorbance of 1 corresponds to 2.6 mg protein A/ml).

8. Rinse the column with 100 ml TST buffer at a speed of 100 ml/h, and subsequently store the column in TST containing 20% ethanol at 4 °C.

4.2 Construction and expression of EGF/TGFα chimeras

DNA constructs encoding the complete nucleotide sequence for human EGF and TGFα are commercially available (R & D systems) or can be made by linking overlapping synthetic oligonucleotides, as described (52), or by the use of PCR techniques (see *Chapter 1*). For expression in *E. coli* bacterial codon use is preferred, although human codon use is also applicable (see *Chapter 1*). Using synthetic oligonucleotides containing the sequence for the factor X_a recognition sequence, the 5′ end of the growth factor sequences can be linked to those for protein A in the pEZZ 18 expression vector. For construction of chimeras of EGF and TGFα in which domains were exchanged bordered by the conserved cysteine residues, a restriction enzyme was chosen which cleaves one of the growth factor genes just 5′ of a cysteine-encoding triplet, and an enzyme cleaving the other growth factor gene just 3′ of this cysteine. By filling in with specific complementary synthetic oligonucleotides with restriction enzyme recognition sites on both ends, the relevant parts of the growth factor genes were linked together again. For making smaller mutations, including exchanges of individual amino acids, synthetic oligonucleotides containing the required mutated sequence were used in the Altered Sites™ II *in vitro* mutagenesis system (Promega Cooperation, Madison, WI).

All chimeras were expressed in the pEZZ 18 expression vector, as described in *Protocol 1*. A representative Western blot of periplasmic material obtained from the protein A–fusion proteins of recombinant human EGF, and three EGF point mutants is shown in *Figure 2*. The protein A part of the fusion proteins was visualized by incubation with goat immunoglobulins linked to horse-radish peroxidase (Nordic Immunological Laboratories). Data show that the fusion proteins run at an apparent molecular size of approximately 20 kDa. Under non-reducing conditions higher molecular weight bands are visible, which are absent after reduction. Most likely these bands represent dimeric and other multimeric forms of these fusion proteins, originated by formation of disulfide bridges between different EGF molecules. Note that the protein A molecule itself contains no cysteine groups. The molecular size of the protein A part of the fusion protein is approximately 14 kDa, but since the vector contains no stop codon immediately following the protein A sequence, a size of 18 kDa is observed for the protein A product in the case in which no growth factor gene is subcloned (lane 5). The protein A molecule derived in this way contains only two IgG binding domains, in contrast to the protein A with five IgG binding domains which is produced by wild-type bacteria (57).

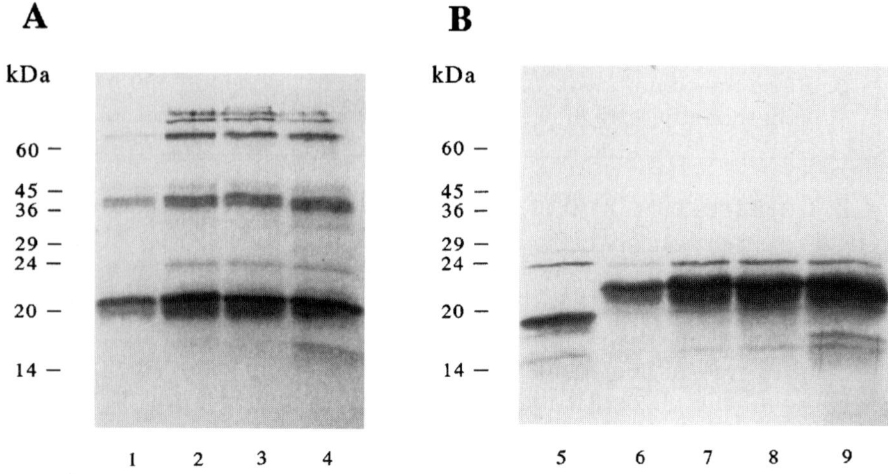

Figure 2. Analysis of protein A–growth factor fusion proteins by SDS–polyacrylamide gel electrophoresis and Western blotting. Aliquots of 10 μl of unpurified periplasm were run on a 12.5% SDS–polyacrylamide gel under (A) non-reducing or (B) reducing conditions. Proteins were transferred to nitrocellulose, and the Western blots were probed with goat immunoglobulins linked to horse-radish peroxidase. Wild-type hEGF, lanes 1 and 6; Q43E, lanes 2 and 7; Y44H, lanes 3 and 8; R45A, lanes 4 and 9; control periplasm (pEZZ 18 without insert), lane 5.

After purification of the protein A–fusion protein on IgG–Sepharose as described in *Protocol 1*, the amount of protein A and EGF-like activity produced can be quantified. For analysis of protein A production, a specific ELISA assay has been established by us using biotin-labelled protein A (Sigma) as labelled ligand. The procedure (61) is described in *Protocol 2*. The amount of EGF-like activity can be quantified by a binding competition assay on the human EGF receptor. This can either be done by competition with biotin-labelled murine (m)EGF (Boehringer Mannheim) for binding to human A431 cells (62), or in a radioreceptor assay by measuring competition with binding of commercially available [^{125}I] mEGF (Amersham International) to HER-14 cells (53). These are NIH3T3 cells transfected wih the human EGF receptor, giving rise to approximately 4.0×10^5 receptors per cell (63). The latter procedure is generally used by us, and is described in *Protocol 3* and worked out in *Figure 3*. Assuming that the protein A–EGF fusion protein has similar binding affinity for the receptor as native EGF, measurement of the molar ratio of protein A and EGF activity gives direct information on the percentage of biologically active growth factor produced. In our hands, the molar ratio of EGF binding activity and protein A production is generally close to 70% for most of the EGF/TGFα chimeras we have studied, indicating that the majority of growth factor molecules produced have the proper conformation.

Protocol 2. ELISA for protein A

Equipment and reagents

- 96-well ELISA plates (Greiner, Cat. No. 655101)
- Eppendorf centrifuge and tubes (1.5 ml)
- Plastic tubes (50 ml, Nunc)
- ELISA plate reader (SLT Lab instruments; filter setting 4)
- DMEM/BSA:Dulbecco's modified Eagle's medium, supplemented with 1% BSA
- PBS/Tween: phosphate-buffered saline (PBS), composed of 137 mM NaCl, 2.7 mM KCl, 0.9 mM $CaCl_2$ 0.5 mM $MgCl_2$, 6.5 mM Na_2HPO_4, 1.5 mM KH_2PO_4 pH 7.4, containing 0.05% Tween 20
- 1 mg/ml biotin–protein A (Sigma) in DMEM/BSA
- 1 mg/ml protein A (Sigma) stock solution in DMEM/BSA
- If a CO_2 incubator is used, DMEM is supplemented with 44 g/litre $NaHCO_3$ to maintain a pH of 7.7. If no CO_2 incubator is available binding studies can also be carried out in DMEM buffered with 15 mM Hepes and 8.8 g/litre $NaHCO_3$ pH 7.7, or in PBS/BSA pH 7.4 with similar efficiency.
- Streptavidin peroxidase (Boehringer) resuspended in 1 ml water
- Rabbit immunoglobulins (Nordic Immunological Laboratories) at 2 μg/ml in PBS
- *E. coli* periplasmic material (see Protocol 1)
- 2 M H_2SO_4
- 30% H_2O_2
- 2 mg/ml OPD (*o*-phenylenediamine, Sigma) in a 0.1 M citric acid, 0.2 M Na_2HPO_4 solution pH 5.0

Method

1. Coat a 96-well ELISA plate overnight at 4 °C with 100 μl per well of the rabbit immunoglobulin solution. Subsequently wash the coated plate twice with 100 μl PBS/Tween, and block non-specific protein binding sites by incubation with 100 μl per well of DMEM/BSA for at least 30 min at 37 °C.
2. Prepare 1 ml of a 1:1000 dilution of the biotin–protein A stock solution in DMEM/BSA in an Eppendorf tube. Take 4 μl the protein A stock solution in an Eppendorf tube and add DMEM/BSA to a final volume of 500 μl, to obtain the protein A standard solution with concentration of 8 μl/ml.
3. Take 25 ml DMEM/BSA in a 50 ml plastic tube, and add 125 μl of the diluted biotin–protein A solution. Take 500 μl *E. coli* periplasmic material in an Eppendorf tube, and add 2.5 μl of the diluted biotin–protein A solution.
4. Make a set of twofold serial dilutions of protein A standard solution with biotin–protein A containing DMEM/BSA, in the following way: pipette 250 μl of protein A standard solution into one well of a separate 96-well plate, and 125 μl of biotin–protein A diluted in DMEM/BSA into 11 subsequent wells. Transfer 125 μl from the protein A-containing well into the second, mix with a pipette, transfer 125 μl of this medium to the third well, and so on. The amount of protein A per 100 μl sample (which is the volume that will be transferred later for the ELISA reaction) therefore ranges from 800 ng in the first, 400 ng in the second, and so on to 0.20 ng in the twelfth well, keeping the final biotin–protein A concentration constant at 50 ng/ml.

Protocol 2. *Continued*

5. Make a similar set of dilutions of the *E. coli* periplasmic material. Pipette 250 μl of this material, as obtained in step 3, in the separate 96-well plate, and 125 μl of biotin–protein A in DMEM/BSA in 11 subsequent wells. Transfer 125 μl periplasmic material to the second well, mix, transfer 125 μl to the third well, and so on. The amount of protein A per 100 μl sample in each well will therefore correspond to 100, 50, 25, and so on to 0.024 μl periplasmic material, keeping again the biotin–Protein A concentration fixed at 50 ng/ml.
6. Remove the DMEM/BSA from the immunoglobulin-coated wells, and pipette 100 μl of the serially diluted standard or samples into the wells. All operations in steps 4–6 are carried out in duplicate. Incubate for 1 h at 37 °C in a 95% air/5% CO_2 atmosphere.
7. Wash the plates twice with 100 μl PBS/Tween. Pipette 5 μl 30% H_2O_2 to 5 ml OPD solution. Add 50 μl per well of this solution, and incubate for 1 min in the dark. Stop the reaction by addition of 50 μl 2 M H_2SO_4. Measure the absorbance at 492 nm using an ELISA reader.
8. Plot in one graph the absorbance as a function of the amount of protein A standard added per well. Plot in a second graph the absorbance as a function of the number of microlitres of periplasmic material added per well. From the first curve determine how many nanograms of proteins A are required for 50% competition of biotin–protein A binding, from the second curve how many microlitres of periplasmic material are required for 50% binding competition. The ratio of these two numbers determines how many protein A equivalents are present per microlitre periplasmic materials (protein A has an approximate size of 30 kDa).

Protocol 3. Radioreceptor assay for EGF activity

Equipment and reagents

- Gamma counter (Packard Cobra II)
- 24-well tissue culture plates (Nunc)
- Eppendorf centrifuge and tubes (1.5 ml)
- Eppendorf centrifuge and tubes (1.5 ml)
- 37 °C incubator (5% CO_2)
- Shaker (G_2 Gyratory; New Brunswick Scientific)
- HER-14 cells: NIH3T3 cells transfected with the human EGF receptor (obtained from Dr J. Schlessinger, New York University)
- 10 μg/ml murine EGF (Collaborative Research)
- DMEM-Bic10% NCS: bicarbonate-buffered DMEM, supplemented with 10% newborn calf serum (Hyclone)
- DMEM-BH10% NCS: the above medium now buffered with 15 mM Hepes and 8.8 g/litre $NaHCO_3$ pH 7.7
- 1% Triton X-100 in H_2O
- 0.5 μg/ml ^{125}I murine EGF (Amersham)
- *E. coli* periplasmic material (see *Protocol 1*)
- PBS (see Protocol 2)
- PBS containing 0.1% BSA

Method

1. Plate 1.0×10^5 trypsinized HER-14 cells per well of a 24-well plate, in 1 ml DMEM-Bic/10% NCS, and incubate dishes for two days at 37 °C in 95% air/5% CO_2.
2. Remove the growth medium and incubate the cells in 100 μl DMEM-BH/10% NCS. Serum-containing media are used to avoid non-specific binding, but experiments can be carried out with similar efficiency in PBS/BSA.
3. Add 10μl of ^{125}I EGF to 2.5 ml DMEM-BH/10% NCS, giving rise to a concentration of labelled ligand of 2 ng/ml.
4. Add 5 μl unlabelled EGF to 250 μl of DMEM-BH/10% NCS in an Eppendorf tube, and 125 μl of DMEM-BH/10% NCS in each of eight additional tubes. Make a set of twofold serial dilutions by transferring 125 μl of the EGF-containing solution from the first tube to the second, and after mixing, 125 μl of the second tube to the third, and so on. Add to each of the tubes a similar volume (125 μl) of radiolabelled EGF solution, obtained in step 3. The amount of unlabelled EGF per 100 μl sample (which will be transferred later for the binding assay) therefore ranges from 10 ng in the first tube, to 5 ng in the second tube, and so on down to 0.078 ng in the eighth tube. The concentration of ^{125}I EGF is kept a constant at 0.1 ng/100 μl.
5. Make a similar set of serial dilutions of *E. coli* periplasmic material, by pipetting 250 μl of material into an Eppendorf tube, and 125 μl of DMEM-BH/10% NCS into eight additional tubes. Make serial dilutions in a similar way by pipetting 125 μl periplasmic material to the second tube, and so on. To each of the tubes add 125 μl of radiolabelled EGF solution. The amount of periplasmic material per 100 μl sample will therefore range from 50 μl in the first tube, to 25 μl in the second, down to 0.39 μl in the eighth tube.
6. Remove the medium from the cells, and transfer in duplicate 100 μl of radiolabelled samples to each well. Also include samples with only labelled EGF at similar concentration. Incubate for 2 h at room temperature, with careful shaking at approx. 20 rounds/min.
7. Remove the radiolabelled medium, rinse the cells twice with PBS/BSA, and once with PBS. To the dry wells add 250 μl 1% Triton X-100, and incubate with shaking for 60 min at room temperature. Transfer the extracts to vials for γ-counting.
8. Plot in one graph the per cent binding of labelled EGF as a function of the amount of unlabelled EGF added, and in a second one the per cent binding as a function of the amount of periplasmic material added. From the amount added to obtain 50% binding competition, it is derived how many EGF binding equivalents are present per microlitre of

Protocol 3. *Continued*

periplasmic material. An example of this calculation method is given in *Figure 3* and corresponding legend. From the data obtained from the EGF binding and protein A assays, the molar ratio of protein A and EGF activity in the fusion protein can be calculated (M_r EGF: 6 kDa), which indicates the percentage of EGF present in an active conformation.

After purification of the fusion protein on IgG–Sepharose according to *Protocol 1*, the protein A part is subsequently cleaved off by treatment with factor X_a. This enzyme is commercially available, or can be isolated from human plasma and subsequently activated by treatment with Russel's viper venom (58). We prefer to couple factor X_a to CNBr-activated Sepharose, and to incubate the protein A-fusion protein for approximately eight hours at room temperature with the column-bound enzyme.

After an additional run on IgG-Sepharose to remove protein A and uncleaved fusion protein, a sample of pure EGF-like growth factor is obtained, which may however still contain disulfide bridge mismatches and multimeric forms. These inactive forms are separated from active growth factor by reverse-phase HPLC using a 15 × 0.39 cm Delta-Pak C18 column (Waters Corporation). Elution is carried ut by applying a linear gradient of acetonitrile in 0.1% trifluoroacetic acid, at a flow rate of 1 ml/min. *Figure 4* shows a representative example of purification of a human EGF point mutant. A comparison of the HPLC elution pattern (A) and an analysis of EGF receptor binding activity in each fraction (B) shows that the major protein peak (I) contains biological activity, as do two smaller peaks (IIa and IIb), which are most likely N or C terminally truncated forms with full mitogenic activity. The other major peaks in the elution pattern appear to be inactive. The fractions containing peak (I) are collected and the total amount of EGF-like binding activity is determined in a radioreceptor assay (*Protocol 3*). The amount of protein present is judged from the surface area of peak (1) in the HPLC elution pattern, assuming that absorbance at 229 nm, which is mainly caused by the presence of peptide bonds, is proportional to the amount of protein present. Calibration between absorbance at 229 nm and protein content was achieved by running EGF from murine salivary glands (Collaborative Research), which has a known receptor binding activity per amount of protein, over the same column (53).

5. Characterization of EGF/TGFα domain-exchange mutants

5.1 Introduction

In order to investigate which domains in EGF and TGFα are involved in high affinity receptor binding, we have initially made the ten EGF/TGFα chimeric

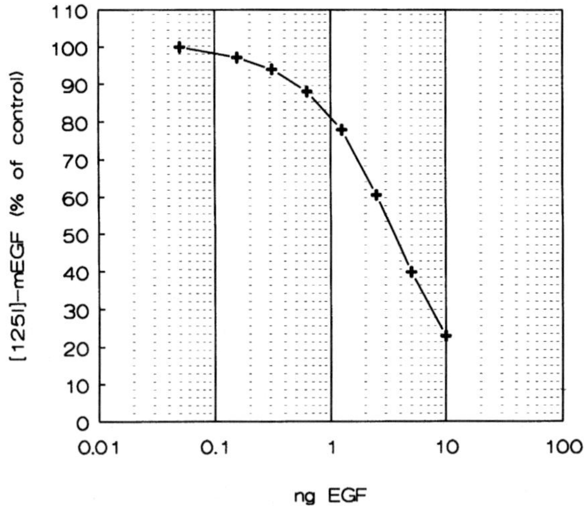

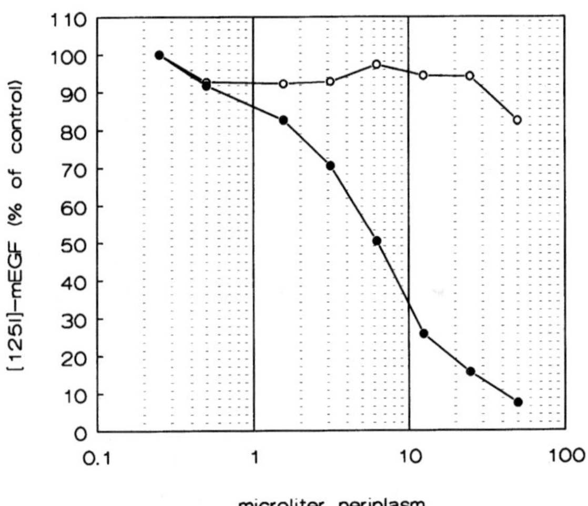

Figure 3. Calculation of EGF binding activity in *E. coli* periplasmic material. In the top panel binding of 0.1 ng [^{125}I]EGF to HER-14 cells is shown, in the presence of increasing amounts of unlabelled EGF, as described in detail in *Protocol 3* (incubation volume of 100 μl). From this graph it is derived that 50% binding competition is obtained upon addition of 3.3 ng unlabelled EGF. In the bottom panel, binding of a similar amount of [^{125}I]EGF to these cells is shown, now in the presence of increasing amounts of periplasmic material from *E. coli*, encoding either protein A alone (○-○) or protein A–EGF fusion protein (●-●). In the latter case, 50% binding competition is obtained upon addition of 6.2 μl of periplasmic material. From a combination of these data, it is derived that 1 μl periplasmic materials contains 3.3/6.2 = 0.53 ng EGF binding equivalents. Assuming a similar binding affinity of EGF and protein A–EGF for the EGF receptor, this implies that the concentration of biologically active protein A–EGF in the periplasmic material equals 88 nM (M_r EGF:6kDa).

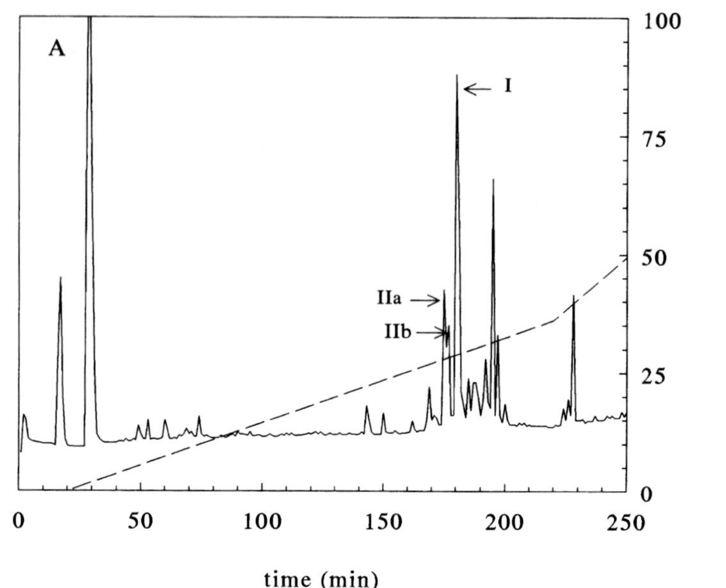

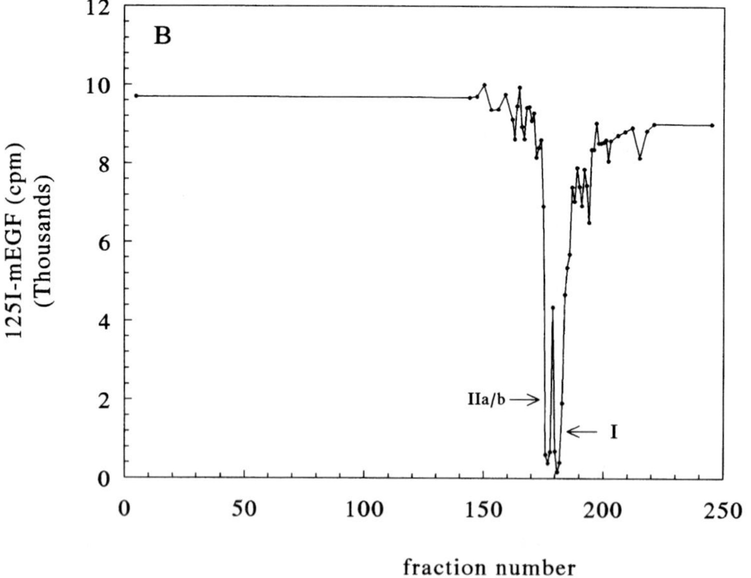

Figure 4. (A) RP-HPLC chromatogram of the hEGF point mutant Q43E and (B) biological activity in the RP-HPLC fractions. Elution was carried out with a linear gradient of CH_3CN in 0.1% trifluoroacetate at a flow rate of 1 ml/min. Biological activity present in the column fractions was determined in a binding competition assay with [^{125}I]mEGF on HER-14 cells (*Protocol 3*).

molecules presented in *Figure 5*. For these chimeric growth factors we have used a nomenclature, in which, e.g. E3T is a growth factor with EGF sequences N terminal of the third cysteine residue, and TGFα sequences C terminal of this residue (52). All growth factors were expressed and HPLC purified, as described above. First of all the affinity of these chimeras for the human EGF receptor was determined by measuring the concentration of chimera, calibrated as explained above, required for half-maximum inhibition of binding of [^{125}I] mEGF to HER-14 cells. These studies demonstrated that within a factor of two all chimeras had similar binding affinity for the human EGF receptor to that of the wild-type growth factors EGF and TGFα(53).

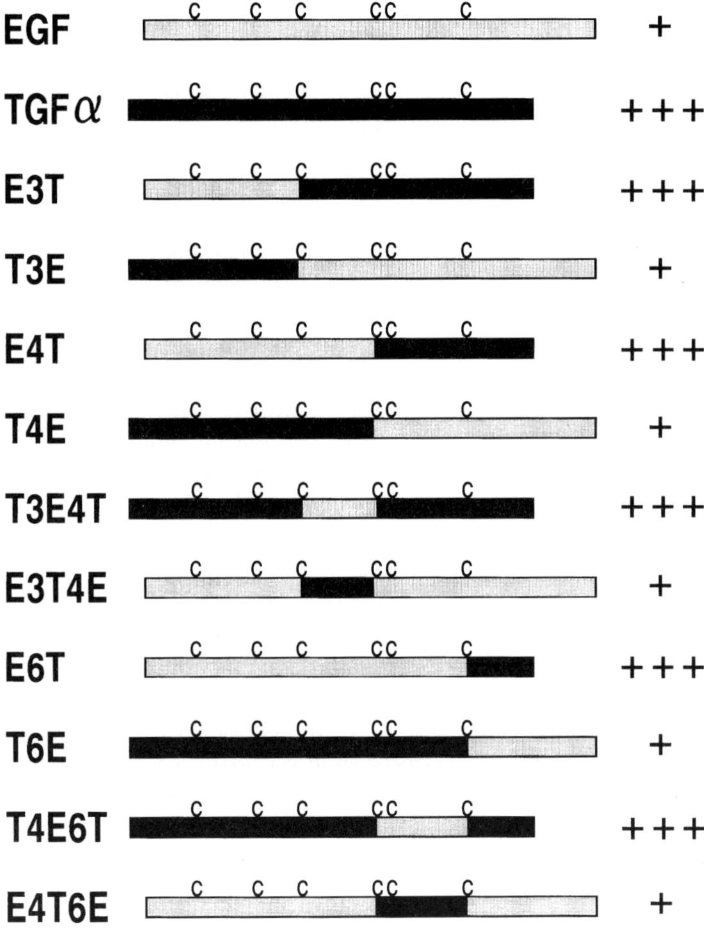

Figure 5. Survey of EGF/TGFα domain-exchange mutants. Domains exchanged were bordered by the indicated conserved cysteine residues (C). Growth factor chimeras with high affinity for the chicken EGF receptor are listed as (+++), with low affinity as (+).

Based on the observation that the chicken EGF receptor binds TGFα with high and EGF with a 100-fold lower affinity, we have subsequently determined the binding affinity of these various chimeras for CER-109 cells (NIH3T3 cells transfected with the chicken EGF receptors; a gift from Dr J. Schlessinger) (30). The data of *Figure 5* show that the mutants E3T, E4T, T3E4T, E6T, and T4E6T have high affinity for the chicken EGF receptor, similar to that of TGFα, while T3E, T4E, E3T4E, T6E, and E4T6E have low binding affinity, close to that of EGF. These data show that all ligands with a linear C terminal domain of TGFα have high affinity for the chicken EGF receptor, while all ligands with EGF sequences in this domain have a low affinity for this receptor. Since the affinity of all these ligands is equally high for the human EGF receptor, most likely sequences are present in the C terminal region of EGF which prevent high affinity binding to the chicken, but not to the human EGF receptor.

After this first approach of narrowing down the region relevant for this affinity difference, we have made an additional set of mutants by exchanging subdomains or individual amino acids between the linear C terminal regions of EGF and TGFα. These mutants are indicated in *Figure 6*, in which the C terminal domain has been subdivided into a domain N terminal and a domain C terminal to the conserved leucine 47 in EGF. Again similar affinities were found for all ligands for binding to the human EGF receptor. When measuring the binding affinity for the chicken EGF receptor, a more complex picture emerged. The presence of TGFα sequences in the linear C terminal region of the molecule was generally sufficient to give intermediate or high binding affinity to the chicken EGF receptor, as demonstrated for the mutants E6ET, E6TE, T6ET, and T6TE. This demonstrates that no single amino acid can be identified which is responsible for the low affinity binding of EGF to the chicken EGF receptor. On the other hand, exchange of single amino acids between EGF and TGFα may be sufficient to make EGF a high affinity ligand for the chicken EGF receptor, as illustrated by the mutant EGF–R45A. Other single amino acid exchanges (EGF–Q43E and Y44H) did not affect the binding affinity of EGF. Finally, truncation of the linear C terminal tail of EGF to a length similar to that of TGFα does also not affect binding affinity, indicating that steric hindrance of the long C terminal tail of EGF most likely does not affect receptor binding. Interestingly, EGF50 appears to have higher binding affinity for the chicken EGF receptor than both EGF48 and intact EGF (53).

The observation that exchange of a single amino acid in EGF (R45A) is sufficient to establish a high affinity ligand for the chicken EGF receptor, strongly indicates that this region of the molecule is directly involved in receptor binding. Mutation of this positively charged amino acid into a neutral one always results in high affinity binding to this receptor, irrespective of the other sequence in the molecules. However, some mutants, including E6ET, T6ET, and EGF50 have high or intermediate affinity in spite of the positive charge at amino acid 45. These ligands all have a short C terminal tail without

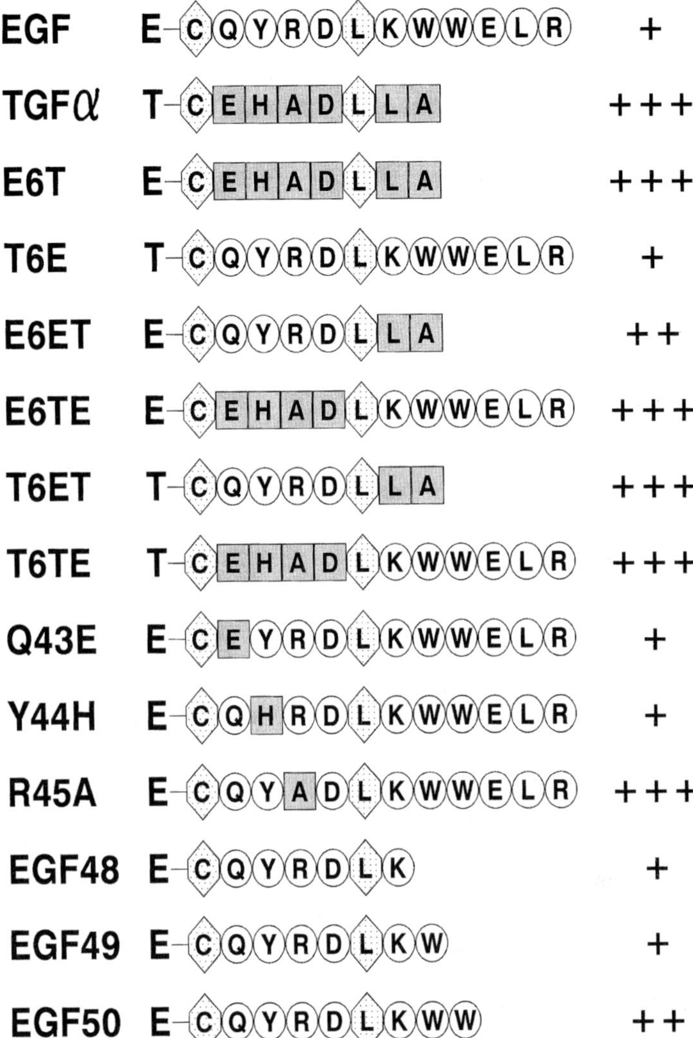

Figure 6. Mutants of EGF/TGFα in the linear C terminal region. The conserved sixth cysteine (C42 in EGF) and leucine (L47 in EGF) have been indicated in diamonds, EGF sequences in circles, and TGFα sequences in squares. The sequence N terminal of the sixth cysteine is indicated by either E (EGF sequence) or T (TGFα sequence). High affinity of the mutants for the chicken EGF receptor is indicated by (+++), and low affinity by (+).

a positive charge at the last or last but one position. This strongly suggests that positive charges in the C terminal region of EGF interfere with high affinity binding to the chicken EGF receptor, because of charge repulsion between them and similarly charged amino acids in the ligand binding domain of the receptor.

It has been well-established that the so-called domain III of the EGF receptor contains the ligand binding site. Using cross-linking studies with an antibody that competes with EGF for receptor binding, it has been established that amino acids 352–366 of the human EGF receptor are in direct contact with the ligand (51). *Figure 7* shows a comparison of the sequences of the human (HER), murine (MER), and chicken (CER) EGF receptor in this region of the molecule, including the charges of the side chains of the amino acids. The amino acids in MER and CER which are not conserved with HER have been indicated in grey (data obtained from ref. 64). These data show that HER and CER differ by four amino acids in this region; studies by Brown *et al.* (65) have shown, however, that substitution of these four amino acids of the chicken EGF receptor into the human receptor does not impair the high affinity binding of EGF to this receptor. In addition to these four amino acids, these receptors also differ by two positively charged lysines just outside the putative ligand binding domain, which are present in CER but not in HER (see *Figure 7*). In the case of the murine EGF receptor (MER) a single arginine is present in this position, and EGF is known to be a high affinity ligand for the murine receptor. Therefore we suggest that the combination of two

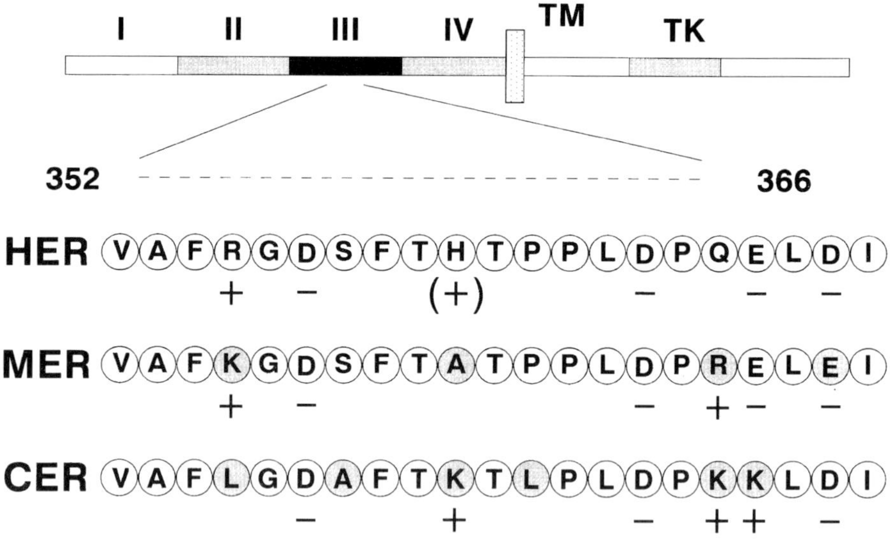

Figure 7. Conservation of the putative EGF binding domain in the human (HER), mouse (MER), and chicken (CER) EGF receptors. The putative ligand binding domain identified by Wu *et al.* (51) corresponds to amino acids 352–366 (horizontal dashed line) from the so-called domain III of the human EGF receptor. Domains II and IV correspond to cysteine-rich domains in the extracellular region of the receptor, TM (spanning the dotted area) stands for transmembrane domains, and TK for tyrosine kinase domain of the receptor. Amino acids in grey in MER and CER are not conserved in HER. In addition, the charge of amino acid side chains has been indicated (data from ref. 64).

positively charged amino acids in CER prevents high affinity binding of EGF due to charge repulsion with R45, and maybe also other positively charged amino acids further C terminal in EGF. Experimental data gained by mutating these two lysines to the corresponding amino acids in the human receptor should enable us to test this hypothesis.

5.2 Functional studies on EGF/TGFα chimeras

Further studies have shown that the above EGF/TGFα chimeras cannot only be used for determining receptor binding domains in these growth factors, but also for studying signal transduction by these factors. Preliminary data show that a number of the chimeras described above are more potent in growth stimulation of human cells than the native growth factors EGF and TGFα, in spite of the fact that all ligands have similar binding affinity for the human receptor. This superagonistic behaviour is associated with higher association and dissociation rate constants for these receptors, increased ability to phosphorylate MAP kinase, and enhanced intracellular signalling of the receptors after ligand-induced internalization (Lenferink *et al.*; unpublished data). Current experiments are aimed at determining which characteristics of these superagonists correlate best with their enhanced mitogenic activity, which may have important consequences for our view on growth factor-induced receptor activation. In conclusion, the chimeric approach described here has allowed us to identify receptor binding domains in EGF-like molecules by testing only a relatively limited number of mutant growth factors. Moreover, these chimeras appear to be of great value for testing EGF receptor-mediated signalling.

Acknowledgements

The current research is supported by the Dutch Cancer Society (KWF).

References

1. Kaushansky, K. (1992). *Proteins: Structure Function Genetics*, **12**, 1.
2. Dorssers, L. C. J., Burger, H., Wagemaker, G., and De Koning, J. P. (1994). *Growth Factors*, **11**, 93.
3. Bonsch, D., Kammer, W., Lischke, A., and Friedrich, K. (1995). *J. Biol. Chem.*, **270**, 8452.
4. McKenzie, A. N., Barry, S. C., Strath, M., and Sanderson, C. J. (1991). *EMBO J.*, **10**, 1193.
5. Van Dam, M., Müllberg, J., Schooltink, H., Stoyan, T., Brakenhoff, J. P. J., Graeve, L., *et al.* (1993). *J. Biol. Chem.*, **268**, 15285.
6. Fiorillo, M. T., Cabibbo, A., Iacopetti, P., Fattori, E., and Ciliberto, G. (1992). *Eur. J. Immunol.*, **22**, 2609.

7. Owczarek, C. M., Layton, M. J., Metcalf, D., Clark, R., Gough, N. M., and Nicola, N. A. (1995). *Ann. N. Y. Acad. Sci.*, **21,** 165.
8. Layton, M. J., Lock, P., Metcalf, D., and Nicola, N. A. (1994). *J. Biol. Chem.*, **269,** 17048.
9. Heinrich, J. N., O'Rourke, E. C., Chen, L., Gray, H., Dorfman, K. S., and Bravo, R. (1994). *Mol. Cell. Biol.*, **14,** 2849.
10. Gladue, R. P., Laguerre, A. M., Magna, H. A., Carroll, L. A., O'Donnell, M., Changelian, P. S., *et al.* (1994). *Cytokine*, **6,** 318.
11. Heldin, C. H. and Ostman, A. (1996). *Cytokine Growth Factor Rev.*, **7,** 3.
12. Wells, T. N., Power, C. A., Lusti-Narasimhan, M., Hoogewerf, A. J., Cooke, R. M., Chung, C. W., *et al.* (1996). *J. Leuk. Biol.*, **59,** 53.
13. Kruse, N., Shen, B. J., Arnold, S., Tony, H. P., Müller, T., and Sebald, W. (1993). *EMBO J.*, **12,** 5121.
14. Savino, R., Lahm, A., Salvati, A. L., Ciapponi, L., Sporeno, E., Altamura, S., *et al.*. (1994). *EMBO J.*, **13,** 1357.
15. Ehlers, M., De Hon, F. D., Bos, H. R., Horsten, U., Kurapkat, G., Van de Leur, H. S. *et al.* (1995). *J. Biol. Chem.*, **270,** 8158.
16. Hannum, C. H., Wilcox, C. J., Arend, W. P., Joslin, F. G., Dripps, D. J., Heimdal, P. L. *et al.* (1990). *Nature,* **343,** 336.
17. Carter, D. B., Deibel, M. R., Dunn, C. J., Tomich, C. S., Laborde, A. L., Slightom, J. L., *et al.* (1990). *Nature,* **344,** 633.
18. Webb, D. J., Atkins, T. L., Crookston, K. P., Burmester, J. K., Qian, S. W., and Gonias, S. L. (1994). *J. Biol. Chem.*, **269,** 30402.
19. Qian, S. W., Burmester, J. K., Merwin, J. R., Madri, J. A., Sporn, M. B., and Roberts, A. B. (1992). *Proc. Natl. Acad. Sci. USA*, **89,** 6290.
20. Kristensen, C., Andersen, A. S., Hach, M., Wiberg, F. C., Schaffer, L., and Kjeldsen, T. (1995). *Biochem. J.*, **305,** 981.
21. Schaffer, L., Kjeldsen, T., Andersen, A. S., Wiberg, F. C., Larsen, U. D., Cara, J. F., *et al.* (1993). *J. Biol. Chem.*, **268,** 3044.
22. Kullanders, K. and Ebendal, T. (1994). *J. Neurosci. Res.*, **39,** 195.
23. Imamura, F., Friedman, S. A., Gamble, S., Tokita, Y., Opalenik, S. R., Thompson, J. A., *et al.* (1995). *Biochim. Biophys. Acta*, **1266,** 124.
24. Heidaran, M. A., Mahadevan, D., and LaRochelle, W. J. (1995). *FASEB J.*, **9,** 140.
25. LaRochelle, W. J., Giese, N., May-Siroff, M., Robbins, K. C., and Aaronson, S. A. (1990). *Science*, **248,** 1541.
26. Ostman, A., Andersson, M., Hellman, U., and Heldin, C. H. (1991). *J. Biol. Chem.*, **266,** 10073.
27. Kreitman, R. J. and Pastan, I. (1994). *Adv. Pharmacol.*, **28,** 193.
28. Cohen, S. (1962). *J. Biol. Chem.*, **237,** 1555.
29. Carpenter, G. and Wahl, M. I. (1991). In *Peptide growth factors and their receptors* (ed. M. B. Sporn and A. B. Roberts), Vol. 1, pp. 69–172. Springer–Verlag, New York.
30. Lax, I., Bellot, F., Howk, R., Ullrich, A., Givol, D., and Schlessinger, J. (1989). *EMBO J.*, **8,** 421.
31. Gullick, W. J. (1994). *Eur. J. Cancer*, **30A,** 2186.
32. Carraway, K. L. and Cerione, R. A. (1993). *J. Biol. Chem.*, **268,** 23860.
33. Lee, D. C., Fenton, S. E., Berkowitz, E. A., and Hissong, M. A. (1995). *Pharmacol. Rev.*, **47,** 51.

34. Groenen, L. C., Nice, E. C., and Burgess, A. W. (1994). *Growth Factors*, **11,** 235.
35. Kimura, H. (1993). *Proc. Natl. Acad. Sci. USA*, **90,** 2165.
36. Mitamura, T., Higashiyama, S., Taniguchi, N., Klagsbrun, M., and Mekada, E. (1995). *J. Biol. Chem.*, **270,** 1015.
37. Campbell, I. D., Cooke, R. M., Baron, M., Harvey, T. S., and Tappin, M. J. (1989). *Prog. Growth Factor Res.*, **1,** 13.
38. Carraway, K. L. and Cantley, L. C. (1994). *Cell*, **78,** 5.
39. Peles, E. and Yarden, Y. (1993). *BioEssays,* ***15,*** 815.
40. Sheng, Z., Wu, K., Carraway, K. L., and Fregien, N. (1992). *J. Biol. Chem.*, **267,** 16341.
41. Beerli, R. R. and Hynes, N. E. (1996). *J. Biol. Chem.*, **271,** 6071.
42. Riese, D. J., Bermingham, Y., Van Raaij, T. M., Buckley, S., Plowman, G. D., and Stern, D. F. (1996). *Oncogene*, **12,** 345.
43. Earp, H. S., Dawson, T. L., Li, X., and Yu, H. (1995). *Breast Cancer Res. Treatm.*, **35,** 115.
44. Hommel, U., Harvey, T. S., Driscoll, P. C., and Campbell, I. D. (1992). *J. Mol. Biol.*, **227,** 271.
45. Walker, F., Nice, E., Fabri, L., Moy, F. J., Liu, F. J., Wu, R., *et al.* (1990). *Biochemistry*, **29,** 10635.
46. Campion, S. R. and Niyogi, S. L. (1994). *Prog. Nucleic Acid Res. Mol. Biol.*, **49,** 353.
47. Komoriya, A., Hortsch, M., Meyers, C., Smith, J., Kanety, H., and Schlessinger, J. (1984). *Proc. Natl. Acad. Sci. USA*, **81,** 1351.
48. Purchio, A. F., Twardzik, D. R., Bruce, A. G., Wizenthal, L., Madisen, L., Ranchalis, J. E., *et al.* (1987). *Gene*, **60,** 175.
49. Richter, A., Drummond, D. R., MacGarvie, J., Puddicombe, S. M., Chamberlin, S. G., and Davies, D. E. (1995). *J. Biol. Chem.*, **270,** 1612.
50. Barbacci, E. G., Guarino, B. C., Stroh, J. G., Singleton, D. H., Rosnack, K. J., Moyer, J. D., *et al.* (1995). *J. Biol. Chem.*, **270,** 9585.
51. Wu, D., Wang, L., Sato, G. H., West, K. A., Harris, W. R., Crabb, J. W., *et al.*(1989). *J. Biol. Chem.*, **264,** 17469.
52. Kramer, R. H., Lenferink, A. E. G., Lammerts van Bueren-Koornneef, I., Van der Meer, A., Van de Poll, M. L. M., and Van Zoelen, E. J. J. (1994). *J. Biol. Chem.*, **269,** 8708.
53. Van de Poll, M. L. M., Lenferink, A. E. G., Van Vugt, M. J. H., Jacobs, J. J. L., Janssen, J. W. H., Joldersma, M., *et al.* (1995). *J. Biol. Chem.*, **270,** 22337.
54. Winkler, M. E., Bringman, T., and Marks, B. J. (1986). *J. Biol. Chem.*, **261,** 13838.
55. Clare, J. J., Romanos, M. A., Rayment, F. B., Rowedder, J. E., Smith, M. A., Payne, M. M., *et al.* (1991). *Gene*, **105,** 205.
56. Brake, A. J., Merryweather, J. P., Coit, D. G., Heberlein, U. A., Masiarz, F. R., Mullenbach, G. T., *et al.* (1984). *Proc. Natl. Acad. Sci. USA*, **81,** 4642.
57. Nilsson, B. and Abrahmsen, L. (1990). In *Methods in Enzymology*, **185**, 144–61.
58. Nagai, K. and Thøgersen, H. C. (1987). In *Methods in Enzymology*, **153**, 461–81.
59. Engel, H., Mottle, H., and Keck, W. (1993). *Protein Expression Purif.*, **3,** 108.
60. Strauch, K. L., Johnson, K., and Beckwith, J. (1989). *J. Bacteriol.*, **171,** 2689.
61. Van Zoelen, E. J. J., Kramer, R. H., Van Reen, M. M. M., Veerkamp, J. H., and Ross, H. A. (1993). *Biochemistry*, **32,** 6275.

62. King, I. C. and Catino, J. J. (1990). *Anal. Biochem.*, **188,** 97.
63. Honegger, A., Dull, T. J., Bellot, F., Van Obberghen, E., Szapary, D., Schmidt, A., *et al.* (1988). *EMBO J.*, **7,** 3045.
64. Avivi, A., Lax, I., Ullrich, A., Schlessinger, J., Givol, D., and Morse, B. (1991). *Oncogene*, **6,** 673.
65. Brown, P. M., Debanne, M. T., Grothe, S., Bergsma, D., Carin, M., Kay, C., *et al.* (1994). *Eur. J. Biochem.*, **225,** 223.

5

Structure–activity relationships of chemokines

IAN CLARK-LEWIS, JENNIFER ANDERSON, PHILIP OWEN, LUAN VO, and JIANG-HONG GONG

1. Introduction

This chapter focuses on structure–activity relationships (SAR) of protein growth and activation factors citing examples from our work on chemokines (1, 2). The emphasis is on general principles and approaches to SAR of secreted and soluble proteins, the primary function of which is to bind specific integral membrane receptors leading to a distinct cellular response.

The study of SAR requires changing the covalent structure of a protein and determining the effects of the changes in terms of function and three-dimensional (3D) structure. The structure could be changed by either chemical modification or by *de novo* synthesis. Compared with chemical modification, the synthesis approach has the advantages that a large repertoire of changes can be introduced into any position and that the covalent structure is defined. Thus, the *de novo* synthesis approach will be the subject of this chapter.

The mutant proteins are usually made by recombinant DNA techniques, but can be made by chemical (peptide) synthesis in the case of chemokines and cytokines. The methods for obtaining the protein will not be discussed here. Rather, the emphasis will be on the key steps in determining SAR, after the crude protein mixture has been obtained. The protocols are from our work with chemically synthesized analogues, but are potentially applicable to proteins from other sources.

The key steps are:

(a) Design of the modified protein.

(b) Generation of the protein product, which includes folding, purification, authentication, and methods for handling of the protein.

(c) Assaying for biological functions and for receptor interaction.

(d) Interpretation of the outcome.

1.1 Chemokines

Chemokines (chemoattractant cytokines) are a structurally related family of proteins with pro-inflammatory functions (3). They are thought to function in the initiation and progression of inflammatory diseases by stimulating the accumulation and activation of myeloid and lymphoid cells. Chemokines lack growth promoting, or apoptosis-inducing properties. Rather they stimulate the migration (chemotaxis) and functional activation (release of mediators) of already fully mature cells that have little capacity for self-renewal. However, they are growth factor-like in that they are inducible, secreted proteins and bind to receptors on target cells resulting in transmission of intracellular signals. Thus the practical approaches to the study of chemokines are similar to those for more typical growth factors.

Despite the similarities to growth factors, the chemokines have some differences. For example, chemokines share considerable sequence similarity, they have a common defined structural fold, and bind to members of the seven transmembrane receptor superfamily. The seven transmembrane receptor class is different in structure and function to the cytokine and tyrosine kinase receptor classes. Therefore interpretation of receptor interactions should be compared to that for other ligands of this receptor family.

Chemokines are about 70–80 amino acids in length, have a M_r around 8000 Da, and usually have four cysteines, which pair to form two intrachain disulfide bonds (1, 2). The human chemokines that have been characterized number about 25 and are divided into two groups according to the relative positions of the first two cysteines. For the CXC chemokines, e.g. interleukin-8 (IL-8),the first two cysteines are separated by one residue, and for the CC chemokines, e.g. monocyte chemoattractant protein (MCP)-1, they are adjacent (3).

1.2 Primary structure: practical considerations

The full amino acid sequence of the naturally occurring and fully active protein must be known definitively. This is particularly important for chemokines as residues close to the amino terminus are usually critical. For example, the actual amino terminal residue of MCP-1 is essential for function (4). When using sequences for proteins that are described in the literature, it is important to realize that often the precursor is described, but the processed functional form is not always known. In addition, the amino terminal residue of the mature protein may have been predicted, rather than determined experimentally and such predictions are not always reliable. Forms that have been further processed after secretion are often found in medium and tissue fluids. Finally, one should be aware of the possibility of post-translational modification, particularly glycosylation as this is the most common covalent modification of growth factors. Even though glycosylation seems to be only rarely important for function, it could affect physical properties such as solubility.

It is essential to produce protein with the native (wild-type) sequence using the same procedure as will be used to prepare any analogues, and to characterize this product to ensure that it has the same physical and functional properties as the molecule derived from natural sources. After this is done, this 'control' preparation will become the standard to which the analogues are compared.

1.3 Importance of the 3D structure

There are two fundamental rules that must be considered in SAR. First, the primary structure determines the folding and 3D structure of the protein (5), and secondly, the 3D structure cannot be predicted from the sequence (6). This means that the analogue will have a different 3D structure from the native protein but this structural difference may not always be detectable by standard techniques. The nature of the difference depends on the individual modification(s).

A knowledge of the 3D structure of the wild-type protein, determined either by X-ray crystallographic or NMR techniques, is a major advantage for the design and interpretation of the experiment. Examination of the structure allows reasonable hypotheses to be proposed and tested. On the other hand, if only sequence information is available, it is difficult to devise analogues that will provide meaningful answers to questions of SAR.

It is often important to be able to interpret the changes to the primary structure in terms of the 3D structure as well as function. The critical question is: does the change affect function directly by affecting binding or indirectly by affecting folding? In the latter case, residue(s) critical for receptor binding could be far removed from the mutated residue, and thus the 3D structure is critical for interpretation of the experiment. The possible effect on 3D structure should be taken into account in designing the experiment so that the outcome can be interpreted. The structure indicates which residues are involved in secondary structure and other interactions, and provides information as to which residues are exposed and which are buried. Nevertheless, knowing the structure may not always mean that the SAR experiments can be interpreted in terms of the structure. For example, in the case of IL-8 the critical regions are substantially disordered in the protein. Presumably these residues assume a defined conformation only upon binding to the receptor.

Comparisons to the sequences of other family members allow comparison of variable and conserved regions with the structural elements. Known structures of related proteins can be used to develop hypotheses for SAR. Although these comparisons assist in the interpretation of the experiment, the ideal is to be able to compare the determined structure of both wild-type and the analogues. This is not always possible as the number of analogues required for SAR, and the rate that they can be produced, greatly exceeds the rate at which they can be analysed structurally.

2. Design: testing an hypothesis

The first step is to develop an hypothesis, e.g. that residue X, which is conserved in related proteins, is functionally important. The more precise and testable the hypothesis, the more likely the design will be a success. Random or *ad hoc* changes to the sequence are not likely to be useful. As a general rule, changes that reduce functional activity require careful interpretation as the effects could be indirect and therefore secondary to the modification.

2.1 Single substitutions

The aim of single substitutions is to determine if a particular residue is sensitive to modification and to be able to make predictions regarding its role in structure or function. Changes that do not affect activity can be as informative as those that do. Often stronger arguments can be made by making multiple substitutions of a given residue. For example, a residue could be substituted by a spectrum of different substituents, e.g. Ala, Leu, Lys, Glu, Ser, and Gln. These substitutions include small and large non-polar, positively and negatively charged, and polar residues. In addition, changes to Cys, Gly, and Pro can permit examination of changes in the secondary and tertiary structure.

2.2 Multiple substitutions

The examination of an entire polypeptide by making four or five analogues with single substitutions at each position is an extremely laborious way of determining SAR. Furthermore, it is possible that the single substitutions might have no effect on function. However, substitutions at multiple positions over a region may reveal critical residues. The binding surface may represent the sum of several interactions whereby the loss of one can be tolerated. Therefore a systematic experiment such as an alanine scan (7) could be a useful first step, but may not identify all the critical elements of the structure. Substitution of pairs of cysteine residues is a classical method of determining the effects of folding.

2.3 Hybrids/chimeras

Hybrids are generated by choosing a structurally related protein to the protein of interest and then 'cutting and pasting' to reveal the activity of the molecule of interest (1). For growth factors, the best approach is to analyse the surface exposed residues while keeping the structure constant. It is much better to start with an inactive, related protein as a scaffold and build in various regions of the molecule of interest, than to start with the molecule of interest and build in regions that inactivate it. In addition the related protein should have most of its identity in the buried regions rather than the solvent exposed regions. Identical residues in the solvent accessible regions, which are the subject of the experiment, cannot be evaluated by this approach.

Proteins usually have surface regions that are solvent exposed and core regions that are solvent inaccessible. For growth factors and chemokines the regions that are critical for receptor interaction and function are generally solvent exposed. When the protein binds to the receptor to form a complex, some of the solvent accessible residues will become substantially solvent inaccessible. The residues in the core region pack closely together and exclude solvent. Changing the non-polar core can affect this packing and therefore the folding and overall structure.

3. Purification, folding, and analysis

3.1 Sample preparation and RP-HPLC analysis

Proteins may be prepared in many ways as reviewed in Chapter 1 and described in other chapters of this volume. The methods described here are appropriate for chemically synthesized chemokines, but they could be adapted to material derived from other sources. The protein product should be precipitated; the method used will depend on the source, volume, and required concentration. Precipitation can be accomplished by salting-out or using organic solvents. Alternatively, instead of precipitation, the clarified crude mixture can be concentrated using molecular filtration methods.

Protocol 1. Preparation of the crude protein material for purification and HPLC analysis of the crude product

Equipment and reagents

- HPLC column: C18 silica 300 Å pore size, 5 μm particle, 250 × 4.6 mm (Vydac 218TP54, The Separations Group)
- Buffer A: 0.1% TFA in water, pass through a 0.2 μm filter (Millipore)
- Buffer B: acetonitrile containing 0.1% TFA, degas under vacuum
- Buffer C: 6 M guanidine–HCl, 0.3 M Tris–HCl pH 8.5
- 0.2 μm filter unit, 150 ml reservoir size (Nalgene)
- HPLC pre-column (Upchurch C130B) packed with C18 silica HPLC packing (Vydac, The Separations Group)
- Analytical scale HPLC system capable of running binary gradients up to a maximal flow rate of 5 ml/min, UV detector, chart recorder, and autosampler (optional)

Method

1. Dissolve the precipitated protein in buffer C containing 10% 2-mercaptoethanol, to make an approx. 30 mg/ml solution.
2. Remove 50 μl for analysis by RP-HPLC and store the remainder at –80 °C.
3. Inject a small sample (e.g. 5 μl)[a] on an analytical RP-HPLC column which has been equilibrated in buffer A.
4. Elute with a gradient up to 60% buffer B over 1 h.

Protocol 1. *Continued*

5. Monitor and record the elution by UV absorbance at 214 nm, with the detector set at 2.0 absorbance units full scale.[b]

[a] The amount required to provide a clear profile depends on the purity and nature of the crude product. It is preferable to increase the loading than to reduce the absorbance scale because of increased drift at low UV wavelengths, absorbance of the acetonitrile, and ease of comparison with future chromatograms.
[b] This chromatogram will reveal most of the proteins in the crude mixture (*Figure 1A*). Retain this chromatogram for future reference as it provides the profile of the unfolded material.

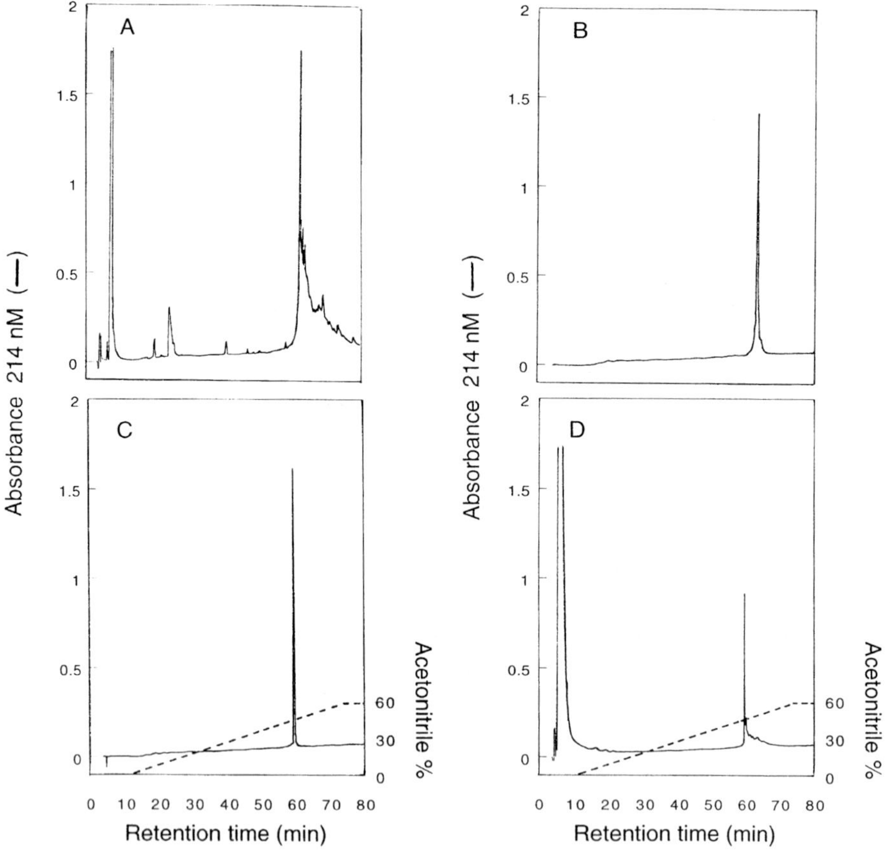

Figure 1. Analytical RP-HPLC profiles of IL-8 at various stages of purification. The sample was loaded on to a C18 silica column 250 × 4.6 mm and eluted using a linear gradient from 0–60% acetonitrile. TFA was kept constant at 0.1%. (A) Unfolded (reduced) crude chemically synthesized material. (B) Unfolded reduced material after preparative HPLC. (C) Material in (B) that has been folded to form IL-8. (D) Final purified IL-8 after semi-preparative HPLC.

3.2 Folding

Proteins exist in a compact, folded form that is unique to each molecule and it is essential that the analogue is present in its optimally folded form. The way a protein folds into its final 3D conformation is determined by its primary structure (amino acid sequence), so most native proteins fold spontaneously. Most growth factors and all known chemokines are stabilized by disulfide bridges. The formation of these disulfides is determined by the folding of the polypeptide chain and hence by the sequence. The function of the disulfides in folding is primarily to prevent unfolding of the protein and thus stabilize the structure.

There are two requirements for optimal *in vitro* folding. First, the protein must be soluble in both the unfolded and folded form. Thus solvent conditions must be selected to fulfil this requirement. Compared with native protein, an analogue that has sequence changes may not fold or may require different conditions. This very much depends on the particular changes that have been made to the primary structure. The second requirement is the addition of a mild oxidant that will convert the SH group to S^-, but not to higher oxidation states. It is essential that the reaction between the oxidant and the SH group be reversible. Examples of suitable oxidants are DMSO, or oxygen that is kept dissolved by vigorous stirring.

Not all analogues of a given protein will fold under the same conditions as the native molecule. Changes in the primary structure affect the solubility and the energetics of the folding process. Therefore it is best to check a number of different conditions on a small scale before committing the entire product to a particular set of conditions.

Protocol 2. Small scale tests to establish optimal folding conditions

1. Weigh out five 500 μg samples of chemokine polypeptide into 5 ml glass tubes. Make 1 mg/ml solutions of the chemokine polypeptide using the following solvents and conditions:
   (a) 1 M guanidine–HCl, 0.1 M Tris pH 8.5, 10% DMSO in a sealed flask, in the dark without stirring.
   (b) 1 M guanidine–HCl, 0.1 M Tris pH 8.5, stir vigorously in air, avoid frothing.
   (c) Water, stir vigorously in air (avoid frothing).
   (d) 10% DMSO in a sealed flask, in the dark without stirring.
   (e) PBS, stir in air.

   After 24 h, acidify, filter, and analyse each sample as for *Protocol 1*.

Protocol 2. *Continued*

2. Compare its chromatogram with that of the reduced material described in *Protocol 1* (*Figure 1A* and *C*).
3. Choose the conditions that give maximum conversion to a single product on RP-HPLC and that elutes earlier than the reduced starting material.[a]

[a] Folded chemokines will always have an earlier retention time, usually by about 3 min, than the reduced form (8). This means that the post-folding chromatogram should reveal a new earlier eluting peak(s). Some intermediate forms, that elute between the folded and unfolded form, may also be present. The level of the reduced, unfolded form indicates the extent of folding. The result should be apparent from the chromatogram.

When the desired conditions have been selected the entire batch of polypeptide can be folded by a scaled up procedure. In the example described in *Protocol 3* the conditions described in *Protocol 2*, step 1(a) were selected for batch folding.

Protocol 3. Large scale folding of partially purified chemokine polypeptide

Reagents

- Buffer D: 1 M guanidine–HCl, 0.1 M Tris pH 8.5, 10% DMSO, filter through a 0.2 μm filter (Nalgene)

Method

1. Dissolve the lyophilized material obtained in *Protocol 4* in buffer D.
2. Keep the sample overnight in the dark and sealed from air.
3. Run a sample on analytical RP-HPLC as in *Protocol 1*.[a]

[a] If the folding is incomplete then more time may be needed. The conversion to the folded form should be monitored until the conversion is complete *Figure 1C*.

3.3 Purification

The purification of synthetic proteins is HPLC based. Reverse-phase HPLC is the main method used for separation, however, ion exchange can also be used. The following strategy is suitable for use with up to 2 g of product:

(a) Purify the reduced protein on a preparative RP-HPLC (25 × 250 mm) column.
(b) Fold.
(c) Purify the folded protein using RP-HPLC on a semi-preparative (10 × 250 mm) column.
(d) Repeat step (c) if required, or use ion exchange (not described).

Note that it is important to avoid excessive dilution both on the column and during elution to prevent losses. RP-HPLC columns have high capacities for protein retention. For optimal separation, loading in the mid-capacity range of the column is best. A guide is: 2 g for 25 × 250 mm column, 100 mg for 10 × 250 mm column, and 5 mg for 4.6 × 250 mm column. Low loading results in poor yields. On the other hand, overloading could cause peak broadening due to solubility problems on elution from the column.

Protocol 4. Preparative scale purification

Equipment

- C18 silica column: 30 nm pore size, 10–15 µm particle size, 22 × 250 mm (Vydac, The Separations Group; 218TP1022)
- Pre-column (see *Protocol 1*)
- Lyophilizer
- HPLC system capable of binary gradients and flow rates up to 30 ml/min, UV detector, and chart recorder

Method

1. Adjust the pH of the crude reduced product to 3.0 using 20% acetic acid, and pass through a 0.2 µm filter.
2. Load on to a preparative column using the solvent inlet line for buffer A.
3. Develop the column using a flow rate of 15 ml/min and a gradient of 0–60% acetonitrile over 200 min, and monitor protein elution at 214 nm. Most polypeptides elute between 20% and 40% acetonitrile.
4. Collect fractions every 1.5 min and analyse them as described in *Protocol 1*. Pool those containing the protein of interest.
5. Lyophilize the pooled sample and analyse the material as described in *Protocol 1* (*Figure 1B*).

Protocol 5. Purification of the folded protein by semi-preparative HPLC

Equipment and reagents

- C18 silica column: 30 nm pore size, 5 µm particle size, 10 × 250 mm (Vydac, The Separations Group; 218TP510)
- Filter (Nalgene)
- Lyophilizer
- HPLC system capable of continuous flow rates of 10 ml/min, with UV detector and chart recorder
- 20% acetic acid in water solution

Method

1. Acidify the sample from *Protocol 3* to pH 3.0 using 20% acetic acid.
2. Filter through a 0.2 µm filter.

Protocol 5. *Continued*

3. Load approx. 100 mg of protein on to the semi-preparative column via the buffer A solvent inlet.[a]
4. Develop the column using a flow rate of 3 ml/min and a gradient of 0–60% acetonitrile over 200 min.
5. Collect fractions every 1.5 min.
6. Run the fractions on analytical RP-HPLC as described in *Protocol 1*.
7. Pool and lyophilize the fractions that are homogeneous. Analyse the pooled fraction by analytical RP-HPLC as described in *Protocol 1* (*Figure 1D*).

[a] Base the amount on the weighed material after folding as described in *Protocol 3*.

3.4 Verification of covalent structure

It is essential to take all reasonable steps to show that the purified material is in fact the molecule that was originally intended. In the case of recombinant methods the DNA should be sequenced. However, whichever method is used proof of the covalent structure of the protein is required. There are various methods for analysing the protein. Neither amino acid analysis nor amino terminal sequencing are adequate, although the latter could identify an incorrect amino acid within the region determined. Edman degradation of overlapping peptides derived from the protein is certainly adequate proof of structure, provided the data is complete. This is a very laborious method and is not without problems. However, determination of the mass of the protein is generally considered adequate verification of the covalent structure. The covalent structure is inferred from the mass determined, and other data are required to conclude that the correct sequence was made, e.g. the DNA sequence. The method has the advantage of speed, high accuracy (within one atomic mass unit), and reliability. Furthermore the entire molecule is analysed without chemical treatment and in the form that it is to be used for the SAR experiments. This means that any change to the sequence or modification to the structure that may have occurred during the synthesis, can be identified. It allows the investigator to distinguish between the disulfide bonded and the reduced unfolded form of a chemokine because of the 2 amu mass difference. Analysis of the spectrum can also give an indication of purity. Thus, even though mass spectroscopy is a verification rather than proof of structure, it is the most powerful method of protein analysis.

Protocol 6. Mass spectrometry

Equipment

- Electrospray mass spectrometer (Perkin-Elmer/SCIEX API-3)
- Harvard injection pump (Harvard Apparatus Canada)

Method

1. Prepare 100 μl of a 1 mg/ml solution of purified chemokine in 2% acetonitrile, 0.5% TFA.
2. Inject the sample at a rate of 5 μl/min using a Harvard pump into the injection port of a pre-calibrated electrospray mass spectrometer.
3. Collect signals from the ionized sample for 1 min.
4. Analyse the data using the *HYPERMASS* program (Perkin-Elmer) to generate the average measured mass, and the standard deviation of the mass measurement.
5. Compare the measured mass obtained with the calculated mass. If the sample is pure, peaks corresponding to only one molecular species should be apparent.

3.5 Handling and storage

After purification the chemokine analogue is lyophilized and stored at –80 °C in glass vials. For use, prepare a stock solution by dissolving a weighed sample of chemokine in water to make a 10 mg/ml solution and store aliquots at –80 °C. Dilute this solution as required.

4. Functional assays

4.1 General considerations

Once the desired analogue has been produced, purified, and characterized to verify its purity and its primary structure, it is ready for functional analysis. This involves using functional assays that are appropriate for the molecule under study. The assays are critical for any SAR experiment. Many growth factors and chemokines have multiple activities and therefore more than one assay may be available. The more functional the analysis, the more complete the SAR. The assays must be reliable, reproducible, and quantitative. The most important parameter is potency, which is usually expressed as the concentration required to give, for example, 50% of the maximal response. The comparison between analogues should be made at a point in the titration curve where there is a linear relationship between concentration and response value. The actual value chosen is not critical and could be 50%, 30%, or other. The curves for different samples should be parallel.

4.2 Cell migration

Chemokines stimulate the migration of cells towards a source. Cells that respond to chemokine migrate from low to high concentrations of chemokine when a concentration gradient is established. This phenomenon is known as chemotaxis. In the example below the chemokine MCP-1 stimulates chemotaxis of a cell line that is maintained *in vitro*.

Protocol 7. Chemotaxis assay for MCP-1

Equipment and reagents

- 48-well Boyden-type microchemotaxis chamber (NeuroProbe)
- Buffer E: RPMI 1640 medium containing 10 mg/ml bovine serum albumin (BSA) and 25 mM Hepes
- Polycarbonate filters (Nucleopore)
- THP-1: a human monocytic cell line that is known to respond to some chemokines (ATCC, Rockville, MA)
- Canco Quik Stain II (Baxter)

Method

1. Culture THP-1 monocytic cells in RPMI 1640 medium supplemented with 10% FCS. Maintain the cells at a logarithmic growth rate.
2. Make dilutions of MCP-1 in buffer E, covering a concentration range from 10^{-10} to 10^{-6} M.
3. Add the test solutions in triplicate to the wells of the bottom tray of the Boyden chamber so that each well is completely full (~ 26 μl/well).
4. Place a polycarbonate filter membrane (5 μm pore size) over the bottom chamber.
5. Place the top chamber over the filter and tighten the top and bottom chambers together.
6. Add THP-1 cells (50 μl of a 10^7 cell/ml suspension in buffer E) to the upper chamber.
7. Incubate the chamber at 37 °C in a 5% CO_2 in air atmosphere for 1–2 h.
8. Remove the filter and scrape the cells from the top side of the filter using the wiper blade that is supplied with the NeuroProbe microchambers.
9. Fix the filter in methanol and stain the attached cells using Canco Quik Stain II.
10. Count the cells that have migrated through the filter under a microscope.
11. Determine the chemotactic index by calculating the ratio of the migrated cells in the presence of MCP-1, to the migrated cells in the absence of MCP-1.

5. Receptor binding

5.1 Iodination of chemokines

It is essential that the chemokine, when labelled, retains full activity and binding affinity for the receptor. Methods that label tyrosines are popular but involve use of oxidants. The oxidation procedure can have detrimental effects on the protein and for this reason, if an oxidative method is to be used, it must be a gentle method, such as lactoperoxidase. Not all chemokines have tyrosine and not all that do can be labelled on tyrosine with retention of receptor binding. There are two potential solutions to this: one is to engineer tyrosines in a non-sensitive part of the peptide structure to allow or enhance subsequent iodination. The other is to use the Bolton–Hunter reagent which involves modification of amino groups (mostly lysine) with an ^{125}I-labelled chemical group. Apart from selection of the method, another important consideration is how 'hot' to make the protein. Many researchers make the mistake of over-labelling and thereby inactivating the protein or causing changes that affect the proteins physical properties as well as function. Highly radiolabelled proteins are also less stable. To summarize, the critical point is to use a gentle labelling procedure to obtain trace labelling with the goal of maximal biological potency rather than maximal radioactivity.

Protocol 8 outlines the method used for labelling the chemokine RANTES, using a lactoperoxidase method which is particularly gentle because it does not require added oxidants. The oxidant that is present in the ^{125}I preparation is sufficient.

Protocol 8. Iodination of the chemokine RANTES using lactoperoxidase

Equipment and reagents

- Sephadex G25 column (PD10, Pharmacia)
- Gamma counter
- Tyrosine: 100 mg/ml in water (saturated solution)
- Buffer F: 0.5 M sodium acetate pH 6.5
- Buffer G: water, 0.2% gelatin, 0.02% sodium azide
- Na[^{125}I] (No. 016303401, ICN)
- Lactoperoxidase (L-8257, Sigma Chemical Co.)

Method

1. Add 1 mCi (3.7 Mbq) Na[^{125}I] to 5 μg RANTES and 1 μg lactoperoxidase in 135 μl buffer F.
2. Allow to react at room temperature for 3 min.
3. Add 135 μl of saturated tyrosine and incubate at room temperature for 1 min to stop the reaction.
4. Separate the ^{125}I-labelled RANTES from other reagents by chromatography on a 1 × 9 cm Sephadex G25 column equilibrated in buffer G. Pool the peak protein-containing fractions that elute between 2.5–4.0 ml.
5. Determine the specific activity by estimating the d.p.m. in the sample, and the recovery of protein.

After labelling, the protein should be analysed by SDS–PAGE. This establishes that label is associated only with molecules of the apparent M_r of the protein of interest. In addition, the labelled preparation should be assayed for biological activity to demonstrate that activity has not been lost. However, it should be pointed out that this does not prove that all the labelled molecules in the preparation are functional because the activity may reside in the unlabelled or partially labelled molecules.

This method resulted in RANTES with a specific activity of around 260 Ci/mmol (9).

Protocol 9. Iodination of MCP-1 using the Bolton–Hunter reagent

Equipment and reagents

- Gamma counter
- Monoiodinated Bolton–Hunter reagent (*N*-succinimidyl-3-(4-hydroxy-3-^{125}I iodophenyl propionate), specific activity 2200 Ci/mmol (Dupont)
- Buffer H: 0.1 M sodium borate pH 8.5
- Buffer I: PBS containing 0.2% gelatin and 0.1% sodium azide

Method

1. Carefully dry 1 mCi of [^{125}I]Bolton–Hunter reagent using nitrogen and place at 4 °C.
2. Add 10 μg chemokine to 10 μl buffer H. Add the solution to the [^{125}I]Bolton–Hunter reagent and react for 20 min at 4 °C.
3. Add 125 μl 0.5 M glycine in buffer H and leave for 10 min at 4 °C to stop the reaction.
4. Separate labelled chemokine from unreacted Bolton–Hunter reagent by chromatography on a Sephadex G25 column equilibrated in buffer I.
5. Determine the specific activity of the ^{125}I-labelled chemokine preparation by measuring the d.p.m. and estimate the protein recovery.
6. Run a sample of the labelled chemokine on SDS–PAGE. The radioactivity should be associated with a protein of the expected size.

5.2 Receptor binding assays

As with other receptor systems, binding kinetics is the most common method of determining the ability of chemokines to bind receptors. Methods based on Scatchard analysis are often used. For SAR, the dissociation constant (K_d) is the most useful parameter.

$$K_d = \frac{[R] \times [L]}{[RL]}$$

where: R = receptor; L = ligand; and RL = receptor:ligand complex.

Binding parameters are measured by competition of the unlabelled chemokine or analogue with the labelled native chemokine for binding sites (receptors) on the cell surface.

5.3 Bindability and kinetics

Before carrying out binding analyses it is essential to determine binding parameters in order to set-up the appropriate experimental conditions. These are the proportion of the labelled preparation that can bind to cells (called the 'bindability') and the time required for binding to reach equilibrium. These are important because Scatchard-type methods involve many assumptions, the most critical of which, for practical purposes, are that saturation of specific binding is achieved and that the time of the reaction is sufficient for equilibrium to be reached.

To determine 'bindability': add 1 μl of [^{125}I]MCP-1 (*Protocol 9*) to a series of tubes containing increasing numbers of cells (between 10^5 and 10^7 cells/tube). Determine cell-associated radioactivity (counts bound) as described in *Protocol 10.* Plot the counts bound versus the number of added cells to determine the maximal bindable counts. Bindability is the amount of radiolabelled protein bound to cells compared with the total added to the tubes. For [^{125}I]MCP-1 labelled by the Bolton–Hunter method the value is around 20%.

To determine the time required for equilibrium binding of MCP-1, add 4 nM [^{125}I]MCP-1 to 5×10^6 cells for various time periods between 0–2 h. Determine the counts specifically bound to the cells as described *Protocol 10.*

For Scatchard kinetic analysis the receptor competition assay must be performed for a time period that allows the specific MCP-1 binding to reach a steady state. The specific binding of MCP-1 achieves a steady state after 15 min.

Protocol 10. Competition binding of unlabelled MCP-1 for labelled MCP-1

Reagents

- Binding buffer (buffer J): RPMI 1640 medium containing 10 mg/ml BSA, 25 mM Hepes, and 0.1% sodium azide

Method

1. For receptor competition assay, add 100 μl of THP-1 cell suspension (5×10^7 cells/ml in buffer J) to each 1.5 ml tube.
2. Make a series of MCP-1 solutions with concentrations of 2×10^{-5}, 6×10^{-6}, 2×10^{-6}, 6×10^{-7}, 2×10^{-7}, 6×10^{-8}, 2×10^{-8}, 6×10^{-9}, 2×10^{-9}, 6×10^{-10}, and 2×10^{-10} M in binding buffer. Add 100 μl of these solutions to duplicate tubes.

Protocol 10. *Continued*

3. Add the same amount of [^{125}I]MCP-1 to each of above tubes to achieve a final concentration of radiolabelled protein of 4×10^{-9} M.
4. Place the tubes at 4°C for 30 min with gently shaking.
5. Centrifuge the cells through a 2:3 (v/v) mixture of diacetylphthalate and dibutylphthalate for 2 sec at maximal speed in an Eppendorf microcentrifuge.
6. Cut off the base of the tube that contains the cell pellet and count the ^{125}I radioactivity present in the pellet to determine cell-associated activity.
7. Calculate the specifically bound c.p.m. by subtracting the non-specifically bound c.p.m. (i.e. the c.p.m. bound in the presence of 100-fold molar excess of unlabelled MCP-1) from the total c.p.m. bound to the cells.
8. Dissociation constants (K_d values) are determined by plotting the data according to Scatchard (10).

References

1. Clark-Lewis, I., Dewald, B., Loetscher, M., Moser, B., and Baggiolini, M. (1994). *J. Biol. Chem.*, **269**, 16075.
2. Clark-Lewis, I., Kim, K.-S., Rajarathnam, K., Gong, J.-H., Dewald, B., Baggiolini, M., *et al.* (1995). *J. Leuk. Biol.*, **57**, 703.
3. Baggiolini, M., Dewald, B., and Moser, B. (1994). *Adv. Immunol.*, **55**, 97.
4. Gong, J.-H. and Clark-Lewis, I. (1995). *J. Exp. Med.*, **181**, 631.
5. Anfinson, C. B. (1973). *Science*, **181**, 223.
6. Richards, F. M. (1991). *Sci. Am.*, 54.
7. Cunningham, B. C. and Wells, J. A. (1989). *Science*, **244**, 1081.
8. Clark-Lewis, I., Moser, B., Walz, A., Baggiolini, M., Scott, G. J., and Aebersold, R. A. (1991). *Biochemistry*, **30**, 3128.
9. Gong, J.-H., Uguccioni, M., Dewald, B., Baggiolini, M., and Clark-Lewis, I. (1996). *J. Biol. Chem.*, **271**, 10521.
10. Munson, P. J. and Rodbard, D. (1980). *Anal. Biochem.*, **107**, 220.

6

Generation and selection of RNA ligands that inhibit the interaction of vascular permeability factor/vascular endothelial growth factor (VPF/VEGF) with its receptors

LOUIS S. GREEN and NEBOJŠA JANJIĆ

1. Introduction to the SELEX process

Storage and transmission of genetic information by nucleic acids is critically dependent on their ability to interact with proteins. The formation of specific protein:nucleic acid complexes is a function of precisely defined three-dimensional structures of both binding partners. Like proteins, nucleic acids are capable of folding into a wide variety of three-dimensional structures. The ability of nucleic acids to form secondary and tertiary structures such as Watson–Crick base pairing, Hoogstein, and other non-canonical base pairing, pseudoknots, and G-quartets is determined by their sequence. Sequence-randomized nucleic acid libraries therefore contain a collection of binding specificities.

The advent of solid phase chemical synthesis of oligonucleotides has facilitated the generation of nucleic acid libraries of enormous sequence diversity by employing an equimolar mixture of the four monomers (e.g. phosphoramidites) at defined positions in the sequence. DNA, RNA, and modified nucleic acids are now commercially available from many suppliers. The SELEX (systematic evolution of ligands by exponential enrichment) process has recently emerged as a powerful method for screening sequence-randomized nucleic acid libraries for rare molecules (aptamers) that bind with high affinity and specificity to a variety of target molecules, by successive rounds of affinity selection and amplification (1, 2). Aptamers that bind to nucleic acid binding proteins, non-nucleic acid binding proteins, amino acids, peptides, carbohydrates, cofactors, and small aromatic molecules have been identified using this process (3, 4), attesting to the multitude of binding specificities present in random libraries. Similar selection/amplification strategies have been used to identify nucleic acids with unique catalytic functions (4).

Detailed protocols associated with SELEX experiments have been recently published (5) and the application of the SELEX process to various molecular targets has been reviewed (3, 4). In this chapter, we will focus on the application of the SELEX process for identifying RNA and modified RNA ligands to vascular permeability factor/vascular endothelial growth factor (VPF/VEGF).

2. Application of the SELEX process to identifying RNA ligands to VPF/VEGF

VPF/VEGF was chosen as a SELEX target because it is a potent inducer of angiogenesis (the growth of new blood vessels) *in vivo* (6). Angiogenesis in adults is rare and normally occurs under tightly regulated conditions. However, in a number of proliferative disease states including various retinopathies, rheumatoid arthritis, psoriasis, and cancer, disregulated angiogenesis contributes to disease progression (7). The notion that angiogenesis antagonists may be useful clinically is now widely accepted and several such agents have entered clinical trials (8, 9).

VPF/VEGF is a disulfide linked homodimer that occurs in four molecular isoforms as a result of alternative splicing of the same mRNA (121, 165, 189, and 206 amino acids) (10, 11). We chose VPF/VEGF–165 as the target of our SELEX experiments since it is the most abundant form. VPF/VEGF–165 can be obtained in lyophilized form, free from carrier protein from R & D Systems. We typically prepare 10–50 μM stock solutions by dissolving VPF/VEGF powder in phosphate-buffered saline (PBS) pH 7.4. VPF/VEGF concentration in stock solutions is determined from the absorbance reading at 280 nm ($\varepsilon_{280} = 1.3 \times 10^4$ a.u. M^{-1} cm^{-1}). After the protein concentration is determined, we generally add 0.01% human serum albumin (HSA) (Sigma Chemical Co.) to the stock solution of VPF/VEGF and store the protein frozen in 10–50 μl aliquots at –20 °C. When stored in this manner, VPF/VEGF retains full mitogenic activity for at least six months.

2.1 Random RNA libraries

The starting point of any SELEX experiment is the generation of sequence-randomized nucleic acid libraries. Since the shape repertoire of double-stranded nucleic acids is limited, we have focused entirely on selection of aptamers from single-stranded libraries. In RNA SELEX experiments, the starting RNA library is prepared by *in vitro* transcription from synthetic double-stranded DNA templates. The randomized region is flanked by defined sequence regions that contain PCR primer annealing sites, a primer annealing site for cDNA synthesis, a T7 RNA polymerase promoter region, and restriction enzyme sites that allow cloning into vectors. The restriction sites can alternatively be added in a separate amplification step immediately prior to cloning. In DNA SELEX experiments, the constant regions only need to contain the PCR primer annealing sites. One of the PCR primers usually contains a biotin group(s) at the 5′ end which allows easy strand separation (5, 12).

The size of the randomized region determines the maximal number of different molecules in the library, i.e. 4^n, where n is the number of randomized positions. Randomized regions as short as 8 nucleotides (1) to as long as 220 nucleotides (13) have been used. We have preferred to initiate our SELEX experiments with contiguous randomized regions of 30–50 nucleotides (14–17). For most protein targets, random regions in this size range have been adequate (4). More importantly, because chemical synthesis becomes more difficult with increasing nucleic acid length, it is generally our goal to identify short, high affinity aptamers. For practical reasons, we have typically performed initial selections with 0.2–2 nmol (10^{14} to 10^{15} molecules) of RNA. It should be emphasized that for a random region of 30 nucleotides, this represents a small fraction of all possible sequences ($4^{30} = 1.2 \times 10^{18}$).

Following *in vitro* transcription, the sequence-randomized RNA for initial selections is purified by polyacrylamide gel electrophoresis using conventional procedures (5). Prior to the addition of RNA to VPF/VEGF solutions in affinity selections or in binding experiments, we prefer to heat the RNA in PBS at 90–95 °C for 1–2 min and cool it on ice to minimize the formation of high molecular weight aggregates (*Protocol 1*).

Protocol 1. Iterative affinity selection of randomized RNA libraries

Equipment and reagents

- Nitrocellulose filters: type HA, 0.45 μm pore, 25 mm diameter (Millipore)
- Cellulose acetate syringe filters: 0.22 μm pore, 25 mm diameter (Millipore)
- Binding buffer, Tris-buffered saline (TBS): 10 mM Tris–HCl, 137 mM NaCl, 2.7 mM KCl pH 7.4, containing 1 mM $MgCl_2$, 1 mM $CaCl_2$, 0.01% human serum albumin (HSA)
- Synthetic randomized DNA templates and PCR primers
- 10 μM VPF/VEGF in phosphate-buffered saline (PBS): recombinant human 165 amino acid isoform (R & D Systems)
- 20 mg/ml glycogen (Boehringer Mannheim)
- Bromophenol blue and xylene cyanol FF dyes (Sigma)

A. *Gel purification of RNA*

1. Run the ^{32}P-labelled RNA obtained by *in vitro* transcription from the synthetic randomized double-stranded DNA template for round one, or from the previous round of selection (from part B, step 14), on an 8% denaturing polyacrylamide gel. Use bromophenol blue and xylene cyanol dyes as approximate size markers.
2. Expose the Saran wrapped gel to X-ray film.
3. Cut out a gel slice with the full-length transcript band, using the developed film as a template.

Protocol 1. *Continued*

4. Place the gel slice in a 1.5 ml microcentrifuge tube and freeze on dry ice.
5. Thaw and add 300 µl of 2 mM EDTA at room temperature.
6. Crush the slice against the bottom of the tube with the plunger from a 1 ml sterile plastic syringe until a uniform slurry is obtained.
7. Add 100 µl of 3 M sodium acetate pH 5.2 at room temperature.
8. Push the slurry through a cellulose acetate syringe filter (to remove the polyacrylamide bits) into a new 1.5 ml microcentrifuge tube.
9. Add 1 ml of 1:1(v/v) isopropanol:ethanol and precipitate the RNA on dry ice for 20 min.
10. Spin down the RNA at 14000 *g* for 30 min in a microcentrifuge at 4°C.
11. Wash the pellet with 70% ethanol, vacuum or air dry, and resuspend in water.
12. Determine the RNA concentration by absorbance at 260 nm.

B. *Affinity selection of RNA*

1. Prepare three to five VPF/VEGF RNA binding mixtures in binding buffer, as follows. For the first (most concentrated) binding mixture use the VPF/VEGF concentration that yielded 1–10% RNA binding in the previous round,[a] using successive threefold dilutions for the other binding mixtures. Add heat denatured RNA to give a five- to tenfold molar excess over VPF/VEGF.
2. Incubate the binding mixtures at room temperature for 4 h (or 4°C for 16 h), then at 37°C for 15–30 min.
3. Filter the mixtures (the bound complexes are retained) through nitrocellulose filter discs, and wash with 5 ml of binding buffer without HSA.
4. Measure the ^{32}P radioactivity of the wet filters by Cerenkov counting using a scintillation counter, along with 5–10% of the input RNA spotted on a filter, to determine the per cent of RNA bound.
5. Slice the selected wet filter(s) into small pieces with a scalpel blade and add to a 1.5 ml microcentrifuge tube containing 400 µl of saturated phenol pH 7, and 200 µl of fresh 7 M urea.
6. Vortex, heat to 65°C for 5–10 min, then cool to room temperature.
7. Add 400 µl of chloroform and vortex.
8. Push the nitrocellulose pieces to the bottom of the tube.
9. Centrifuge at 14000 *g* to separate the phases and pipette the aqueous phase (about 400 µl) to a new microcentrifuge tube.

10. Extract with chloroform, then precipitate with 0.1 vol. of 3 M sodium acetate pH 5.2, 10 μg of glycogen as a carrier, and 1 ml of 1:1 (v/v) isopropanol:ethanol on dry ice.
11. Wash the pellet with 70% ethanol and vacuum or air dry.
12. Resuspend the selected RNA pellet in 25 μl of water on ice. Reverse transcribe the selected RNA with the 3′ primer according to ref. 5.
13. Amplify the cDNA with 3′ and 5′ primers by PCR according to ref. 5.
14. Transcribe the resulting double-stranded DNA template with 10–20 μCi of [α-^{32}P]ATP according to ref. 5.
15. Return to part A, step 1 to begin the next round of affinity selection.

[a] Early rounds of selection typically range from 20 nM to 1 μM VPF/VEGF and in late rounds as low as 1 pM. Amounts of VPF/VEGF used are 100–1000 pmol in early rounds and 1–50 pmol in later rounds. The volume of the binding mixture is adjusted to accommodate these amounts and concentrations and may range from 100 μl to 100 ml. See *Protocol 2*, footnote *b*.

2.2 Selection conditions

Typical affinity selection conditions are outlined in *Protocol 1*. Since our goal is to identify ligands that inhibit VPF/VEGF *in vivo*, the affinity selections are done in a buffer at physiological pH, ionic strength, and temperature. Although our initial SELEX experiments targeting VPF/VEGF were done in PBS buffer without divalent metal ions (15), we now recommend that the selection buffers include $\approx$ 1 mM $MgCl_2$ and $CaCl_2$. Ca^{2+} and Mg^{2+} are present in most bodily fluids and their role in the structure and function of many nucleic acids is well known. If Ca^{2+} and Mg^{2+} are absent from the selection buffer, it is possible that the structure of the evolved ligands may be changed by these metal ions *in vivo* in a manner that reduces the binding affinity of the evolved nucleic acid ligands for their target.

The binding affinity of sequence-randomized nucleic acids in the 50–80 nucleotide size range to VPF/VEGF is typically 100–500 nM, as determined by nitrocellulose filter binding. It is important to note that protein:nucleic acid complexes may dissociate to varying extents during washing on nitrocellulose filters so that the actual binding affinity may be higher than the observed. In any event, nucleic acids clearly have a significant intrinsic tendency to bind to VPF/VEGF–165. This is not surprising since VPF/VEGF–165 is known to bind heparin, a sulfated (polyanionic) polysaccharide.

The initial affinity selection is extremely important for the success of a SELEX experiment. This is primarily because sequences that have the highest affinity for the target may only be present as a single copy in the initial pool. These rare sequences are vastly outnumbered by sequences that have lower affinity for the target. To reduce the chance of losing the best ligands before they are amplified, our binding conditions in the first round of SELEX are generally not overly stringent. With VPF/VEGF and similar proteins, we have had good success starting with ≈ 1 nmol of the nucleic acid library and 0.2–0.3 nmol VPF/VEGF in a total volume of 1 ml. The modest molar excess of the RNA over VPF/VEGF (three- to fivefold) at concentrations that exceed the observed dissociation constant ensures that a significant fraction of the input RNA is selected. The length of incubation of the RNA with VPF/VEGF prior to partitioning of the free from bound RNA is another adjustable parameter that should be carefully considered. Longer incubation times increase the likelihood that equilibrium conditions will be attained. At more advanced rounds of SELEX, as the overall affinity of the pool improves, the fraction of ligands with slower dissociation rates increases and the time necessary to reach equilibrium therefore becomes longer. However, protein and/or RNA may be degraded on prolonged incubation at 37°C. With VPF/VEGF, which is a heat and acid stable disulfide linked homodimer, we have used incubation times from 10 min to 16 h (overnight).

The RNA recovered from nitrocellulose filters is reverse transcribed into cDNA and amplified by PCR. The RNA for the next round of affinity selection is prepared by *in vitro* transcription with double-stranded PCR product as the template. These steps have been described in detail recently (5).

In subsequent rounds of affinity selection, stringency can be gradually increased by reducing the concentrations of both RNA and VPF/VEGF. For example, in the second round, starting with the initial selection conditions (e.g. 1 μM RNA and 0.2 μM VPF/VEGF), we generally perform several selections with serial dilutions of the original sample covering a concentration range of several orders of magnitude. This allows us to choose the most stringent selection condition while maintaining significant VPF/VEGF-dependent RNA binding. Protein-independent binding of RNA to nitrocellulose filters is generally low at first (≈ 0.01–0.2% of the input), but typically increases with successive rounds and can become an overwhelming problem. Although we have not examined this problem systematically, the background binding to nitrocellulose appears to emerge sooner when the ratio of protein-dependent binding to protein-independent binding in affinity selections is low. Protein-independent binding can be reduced by increasing the wash volume with binding buffer (we have used up to 1 litre), additional filter washing with 0.5 M urea solution, or pre-blocking the filters with a 5–20 ml wash of 0.01% HSA solution (18). Alternatively, a different partitioning method can be used. Gel electrophoresis mobility shift partitioning is well suited for VPF/VEGF SELEX experiments since the mobility of the protein: nucleic acid complexes is considerably lower than that of the free nucleic acids. Importantly, as with nitrocellulose filter partitioning, no protein modification is required.

The progress of SELEX can be monitored by periodic examination of the binding affinity of the enriched pools. With RNA libraries containing 30–50 randomized positions and selecting on average ≈ 1% of the input RNA, we generally observe a significant affinity improvement by SELEX round 5. After 10–13 rounds, the dissociation constants of the affinity enriched pools are in the nanomolar range or lower. At this point, additional rounds of SELEX generally do not result in significant improvement in affinity.

2.3 Consensus sequence and secondary structure identification

The procedures used for cloning and sequencing of ligands from affinity enriched pools have been described in detail recently (5). In many respects, sequences of individual selected ligands are the raw product of a SELEX experiment. Since individual molecules isolated from the affinity enriched pool are functionally related in the sense that they bind VPF/VEGF with high affinity, the sequences in the initially randomized regions can be arranged into families based on primary structure similarities. The primary structure similarity indicates a common secondary structure and ultimately, 3D structure of the ligands. The existence of multiple families, observed in most SELEX experiments, demonstrates that there are many sequence solutions for a given functional property. For example, six different sequence families have been identified in the RNA SELEX experiment targeting VPF/VEGF (15), which is undoubtedly only a fraction of existing (but as yet unidentified) families.

Primary structure similarity among ligands can be detected either by inspection or with the use of multiple sequence alignment algorithms. This is generally straightforward in sequences with a long contiguous conserved segment. In certain cases, however, regions of strong sequence conservation are separated by highly variable segments. The conserved segments may be circularly permuted (*Figure 1A*) in a manner that preserves the consensus secondary structure (*Figure 1B*). In such cases, the identification of *clusters* of conserved sequence patterns is facilitated by a new algorithm developed by Brenda Javornik and Dominic Zichi (19).

The identification of a common secondary structure for a group of related sequences is generally more difficult. Secondary structure predictions based on thermodynamics of RNA folding (20) are of limited value for analysing SELEX sequences. Individual sequences in a family often have multiple possible secondary structures with comparable predicted free energies of folding. In large sequence sets, the consideration of all reasonable suboptimal folding structures can be an overwhelming task. More alarmingly, these analyses do not consider certain well-known RNA motifs such as pseudoknots and G-quartets. Comparative sequence analysis methods, such as the identification of compensatory base changes including base pairing co-variation, are considerably more useful (21, 22). We have recently described a method of simultaneously examining possible secondary structures of all ligands in a

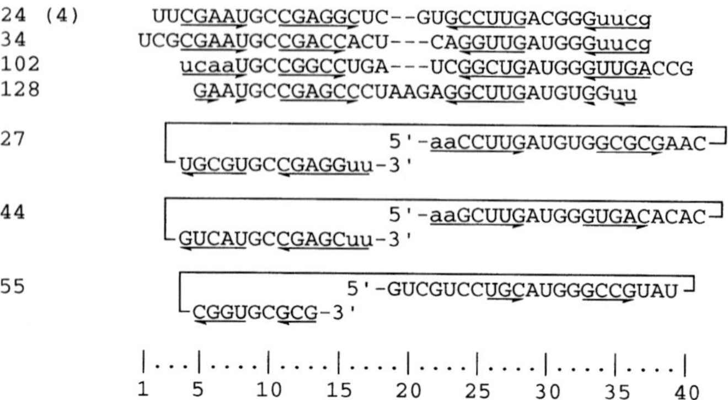

B

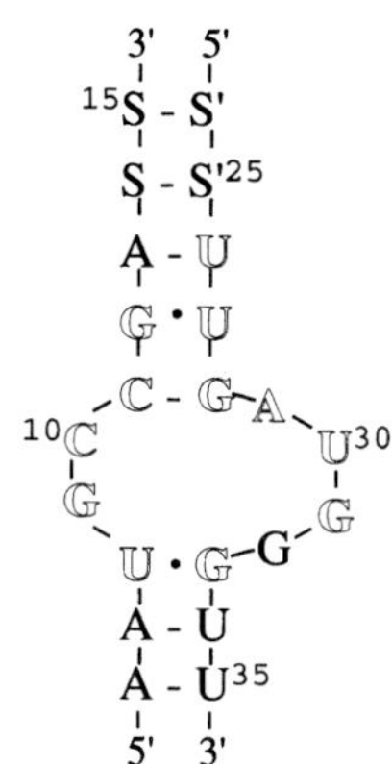

Figure 1. (A) A representative family of RNA ligands to VEGF showing aligned sequences and predicted secondary structures. Underline arrows indicate regions that are expected to be base paired. Lowercase and uppercase letters distinguish nucleotides in the constant and the evolved sequence regions, respectively. The number in parentheses following clone number 24 indicates that this sequence was encountered four times. Numbering (arbitrary) is indicated with the ruler. (B) The predicted secondary structure for the family of ligands shown in (A). Plain text designates positions that occur at > 60% but < 80% frequencies. Positions where individual nucleotides are strongly conserved (frequencies > 80%) are outlined. Nucleotide positions are numbered as they appear in (A). S = G or C, and prime (′) indicates a complementary base. (Adapted from ref. 15 with permission.)

family by using a composite dot matrix representation of aligned sequences (23). The basis for the method is selective visual enhancement of common elements of secondary structure in an aligned set.

The advantages of the above methods in analysing SELEX sequence sets and their limitations have been described recently (19, 23). The programs developed at NeXstar are available on request from Dominic Zichi.

2.4 Minimal sequence determination

In general, the functional activity of SELEX ligands is encoded in a fraction of the entire molecule (including the constant sequence regions). Leads about the consensus primary and secondary structure can be used as a guide for determining the minimal sequence required for activity. For example, highly conserved nucleotides or secondary structure elements (such as base pairs) are probably functionally important. In conjunction with (or in the absence of) such data, information about the minimal sequence requirements can be obtained experimentally by generating RNA ligand fragments and examining the binding properties of the fragments. A method we have used advantageously relies on radiolabelling the ligands at either the 3′ or 5′ ends (for the 5′ or 3′ boundaries, respectively), generating fragments by partial hydrolysis in mildly alkaline buffer and affinity selecting under conditions that demand high affinity binding to the target (15). The affinity selected fragments are then resolved on sequencing gels where the smallest radiolabelled fragment that retains high affinity binding to VPF/VEGF represents the boundary. The combined results from the 5′ and 3′ boundaries define the minimal ligand. This is a highly efficient method of screening the binding affinities of hydrolytic fragments, however, its main disadvantage is the fact that the 5′ and 3′ boundaries are determined sequentially in the context of one (radiolabelled) intact end. The intact end can either augment, diminish, or have no effect on the binding properties of the fragments. The method is described in *Protocol 2* and illustrated in *Figures 2* and *6*.

Protocol 2. Minimal ligand determination

Equipment and reagents

- Nitrocellulose filters (see *Protocol 1*)
- VPF/VEGF (see *Protocol 1*)
- Bacterial alkaline phosphatase (Boehringer Mannheim)
- T4 polynucleotide kinase (New England Biolabs)
- T4 RNA ligase (New England Biolabs)
- [γ-^{32}P]ATP, 3000 Ci/mmol (New England Nuclear)
- [5′-^{32}P]cytidine 3′, 5′-bisphosphate ([^{32}pCp]), 3000 Ci/mmol (New England Nuclear)
- RNase T1 (Boehringer Mannheim)

Method

1. Gel purify a cloned ligand transcript RNA by the method described in *Protocol 1A*.

Protocol 2. *Continued*

2. Dephosphorylate the 5′ end of the ligand with bacterial alkaline phosphatase as described in ref. 5.
3. 5′ end-label about 20 pmol of the ligand using [γ-^{32}P]ATP and T4 polynucleotide kinase for the 3′ boundary determination. 3′ end-label the ligand with [^{32}pCp] and T4 RNA ligase for the 5′ boundary determination (5).
4. Dilute the radiolabelled RNAs at least tenfold into 50 mM sodium carbonate buffer pH 9.0, 100–200 μl final volume, in a 1.5 ml microcentrifuge tube.
5. Subject the RNA to partial alkaline hydrolysis by incubation at 90 °C in a heating block for about 15 min.[a]
6. Transfer the tube containing the hydrolysis reaction mixture to ice, add 0.1 vol. of 3 M sodium acetate pH 5.2, and precipitate with 3 vol. of 1:1 (v/v) isopropanol:ethanol.
7. Spin the tube containing the RNA at 14000 *g* for 30 min in a microcentrifuge.
8. Wash the pellet with 70% ethanol, and vacuum or air dry.
9. Resuspend the pellets in 50 μl water.
10. Set-up three to five binding reactions as follows. Add about 2 pmol aliquots of the partially hydrolysed RNA to binding buffer (*Protocol 1*), with 2 pmol VPF/VEGF at concentrations ranging from about fivefold above to tenfold below the measured K_d,[b] and incubate at 37 °C for 20 min.
11. Filter the binding reactions through nitrocellulose filter discs and wash the discs with 5 ml of binding buffer without HSA.
12. Elute the bound RNA fragments from the filters as in *Protocol 1*B, steps 5–11.
13. Resolve the affinity selected RNA fragments on an 8% denaturing polyacrylamide gel. Include a lane with the unselected alkaline hydrolysate. Include a lane with a partial RNase T1 digest of the full-length ligand to establish the G positions on the gel.

[a] The incubation time is empirically determined. Sharper boundaries usually result if most of the RNA is shorter than full-length.
[b] Total volumes may range from 100 μl to 100 ml. For volumes larger than 10 ml, decrease the HSA in the binding buffer so that it does not exceed 1 mg total since retention of bound complexes on nitrocellulose filters is inhibited by larger amounts of HSA.

2.5 Selection of VPF/VEGF antagonists: inhibition of receptor binding

Affinity screening of all unique molecules from the affinity selected population allows us to identify the best ligands to VPF/VEGF. For certain

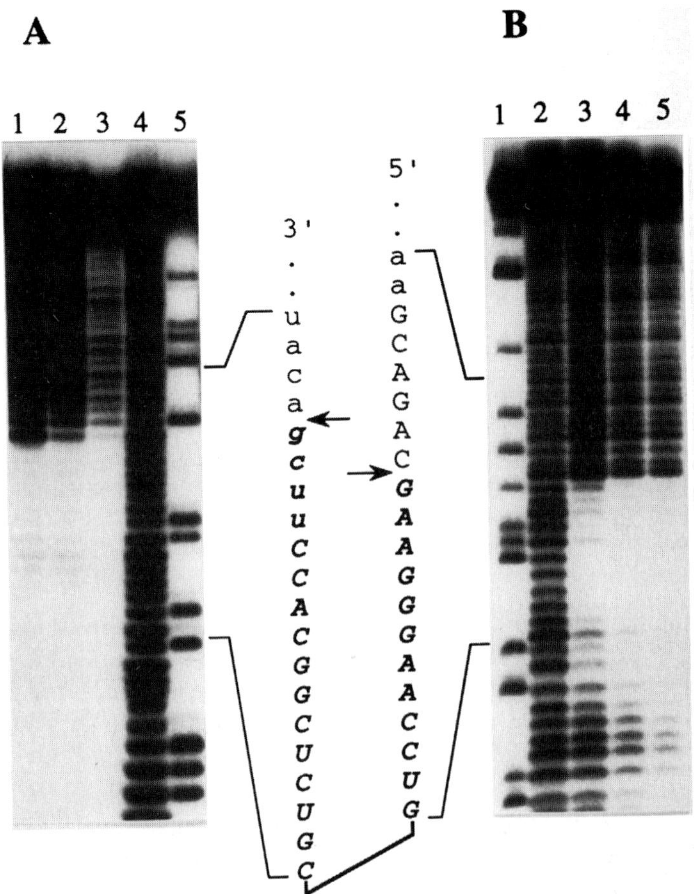

Figure 2. Determination of the 3′ and 5′ boundaries for a representative high affinity VEGF ligand. (A) The 3′ boundary determination showing partially hydrolysed 5′ end-labelled RNA (lane 4), hydrolytic fragments retained on nitrocellulose filters following incubation of the partially hydrolysed RNA with VEGF at 5 nM (lane 1), 0.5 nM (lane 2), and 0.125 nM (lane 3), and a partial digest of the 5′ end-labelled RNA with RNase T1 (lane 5) resolved on an 8% denaturing polyacrylamide gel. (B) The 5′ boundary was determined in an identical manner except that RNA radiolabelled at the 3′ end was used. Shown are RNase T1 digest (lane 1), partial alkaline hydrolysis (lane 2), and hydrolytic fragments retained on nitrocellulose filters following incubation with VEGF at 5 nM (lane 3), 0.5 nM (lane 4), or 0.125 nM (lane 5). *Arrows* indicate the 3′ and the 5′ boundaries that define the minimal ligand (italicized). Lowercase and uppercase letters indicate nucleotides in the constant and the evolved sequence regions, respectively. (Adapted from ref. 15 with permission.)

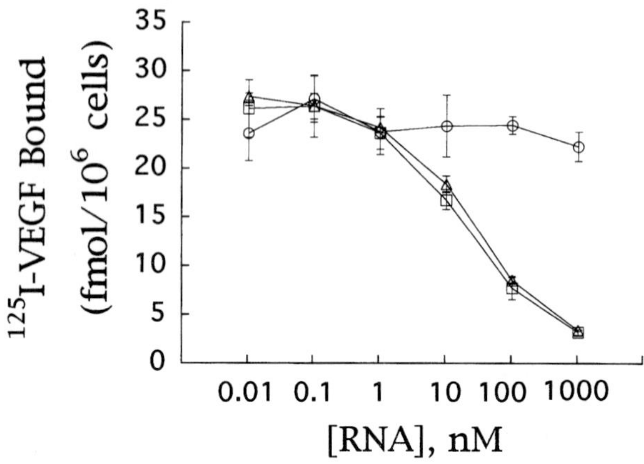

Figure 3. The effect of random RNA (circles) and two representative high affinity RNA ligands 100t (triangles) and 44t (squares) on binding of [^{125}I]VEGF to cell surface receptors as a function of RNA concentration. Data points are the averages ± SD of two to four determinations. (Adapted from ref. 15 with permission.)

diagnostic and affinity separation applications of SELEX-derived ligands, this is a sufficient criterion. However, the main objective of the SELEX experiments described here was to identify potent and specific antagonists of VPF/VEGF. VPF/VEGF exerts its biological functions by binding to cell surface proteins, VPF/VEGF receptors flt-1 and flk-1/KDR. Both receptors belong to the receptor tyrosine kinase class and contain seven immunoglobulin-like domains, a transmembrane domain, and an intracellular tyrosine kinase domain. Binding of VPF/VEGF to the extracellular domain leads to phosphorylation of several tyrosine residues in the intracellular domain and further signal transduction (24).

Receptor binding is the first and necessary step in the VPF/VEGF signal transduction cascade. As the initial screen of the inhibitory potential of our ligands, we chose to examine their ability to inhibit the binding of ^{125}I-labelled VPF/VEGF to receptors expressed on human umbilical vein endothelial cells (HUVEC). HUVEC are commercially available and are known to express both the flt-1 and flk-1/KDR receptors for VPF/VEGF. The assay is rapid and can be readily adapted for screening a large number of ligands. The receptor binding assay is described in *Protocol 3* and illustrated in *Figure 3*.

All high affinity RNA ligands tested (including representative minimal ligands from the six sequence families described in ref. 15) were found to be inhibitors of the binding of VPF/VEGF to its receptors. Furthermore, the binding of any one ligand to VPF/VEGF could be competed with all other ligands and also with heparin. It should be recalled that we were only selecting for high affinity binding to VPF/VEGF without taking special measures to select for functional inhibitors. One possible explanation for these results is that VPF/VEGF–165 has a 'dominant' domain for nucleic acid binding that is in proximity with (or coincident with) its heparin binding and receptor binding domains. It is important to realize, however, that oligonucleotides in the 25–40 nucleotide size range are not small molecules ($M_r \approx 9000$–14 000 Da) and that partial overlap of the binding sites for nucleic acids, heparin, and receptors may be sufficient to produce mutually exclusive binding.

Protocol 3. Receptor binding inhibition assay

Equipment and reagents

- Orbital shaker
- Gamma counter (e.g. Cobra II, Packard)
- [^{125}I]VPF/VEGF (Amersham)
- Human umbilical vein endothelial cells (HUVEC) (Clonetics)
- 24-well cell culture plates (Falcon)
- Endothelial growth medium (EGM) (Clonetics)
- Dulbecco's phosphate-buffered saline (DPBS) (Gibco) containing 1 mM $MgCl_2$, 1 mM $CaCl_2$
- 1% Triton X-100

Method

1. Plate HUVEC (population doubling 4–12)[a] in 24-well cell culture plates at a density of 1–2 × 10^5 cells/well in 0.5 ml/well EGM.
2. At confluence, incubate the cells for 2 h at room temperature in 0.5 ml DPBS (serum-free), containing 10 ng/ml ^{125}I-labelled VPF/VEGF, and increasing concentrations of the oligonucleotide ligands (0.1 nM to 1 μM).
3. At the end of the 2 h incubation period aspirate the medium and wash the wells two times with cold DPBS, taking care not to disturb the monolayer.
4. Lyse the cells by adding 0.3 ml 1% Triton X-100 to each well and shaking the plate on an orbital platform for 30 min at room temperature.
5. Transfer the contents of each well to plastic gamma counter tubes. Wash each well with another 0.3 ml aliquot of 1% Triton X-100, transferring the contents into the corresponding tube.
6. Measure radioactivity from the tubes using a gamma counter.

[a] Cells arrive from Clonetics at population doubling 2, and are grown to population doubling 4–12 in EGM before use.

3. Methods for making RNA ligands nuclease resistant

Since RNA is exquisitely sensitive to ribonucleases, the most efficient method for making RNA ligands nuclease resistant is to initiate the SELEX experiment with nuclease-resistant libraries (25). Completely nuclease-resistant libraries compatible with SELEX have not yet been reported. However, there is now ample evidence that RNA modified at the 2′ position of pyrimidine nucleotides by certain functional groups has a dramatically greater stability in serum compared to unmodified RNA (16, 26, 27). Presumably, the success of achieving a large increase in the resistance to nucleases by only protecting the pyrimidine positions is related to the substrate specificity of most serum ribonucleases. At least two such modifications (2′-amino and 2′-fluoro) at the pyrimidine nucleotide positions (*Figure 4*) are also compatible with the enzymatic steps of SELEX. We have reported previously on the use of 2′-aminopyrimidine RNA libraries containing 30 or 50 randomized positions in SELEX experiments targeting VPF/VEGF (17). A representative sequence family that contains members from both SELEX experiments is shown in *Figure 5A*. In these SELEX experiments, as well as in analogous SELEX experiments targeting basic fibroblast growth factor (16), there does not appear to be an advantage in using a longer randomized region. Perhaps more important is the observation that rare winning motifs can be reproducibly identified in independent SELEX experiments. We have indeed observed this with regularity (14, 16, 17). The performance of SELEX with natural RNA and modified (2′-aminopyrimidine) RNA is comparable in the sense that ligands with similar affinities have been identified. However, the aggregate effect of hydroxyl to amino substitution at the 2′ position over the entire molecule on target binding is very significant in that the sequences and consensus secondary structures of ligands obtained in natural RNA (*Figure 1*)

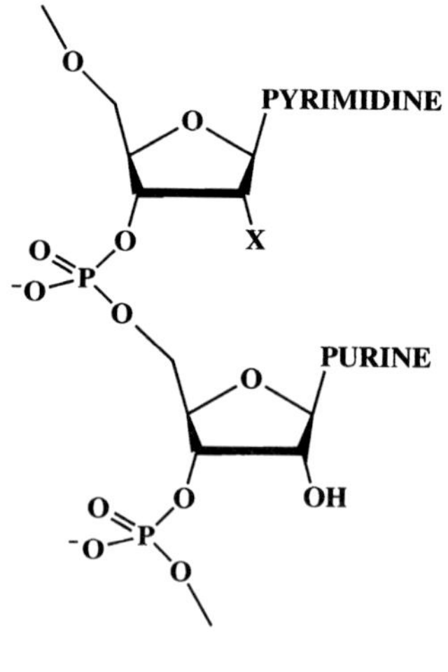

PYRIMIDINE
X
PURINE
OH

$X = NH_2$ or F

Figure 4. Structure of 2′-amino- or 2′-fluoropyrimidine RNA.

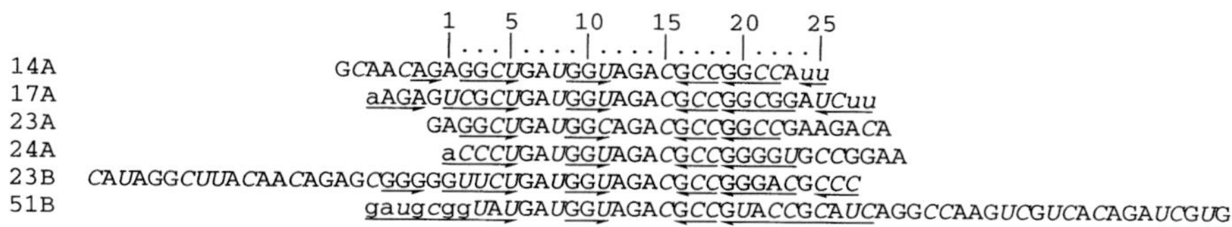

B

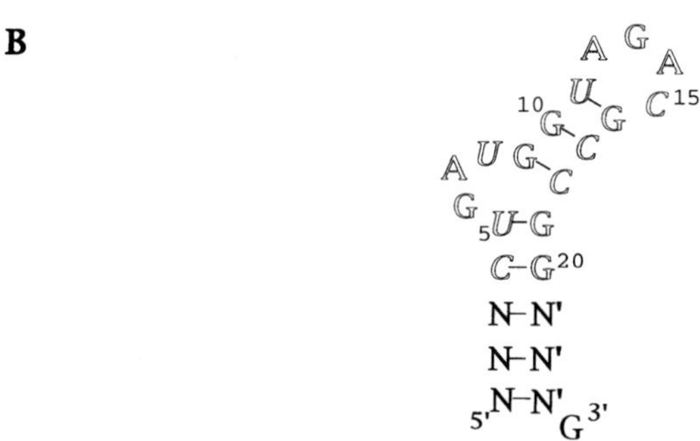

Figure 5. (A) Primary structures of a family of high affinity 2′-aminopyrimidine RNA ligands to VPF/VEGF. Aligned sequences isolated from the 30N and 50N SELEX experiments (designated by letters A and B following the clone number) are shown, depicting nucleotides in the evolved region (uppercase letters); also shown are nucleotides in the constant region (lowercase letters) that participate in the predicted consensus secondary structure formation (underline arrows). The 2′-aminopyrimidines are shown in italic letters. (B) The predicted secondary structure for the family of ligands shown in (A). Positions where individual nucleotides are strongly conserved (frequencies > 80%) are outlined. N–N′ indicates any base pair. (Adapted from ref. 17 with permission.)

and modified RNA (*Figure 5*) SELEX experiments are notably dissimilar. This again is a common finding (14–17).

Minimal sequence determinations with the 2′-aminopyrimidine ligands are done in an analogous manner to that described for the unmodified RNA. However, because of the resistance of phosphodiester bonds adjacent to the 2′-amino groups to alkaline hydrolysis, only the bands that correspond to cleavage at the purine positions are visible and the boundary can be determined only to the nearest purine nucleotide position (*Figure* 6).

Starting with a substantially stabilized minimal 2′-aminopyrimidine RNA ligand (24-mer), we achieved additional stabilization by introducing short caps containing four phosphorothioate linkages at the 3′ and 5′ termini (thereby attenuating the exonucleolytic degradation of the ligand). Compared with the uncapped minimal ligand, which has a half-life in rat urine of 1.4 ± 0.1 h, the capped ligand is significantly more stable with a half-life of 17 ± 2 h. In an attempt to further increase the stability of this ligand, we have examined

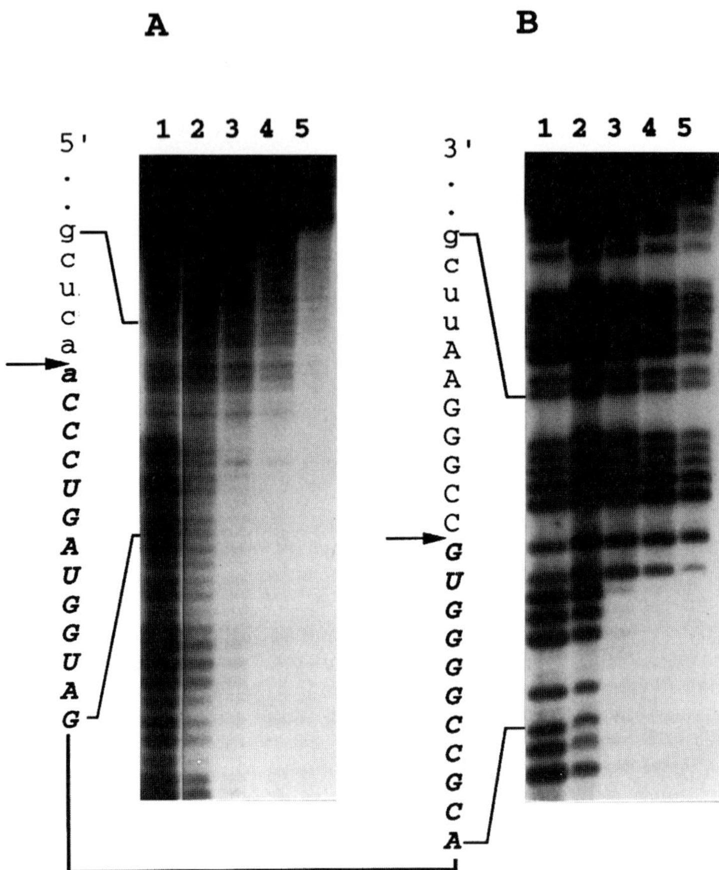

Figure 6. Determination of the minimal sequence requirement for binding to VPF/VEGF of 2′-aminopyrimidine ligand 24A (*Figure 5*). (A) For the 5′ boundary, the partially hydrolysed 3′ labelled ligand (lane 1) was affinity selected with 170 nM (lane 2), 17 nM (lane 3), 1.7 nM (lane 4), or 0.43 nM (lane 5) VPF/VEGF. (B) The 3′ boundary was determined in the same manner with 5′ labelled ligand (same lane assignments as panel A). The absence of pyrimidine nucleotide bands (due to resistance to alkaline hydrolysis of the phosphodiester bonds 3′ to the 2′-aminopyrimidines) allows sequence assignment without external markers. *Arrows* indicate the 5′ and 3′ boundaries that define the minimal ligand. Lowercase and uppercase letters indicate nucleotides in the constant and evolved sequence regions, respectively, and 2′-aminopyrimidines are shown in italicized symbols. (Adapted from ref. 17 with permission.)

the effect of 2′-*O*-methyl (2′-OMe) for 2′-OH substitution at the purine nucleotide positions. The replacement of ribopurine nucleotides by 2′-OMe purine nucleotides at all positions that tolerate such a substitution (10 out of 14 purine positions) extends the half-life of the ligand in rat urine to 131 ± 4 h.

Screening for purine positions that tolerate the 2′-OMe substitution can be done by chemically synthesizing individually substituted ligand variants and examining the effect of the 2′-OMe substitution on binding. The disadvantage of this approach is twofold:

(a) The need to synthesize and screen a large number of oligonucleotides.
(b) Lack of information about the binding properties of multiply 2′-OMe substituted ligands.

Alternatively, one multiply substituted ligand variant can be synthesized, however, in this case, if weaker binding is observed with a particular substitution pattern, further experiments are required to determine the location of substitution-intolerant position(s).

The effect of 2′-OMe substitution on binding can be examined at multiple positions simultaneously by chemically synthesizing ligands in which defined positions contain a mixture of 2′-OH and 2′-OMe nucleotides. Analogously to the situation observed with the 2′-amino substitution (and for the same reason), the phosphodiester bonds adjacent to the 2′-OMe groups are resistant to alkaline hydrolysis. Partially 2′-OMe substituted positions will therefore give rise to fainter alkaline hydrolysis bands on sequencing gels compared with the unsubstituted (all 2′-OH) positions. If the 2′-OH group is preferred at a certain partially substituted position, the alkaline hydrolysis band will become darker following affinity selection. Conversely, if the 2′-OMe group is preferred, the band will become fainter following affinity selection (*Figure 7A*). With the minimal 2′-aminopyrimidine ligand to VPF/VEGF, we have found it advantageous to introduce mixed 2′-OH/2′-OMe groups at three or four purine positions, with the remainder of ribopurines being unmodified. More extensively substituted ligands can also be used, however, we have observed in such cases a significant overall reduction in the binding affinity to VPF/VEGF as well as a smaller difference at all partially substituted positions between the unselected and selected bands. Splitting the partial 2′-OMe substitution among several ligands also allows for the effect of the substitution at a given position to be analysed in quantitative terms.

In the case of the minimal 2′-aminopyrimidine VPF/VEGF ligand, the partial 2′-OMe substitution was distributed among four ligand pools, so that each unsubstituted (all 2′-OH) purine position occurred three times. The mean of the three unsubstituted normalized band intensity ratios for the unselected versus the selected lanes can be obtained for each purine position (*Figure 7B*). These values provide a standard against which the effect of the 2′-OMe substitution on binding can be quantitated. The upward curvature in this plot results from the difference in the extent of partial alkaline hydrolyses in the unselected and the selected lanes. Normalized band intensity ratios for the selected versus the unselected lanes for the partially 2′-OMe substituted positions that are significantly above or below the curve fit line reveal positions where the 2′-OMe substitution reduces (G6, A7, G10, A14) or enhances (A12, G16) binding affinity . The method is outlined in *Protocol 4* and results are shown in *Figure 7*.

Protocol 4. 2′-*O*-methylpurine substitution binding interference in 2′-amino- or 2′-fluoropyrimidine minimal ligands

Equipment and reagents

- Nitrocellulose filters (see *Protocol 1*)
- Vacuum gel drier
- Phosphorimager (e.g. BAS 1000, Fujix) or film densitometer (e.g. Personal Densitometer, Molecular Dynamics)
- [γ-^{32}P]ATP, 3000 Ci/mmol (New England Nuclear)
- VPF/VEGF (see *Protocol 1*)
- Chemically synthesized minimal ligands with a mixture[a] of 2′-hydroxyl and 2′-*O*-methyl (2′-OMe) phosphoramidites at defined purine positions
- T4 polynucleotide kinase (New England Biolabs)
- Binding buffer (see *Protocol 1*)
- Silicone oil (d = 0.963, Sigma)

Method

1. 5′ end-label about 20 pmols of the partially 2′-OMe substituted ligand(s) with [γ-^{32}P]ATP and T4 polynucleotide kinase.
2. Affinity select the ligand populations by nitrocellulose filter partitioning as in *Protocol 2*, steps 10 and 11.
3. Elute the selected ligands from the filters as in *Protocol 1*B, steps 5–11. Also, elute the unselected ligand (the same amount of RNA simply spotted on a filter, without affinity selection) from the control filter.
4. Resuspend the dried selected ligands (and unselected control) in 20 μl of 50 mM sodium carbonate buffer pH 9.0, and overlay with 20 μl silicone oil to prevent evaporation.
5. Subject the ligands to partial alkaline hydrolysis as in *Protocol 2*, step 5. Stop (slow) the hydrolyses by placing the tubes on ice.
6. Resolve the radiolabelled ligand fragments on a 20% denaturing polyacrylamide gel and detect by exposing the vacuum dried gel to a phosphorimaging plate or to X-ray film.
7. Quantitate changes in mixed substitution purine band intensities relative to the unselected bands by image analysis or film densitometry.

[a] The ratio of the mixture of phosphoramidites in the synthesis is not the same as the ratio of substituted bases in the oligo product because of a higher coupling efficiency for the 2′-*O*-methyl phosphoramidites. We have used ratios of from 2:1 to 1:2, 2′-OH:2′-OMe purine phosphoramidites in syntheses for this purpose.

It is worth noting that the 2′-*O*-methyl substitution at all purine positions except G6, A7, G10, and A14 (*Figure 7C*) leads to a significant (≈ 17-fold) *increase* in binding affinity to VPF/VEGF but not to other heparin binding proteins. Therefore, in addition to enhancing nuclease resistance, affinity as well as specificity of RNA ligands for their targets can be improved by these and related modifications (17).

4. Clinical applications of VPF/VEGF antagonists

Angiogenesis is a complex process controlled by both positive and negative stimuli (8, 9). Among positive regulators of angiogenesis VPF/VEGF is unique in several ways. As a secreted protein produced by a variety of cell types, VPF/VEGF acts selectively on endothelial cells to induce cell proliferation, cell migration, and the expression of proteolytic enzymes and their inhibitors (28–30). VPF/VEGF also potently induces a transient increase in permeability of blood vessels to macromolecules. The leakage of plasma proteins into the extravascular space is believed to assist the growth of new blood vessels by supplying a suitable matrix for endothelial cell attachment and migration (31). In this context it is of interest to recall that it was the presence of fibrin in the extravascular space observed in histological sections of tumour tissues that led to the initial discovery of VPF/VEGF (32). Tumour hypoxia, which is associated with inadequate blood supply, induces the expression of VPF/VEGF, but not of the other angiogenic growth factors (33, 34).

VPF/VEGF is essential for proper formation of blood vessels during development. Homozygous deletion of genes encoding either flt-1 or flk-1/KDR tyrosine kinase receptors (35, 36) or heterozygous deletion of the VPF/VEGF gene (37, 38) results in embryonic lethality. There is now direct evidence that VPF/VEGF is also involved in pathological angiogenesis. Most tumour cells produce and secrete VPF/VEGF that then acts on adjacent endothelial cells in a paracrine manner to induce angiogenesis (31). Interference with VPF/VEGF signalling by neutralizing anti-VPF/VEGF antibodies (39), or through the expression of dominant-negative flk-1 receptor (40, 41) results in inhibition of tumour growth in mice. In addition to inhibiting the growth of primary tumours, VPF/VEGF antagonists have been shown more recently to be capable of dramatically reducing the incidence of tumour metastases (42, 43). From a clinical perspective, this is a more relevant observation. In addition to cancer, VPF/VEGF antagonists may be useful in the treatment of ocular angiogenesis disorders, rheumatoid arthritis, and psoriasis (8, 9, 24).

As the role of VPF/VEGF in pathological angiogenesis becomes more firmly established, the incentive to develop potent and specific VPF/VEGF antagonist becomes greater. VPF/VEGF antagonists reported to date could be classified in the following categories:

(a) Agents that interfere with the binding of VPF/VEGF to its cell surface receptors—antibodies (39), extracellular VPF/VEGF receptor domains (44), and aptamers (15, 17).

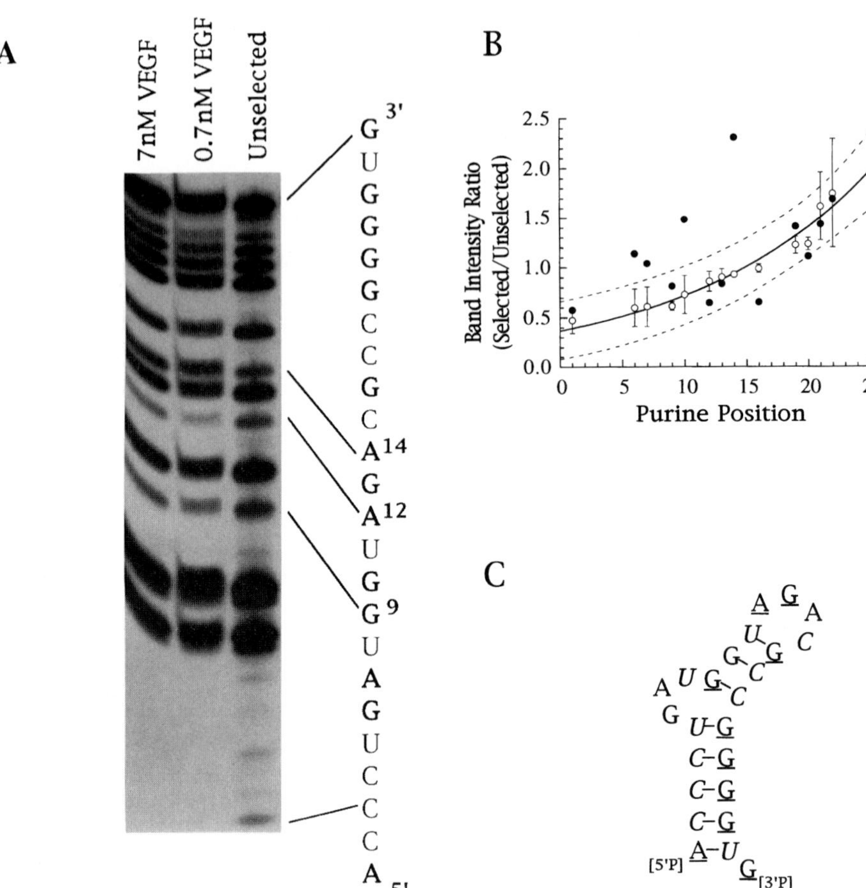

Figure 7. 2′-*O*-methylpurine nucleoside interference analysis. (A) A minimal 2′-aminopyrimidine VPF/VEGF ligand containing 24 nucleotides (shown on the *right*) was chemically synthesized to contain a mixture of 2′-*O*-methyl (2′-OMe) and 2′-OH purine nucleosides at positions 9, 12, and 14, using a 2:1 molar ratio of 2′-OMe:2′-OH (protected as the *t*-butyldimethylsilyl ether) phosphoramidites. Other purine nucleosides were unmodified (2′-OH) and all pyrimidine nucleosides were 2′-NH_2 modified. All purine nucleosides are shown in boldface symbols. The unselected lane shows the partial alkaline hydrolysis pattern of the 5′ ^{32}P end-labelled ligand population prior to affinity selection resolved on a 20% polyacrylamide denaturing gel. The phosphodiester bonds 3′ to the 2′-aminopyrimidines are essentially resistant to alkaline hydrolysis resulting in the absence of bands at those positions. Partial 2′-OMe substitution at the purine positions 9, 12, and 14 is reflected in fainter bands compared to those of the neighbouring (all 2′-OH) purines. The first two lanes from the left show the partial alkaline hydrolysis pattern following affinity selection with at 7 nM or 0.7 nM protein. The preference for the 2′-OMe or the 2′-OH group at the three substituted positions is deduced from the differences in the relative band intensities in the affinity selected versus the unselected ligand pools. For example, the preference for the 2′-OH group at position 14 is deduced from the ratio of A14 to G13 band intensities which is greater in the affinity selected lanes than in the

(b) Agents that inhibit the tyrosine kinase activity of VPF/VEGF receptors—VPF/VEGF-selective tyrphostins (45).

(c) Agents that target VPF/VEGF mRNA and interfere with VPF/VEGF protein synthesis—antisense (46) and ribozyme (47) oligonucleotides.

The relative efficacy of these antagonists and the practicality of their use in clinical settings remains to be established. It is reasonable to expect that anti-VPF/VEGF therapy would need to be delivered for a prolonged period of months to years. Antibodies and VPF/VEGF receptor constructs have very good affinities, specificities, and plasma half-lives, however, their expected immunogenicity (even for humanized antibodies) may limit the duration of treatment. Tyrphostins have the advantage of being small molecules that are relatively easy to manufacture, however, the issues of lower potency and specificity, higher toxicity, and even adequate solubility will need to be addressed. For antisense and ribozyme oligonucleotides, which owe their selectivity to their ability to preferentially hybridize to VPF/VEGF or VPF/VEGF receptor mRNA, the main obstacle is delivery to the cell interior. For nucleic acids, this is an inefficient and poorly understood process (48). Importantly, because nucleic acids tend to bind to certain proteins (including VPF/VEGF) with appreciable affinity, the possibility of observing inhibition through the 'aptamer effect' always needs to be considered (49).

Nuclease-resistant aptamer antagonists of VPF/VEGF offer several unique advantages. Their affinity and specificity is comparable to that of antibodies and receptors while their immunogenicity and toxicity is expected to be low. Intracellular delivery for the aptamers described here is obviously not required. In addition, aptamers in the size range of 20–30 nucleotides (6–10 kDa) are considerably smaller than antibodies which may promote better tissue penetration and diffusability. Like antisense and ribozyme oligonucleotides,

unselected lane. Similarly, the preference for the 2′-OMe group at position 12 is deduced from the ratio of A12 to G13 band intensities which is lower in the affinity selected lanes than in the unselected lane. Band intensities were quantitated by PhosphorImager analysis and the data were normalized for differences between lanes and the extent of partial alkaline hydrolysis (see panel B). (B) Graphical representations of the normalized Phosphorimager data for determination of the 2′-OMe purine nucleoside substitution pattern. The mean of the three normalized band intensity ratios (selected/unselected lanes) for the unsubstituted (exclusively 2′-OH) purine positions is shown in open circles, with standard deviations. A smooth curve is drawn through the data points to guide the eye ($y = 0.403e^{0.609\times}$). The equivalent ratio for each of the partially substituted purine nucleoside positions is shown as filled circles. Ratios for the substituted positions that are signficantly above or below the curve fit line reveal positions where the 2′-OMe substitution reduces (G6, A7, G10, A14) or enhances binding affinity, respectively. The 95% confidence intervals are shown in dashed lines. (C) The proposed secondary structure of the 2′-OMe substituted (underlined positions) and capped 2′-aminopyrimidine RNA ligand. 2′-aminopyrimidine nucleotides are shown in italic letters. [5′P] represents the 5′ cap, d(TsTsTsTs), where s represents the internucleoside phosphorothioate linkage, and [3′P] represents the 3′ cap, d(TsTsTsTsT). (Adapted from ref. 17 with permission.)

aptamers are synthesized chemically. Although the cost of manufacture of oligonucleotides on a large scale is still considerable, the trend toward more efficient and less expensive synthetic methods continues. Importantly, the ease of introducing modifications in aptamers at defined positions provides an opportunity to readily modulate their functional properties (e.g. pharmacokinetics and biodistribution). As we enter a phase in which VPF/VEGF and other aptamers are being tested in a number of animal efficacy models with encouraging preliminary results, the dream of having a robust technology that rapidly produces drug candidates is coming closer to reality.

References

1. Tuerk, C. and Gold, L. (1990). *Science*, **249**, 505.
2. Ellington, A. and Szostak, J. (1990). *Nature*, **346**, 818.
3. Uphoff, K. W., Bell, S. D., and Ellington, A. D. (1996). *Curr. Opin. Struct. Biol.*, **6**, 281.
4. Gold, L., Polisky, B., Uhlenbeck, O. C., and Yarus, M. (1995). *Annu. Rev. Biochem.*, **64**, 763.
5. Fitzwater, T. and Polisky, B. (1996). In *Methods in enzymology* (ed. J. N. Abelson), Vol. 267, pp. 275–301. Academic Press, London.
6. Leung, D. W., Cachianes, G., Kuang, W.-J., and Goeddel, D. V. (1989). *Science*, **246**, 1306.
7. Folkman, J. and Klagsbrun, M. (1987). *Science*, **235**, 442.
8. Folkman, J. (1995). *Nat. Med.*, **1**, 27.
9. Hanahan, D. and Folkman, J. (1996). *Cell*, **86**, 353.
10. Tischer, E., Mitchell, R., Hartman, T., Silva, M., Gospodarowicz, D., Fiddes, J. C., *et al.* (1991). *J. Biol. Chem.*, **266**, 11947.
11. Houck, K. A., Ferrara, N., Winer, J., Cachianes, G., Li, B., and Leung, D. W. (1991). *Mol. Endocrinol.*, **5**, 1806.
12. Pagratis, N. (1996). *Nucleic Acids Res.*, **24**, 3645.
13. Bartel, D. P. and Szostak, J. W. (1993). *Science*, **261**, 1411.
14. Jellinek, D., Lynott, C. K., Rifkin, D. B., and Janjić, N. (1993). *Proc. Natl. Acad. Sci. USA*, **90**, 11227.
15. Jellinek, D., Green, L. S., Bell, C., and Janjić, N. (1994). *Biochemistry*, **33**, 10450.
16. Jellinek, D., Green, L. S., Bell, C., Lynott, C. K., Gill, N., Vargeese, C., *et al.* (1995). *Biochemistry*, **34**, 11363.
17. Green, L. S., Jellinek, D., Bell, C., Beebe, L. A., Feistner, B. D., Gill, S. C., *et al.* (1995). *Chem. Biol.*, **2**, 683.
18. Greslin, A. Unpublished results.
19. Davis, J. P., Janjic, N., Javornik, B. E., and Zichi, D. A. (1996). In *Methods in enzymology* (ed. J. N. Abelson), Vol. 267, pp. 302–14. Academic Press, London.
20. Jaeger, J. A., Turner, D. H., and Zucker, M. (1990). In *Methods in enzymology* (ed. R. F. Doolittle), Vol. 183, pp. 281–306. Academic Press, London.
21. Gutell, R. R., Power, A., Hertz, G. Z., Putz, E. J., and Stormo, G. D. (1992). *Nucleic Acids Res.*, **20**, 5785.
22. Woese, C. R. and Pace, N. R. (1993). In *RNA world* (ed. R. F. Gespeland and J. F. Atkins), pp. 91–117. Cold Spring Harbor Laboratory Press, Plainview, NY.

23. Davis, J. P., Janjić, N., Pribnow, D., and Zichi, D. A. (1995). *Nucleic Acids Res.*, **23**, 4471.
24. Brown, L. F., Detmar, M., Claffey, K., Nagy, J. A., Feng, D., Dvorak, A. M., *et al.* (1997). In *Regulation of angiogenesis* (ed. I. D. Goldberg and E. M. Rosen), pp. 233–69. Birkhauser, Basel.
25. Gold, L. (1995). *J. Biol. Chem.*, **270**, 13581.
26. Pieken, W. A., Olsen, D. B., Benseler, F., Aurup, H., and Eckstein, F. (1991). *Science*, **253**, 314.
27. Lin, Y., Qiu, Q., Gill, S. C., and Jayasena, S. (1994). *Nucleic Acids Res.*, **22**, 5229.
28. Ferrara, N., Houck, K., Jakeman, L., and Leung, D. W. (1992). *Endocrinol. Rev.*, **13**, 18.
29. Pepper, M. S., Ferrara, N., Orci, L., and Montesano, R. (1991). *Biochem. Biophys. Res. Commun.*, **181**, 902.
30. Unemori, E., Ferrara, N., Bauer, E. A., and Amento, E. P. (1992). *J. Cell. Physiol.*, **153**, 557.
31. Dvorak, H. F., Brown, L. F., Detmar, M., and Dvorak, A. M. (1995). *Am. J. Pathol.*, **146**, 1029.
32. Dvorak, H. F., Orenstein, N. S., Carvalho, A. C., Churchill, W. H., Dvorak, A. M., Galli, S. J., *et al.* (1979). *J. Immunol.*, **122**, 166.
33. Shweiki, D., Itin, A., Soffer, D., and Keshet, E. (1992). *Nature*, **359**, 843.
34. Levy, A. P., Levy, N. S., and Goldberg, M. A. (1996). *J. Biol. Chem.*, **271**, 2746.
35. Fong, G.-H., Rossant, J., Gersenstein, M., and Breitman, M. L. (1995). *Nature*, **376**, 66.
36. Shalaby, F., Rossant, J., Yamaguchi, T. P., Gertsenstein, M., Wu, X.-F., Breitman, M. L., *et al.* (1995). *Nature*, **376**, 62.
37. Carmeliet, P., Ferreira, V., Breier, G., Pollefeyt, S., Kieckens, L., Gertsenstein, M., *et al.* (1996). *Nature*, **380**, 435.
38. Ferrara, N., Carver-Moore, K., Chen, H., Dowd, M., Lu, L., O'Shea, K. S., *et al.* (1996). *Nature*, **380**, 439.
39. Kim, K. J., Li, B., Winer, J., Armanini, M., Gillett, N., Phillips, H. S., *et al.* (1993). *Nature*, **362**, 841.
40. Millauer, B., Shawver, L. K., Plate, K. H., Risau, W., and Ullrich, A. (1994). *Nature*, **367**, 576.
41. Millauer, B., Longhi, M. P., Plate, K. H., Shawer, L. K., Risau, W., Ullrich, A., *et al.* (1996). *Cancer Res.*, **56**, 1615.
42. Claffey, K. P., Brown, L. F., del Aguila, L. F., Tognazzi, K., Yeo, K.-T., Manseau, E. J., *et al.* (1996). *Cancer Res.*, **56**, 172.
43. Melnyk, O., Zimmerman, M., Kim, K. J., and Shuman, M. A. (1996). In *87th annual meeting of the American association of cancer research*, Vol. 37, p. 62. American Association of Cancer Research, Washington, DC.
44. Kendall, R. L. and Thomas, K. A. (1993). *Proc. Natl. Acad. Sci. USA*, **90**, 10705.
45. Strawn, L. M., McMahon, G., App, H., Schreck, R., Kuchler, W. R., Longhi, M. P., *et al.* (1996). *Cancer Res.*, **56**, 3540.
46. Robinson, G. S., Pierce, E. A., Rook, S. L., Foley, E., Webb, R., and Smith, L. E. H. (1996). *Proc. Natl. Acad. Sci. USA*, **93**, 4851.
47. Cushman, C., Escobendo, J., Parry, T. J., Kisich, K. O., Richardson, M. L., Speirer,

K. S., *et al.* (1996). Poster presented at *Angiogenesis inhibitors and other novel therapeutic strategies for ocular diseases of neovascularization meeting* (abstract). International Business Communications, Boston, MA.

48. Stein, C. A. and Cheng, Y.-C. (1993). *Science*, **261**, 1004.
49. Burgess, T. L., Fisher, E. F., Ross, S. L., Bready, J. V., Qian, Y.-X., Bayewitch, L. A., *et al.* (1995). *Proc. Natl. Acad. Sci. USA*, **92**, 4051.

7

Immunolocalization and RT-PCR for the detection and quantification of growth factor and receptor gene expression

SIMON R. MYERS and HARSHAD A. NAVSARIA

1. Introduction

In many research fields that impact directly on clinical medicine (e.g. wound healing) a new emphasis on the examination of human tissue has emerged facilitated by the availability of more sophisticated laboratory techniques. This immediately defines the limiting factor in choosing an assay as the volume of human tissue available. To investigate human growth factor and receptor gene expression in any depth, assessment of both protein and messenger RNA (mRNA) for ligand and receptor must be made, in terms both of quantity and of distribution, on small tissue samples. 6 mm punch biopsy wounds of skin for example can be sutured directly, or converted into a small ellipse. In the former case the sutures tend to pull out to leave a pale, circular scar, and in the latter a linear scar of < 1 cm length remains. A 6 mm punch biopsy of skin provides around 60 mg wet weight of tissue. For protocols that require repeated biopsy of an individual, 3 mm punch biopsies are preferable. These can be left to heal by secondary intention, leaving a minimal 'pit' scar. Note: researchers must meet all local and/or national ethical requirements before taking biopsies for experimental purposes from human subjects.

1.1 Assays of protein expression in tissues

A variety of techniques may be used to assay growth factor or receptor expression in tissues:

- immunohistochemistry
- enzyme-linked immunosorbent assay (ELISA)
- Western blotting

- radioimmunoassay (RIA)
- ligand binding studies

ELISA, Western blotting, and RIA require extraction of protein from the tissue specimen (1). Protein extraction on its own will, by definition, destroy tissue morphology and the opportunity to assess protein distribution. The volume of tissue required is considerably more than for semi-quantitative immunohistochemistry because of the inevitable protein antigen loss during extraction (a complete 6 mm punch biopsy as compared to 3×5 μm sections staggered throughout a tissue block). Autoradiographic ligand binding studies will provide specific data on unoccupied receptor levels and distribution in tissue, but are more time-consuming than immunohistochemistry, and involve the use of radioisotopes.

Immunohistochemistry enables both an objective semi-quantification of protein levels and the determination of protein distribution from a very small volume of tissue. The functional relevance of protein ligand and receptor information will depend very much on the epitope specificity of the antibodies used (e.g. differential staining with the LC and CC antibodies raised against the same amino terminus peptide sequence of transforming growth factor-β1) (2). The intensity of immunohistochemical staining can be semi-quantified by histochemical scoring, i.e. assigning to each section a pre-arranged number of 'plusses' or 'minuses' in a blinded fashion by at least two independent observers. Computer image analysis systems have been developed to increase the sensitivity of semi-quantification. Such systems require great care in controlling every step from tissue sectioning onwards.

1.2 Assays of mRNA expression in tissues

Levels of mRNA may be semi-quantified by a number of methods which include:

- RT-PCR
- Northern blotting
- RNase protection assay

Northern blot analysis involves hybridization of extracted mRNA with a labelled probe. In an RNase protection assay a labelled RNA probe is hybridized to target mRNA, and any single-stranded, unhybridized probe degraded with RNase (see Chapter 8). *In situ* hybridization will demonstrate mRNA distribution and may be semi-quantified by computer image analysis, but is not covered here (ref. 3, and see Chapter 8). Compared with these more traditional techniques RT-PCR offers higher levels of specificity and sensitivity (4). If message is being semi-quantified by RT-PCR, then one approach to the determination of message distribution is to use non-radiolabelled *in situ* RT-PCR. The methodology is similar, no probes or isotope are necessary, but a PCR block that will evenly heat microscope slides is required (e.g. Hybaid Omnislide system).

2. Immunolocalization of growth factor and receptor protein

2.1 Introduction

Immunohistochemistry is widely used for the localization in two dimensions of protein expression in tissues and numerous protocols are available in the literature. Using the available antibodies to cytokines and growth factors, it is certainly true in skin that different tissue preparations result in quite different apparent distributions of antigen (e.g. dependence of discrimination between active and latent forms of TGFβ on tissue preparation) (5). This makes the control steps discussed in Section 2.5 especially important. Immunohistochemistry with receptor antibodies is better described, and often includes an acetone fixation step (6). The protocol described here is that devised for immunohistochemistry of growth factor and receptor expression in snap-frozen human skin and may therefore not be appropriate to other tissues prepared in alternative ways.

2.2 Tissue preparation

Antibodies whose epitope specificity and staining optimization are well characterized are ideal, but not always available. Starting from scratch requires preparation of the tissue in question in various ways with flash-frozen tissue as the first option. This strategy gives the best chance of preservation of structure and antigen sites, and therefore of appropriate staining. Deciding whether any staining observed is real or artefactual depends on adequate controls, and particularly on pre-absorption. *Table 1* lists a range of commercial growth factor and cytokine antibodies that have been optimized for use with frozen sections of skin.

Protocol 1. Snap-freezing, blocking, and sectioning of frozen tissue for immunohistochemistry

Equipment and reagents

- Cryotubes (Nunc)
- Small vacuum flask half-filled with liquid nitrogen
- Plastic beaker (200 ml) fitted with a string handle and half-filled with *n*-hexane
- Cork mounting discs
- Oblong plastic tray with room for the beaker, surrounding dry ice piled to fluid level of *n*-hexane, and 200–300 ml of industrial methylated spirit (IMS)
- Tissue support, e.g. Cryo-M-Bed (Bright)

Method

1. Place tissue biopsy directly into a labelled cryotube, cap the tube, and drop it into a liquid nitrogen filled vacuum flask.

Protocol 1. *Continued*

2. Store the cryotube at –70 °C (freezer) or –190 °C (liquid N_2) until blocked, or block directly.
3. Take cryotube from –70 °C freezer or liquid N_2, and place on dry ice until required.
4. Remove tissue from cryotube and orientate on a cork disc over a drop of Cryo-M-Bed.[a]
5. Cover biopsy with a few more drops of Cryo-M-Bed and dip for a couple of seconds at a time into *n*-hexane, until all the Cryo-M-Bed is opaque with no air bubbles.
6. Either store at –70 °C, or section directly. Cryostat sections of 5 μm are ideally prepared for staining immediately, but may be stored at –70 °C.

[a] Orientation at this stage is vital, to ensure that the plane required is parallel to the cork base and therefore the cryostat blade.

2.3 Storage and dilution of antibodies

Antibodies are stored in undiluted aliquots of a small volume (often around 10 μl) to avoid repeated freeze–thaw cycles. Antibody titrations are required in both positive and negative control tissue, prepared in various ways, and using different staining protocols until an optimal system is established. Taking skin growth factor staining as an example, active psoriatic skin, which exhibits a hyperproliferative or 'activated' keratinocyte phenotype, may be used as a positive control in wound healing studies (7). Normal skin may be used as a negative control, although a certain level of constitutive expression is usual, perhaps a result of the insult of biopsy retrieval and processing.

2.4 Detection methods

The detection sytem described below (see *Protocol 3*) is a sensitive three-layer system required to detect small quantities of peptides. The primary antibody binds to the antigen in question. The second layer is a biotinylated antibody raised to the species immunoglobulin of the primary antibody. The third layer creates streptavidin–biotin–enzyme complexes which react with an enzyme substrate chromogen. Many laboratories use diaminobenzidine (DAB) which gives a permanent brown colour in reaction with the bound peroxidase molecules. We used 3-amino-9-ethyl-carbazole (AEC) which again reacts with the peroxidase molecules, this time to give a particulate red end-product. This is not as permanent as DAB (it is also alcohol soluble so slides are not dehydrated, cleared in xylene, or mounted in DPX as for DAB), but it does give a cleaner reaction, less dermal background, and also has the advantage of being a different colour from any melanin deposits in skin. Immunofluorescence becomes a nuisance when semi-quantification of stain-

Table 1. Examples of commercial antibodies used for immunohistochemistry on frozen skin sections

Growth factor (Antigen)	Antibody dilution	Species antibody raised in	Commercial source	Pattern and timing of expression[a]
TGFα (Mature human recombinant TGFα)	1/100	Sheep polyclonal	Biogenesis 9129–8035	N - panepidermal P/W - panepidermal
EGF-R1 (A431 human epidermal carcinoma cells)	1/25	Mouse monoclonal	DAKO M 0886	N - panepidermal P - panepidermal
Interleukin-1β (human recombinant IL-1β)	1/40	Goat polyclonal	R & D Systems AB-201-NA	N - panepidermal P - panepidermal W - early basal, late suprabasal
Interleukin-6 (human recombinant IL-6)	1/30	Goat polyclonal	R & D Systems AB-206-NA	N - basal P - panepidermal W - basal
Keratinocyte growth factor (human recombinant FGF-7)	1/30	Mouse monoclonal	R & D Systems MAB251	N - absent P - suprabasal W - early fibroblast, late suprabasal[b]
TGFβ1 (human recombinant TGFβ1 latency-associated peptide)	1/100	Goat polyclonal	R & D Systems AB-246-PB	N - basal P - basal and epibasal W - panepidermal
TGFβ3 (recombinant chicken TGFβ3)	1/20	Goat polyclonal	R & D Systems AB-244-NA	W - early - panepidermal, late suprabasal

[a] N = expression in epidermis of normal skin, P = expression in active psoriasis, W = expression in a suction blister model of superficial cutaneous injury.
[b] KGF expression in wounded epidermis may represent detection of protein bound to its receptor on keratinocytes.

ing is planned, because the staining fades relatively rapidly and variably, so that analysis must be immediate, or images of sections captured and stored at once (on expensive optical discs) for later processing. In a busy laboratory it also necessitates a fluorescent microscope dedicated to image analysis alone.

2.5 Controls

Non-specific cross-reactivity must be considered carefully and excluded. The ideal situation where a monoclonal antibody has been raised to a peptide sequence, and antibody reactivity can be pre-absorbed completely with antigen excess is not always realized. Classically, optimally diluted antibody is incubated with a dilution series from 10 nmol/ml to 0.001 nmol/ml of antigen before staining positive control tissue. Staining should be incompletely removed in the range 0.01–0.001 nmol/ml (8). Problems are encountered, however, when the tissue section contains not only the ligand antigen, but also the receptor. Staining intensity may increase under these circumstances. One way around this is to bind the antigen to an ELISA plate for the pre-absorption. In this way, pre-absorbing antigen is never presented to the tissue section. However, since there is a limit to the quantity of antigen that can be bound to an ELISA plate well, and since large quantities of antigen may be very expensive or unobtainable, it may not be possible to eliminate staining completely. The following protocol suggest a compromise in which increasing antigen pre-absorption can be shown to be mirrored by decreased immunohistochemical staining (see *Figure 1*).

Protocol 2. Antibody pre-absorption control with bound antigen

Equipment and reagents

- 96-well ELISA plate (Costar 2595)
- PBT buffer: PBS containing 0.1% (w/v) BSA and 0.05% (v/v) Tween
- PBS–Tween: PBS containing 0.05% (v/v) Tween
- Antigen stock solution prepared at 10 nmol/ml in manufacturer's recommended solvent or in PBS
- ELISA plate reader (e.g. Anthos reader 2001, Anthos Labtec Instruments)

Method

1. Pipette serial dilutions of antigen from stock solution (10 nmol/ml to 0.01 nmol/ml) into separate wells of a 96-well ELISA plate in a volume of 100 μl diluent.
2. Incubate plate at 37 °C overnight.
3. Tap the supernatant from the inverted plate, and flood the wells with PBT buffer at room temperature for 1 h.
4. Flood the plate with PBS–Tween and leave for a few minutes.
5. Wash the plate with distilled water.
6. Add 100 μl of optimally diluted antibody to each well and incubate at room temperature for 1 h.
7. For each antigen dilution, pipette an aliquot of supernatant antibody (20–50 μl) onto an air dried section of control tissue.
8. Stain sections in the usual way, and measure the intensity by image analysis (see *Protocol 4*).
9. Once the wells have been developed with AEC, redissolve the particulate chromogen staining with absolute alcohol and measure the absorbance on an ELISA plate reader at 450–500 nm.

The absence of tissue staining upon removal of the primary antibody, or replacement of the primary antibody with a non-immune antibody, serves as

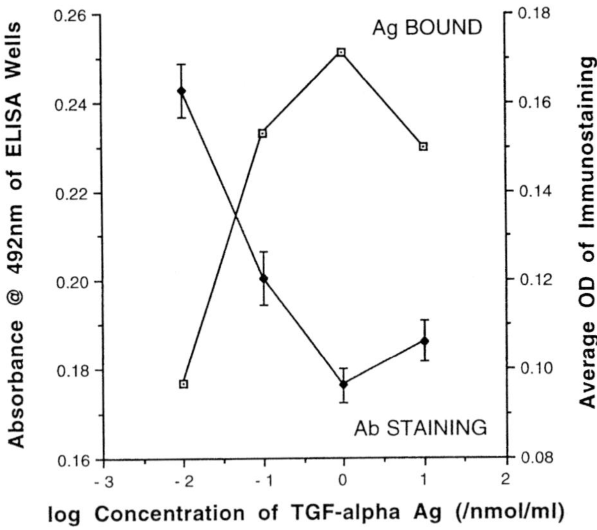

Figure 1. Pre-absorption of a TGFα antibody (Biogenesis 9129–8024). Antigen bound to the ELISA plate varies in a reciprocal fashion with the intensity of antibody staining of positive control (active psoriatic) skin. Only a proportion of the antigen applied to each ELISA well is bound.

a minimal negative control. Similar staining patterns obtained with a range of antibodies strengthens the evidence and Western blotting or immunoprecipitation of the antigen extracted from the tissue in question provide further support.

2.6 Immunohistochemistry

Protocol 3. Three-layer immunoperoxidase staining method for flash-frozen human skin

Equipment and reagents

- Frozen sections of 5 μm thickness cut onto 3-aminopropyltriethoxy-silane (APES, Sigma A-3648) coated glass slides (see *Protocol 3A*)
- Slide chamber lined with filter paper soaked in water—slides should lie flat in this chamber
- Thawed aliquot(s) of primary antibody
- Peroxidase-conjugated streptavidin (1:100, DAKO P0397), or fluorescein isothyocyanate (FITC)-conjugated streptavidin (1:500, Amersham RPN 1232)
- Secondary biotinylated anti-species antibody: e.g. biotinylated rabbit anti-mouse (DAKO E354), biotinylated rabbit anti-goat (DAKO E466)—antibodies should be reconstituted as indicated by the manufacturer
- 3-amino-9-ethyl-carbazole (AEC, Sigma A-5754) detection agent (see *Protocol 3B*)
- Normal rabbit serum (DAKO X902)
- Dilution buffer: manufacturer's recommended buffers, or PBS stabilized with protein[a]

Protocol 3. *Continued*

A. *Preparation of APES coated slides*

1. Rack slides to be coated.
2. Clean slides by washing in acid alcohol (70% methanol, 30% distilled water, 1 ml/100 ml concentrated HCl) for 5 min.
3. Rinse in distilled water for 5 min.
4. Air dry with a fan for 10 min, or until completely dry.
5. Coat slides using a 2% (v/v) solution of APES in acetone for 3 min.
6. Rinse briefly (30 sec) in acetone, and then briefly in distilled water, to wash away excess coating.
7. Air dry at room temperature overnight, or fan dry for 10–15 min.
8. Store in a box protected from light at 4°C for up to one month.

B. *Preparation of the AEC reagent*

1. In a fume-hood, dissolve 60 mg 3-amino-9-ethyl-carbazole (Sigma, A-5754) in 15 ml *N,N*-dimethylformamide (Sigma, D-4254) in a glass container.
2. Add 210 ml of sodium acetate buffer (0.01 M, pH 5.2).
3. Filter and add 225 μl of 33% hydrogen peroxide (BDH, 101284N) as a catalyst.
4. To check that the solution is working, add a couple of drops of reagent to a drop of any left over third layer reagent and examine under a microscope—a peripheral colour reaction should be observed after a couple of minutes. If not, make up fresh AEC reagent before proceeding.

C. *Staining procedure*

1. Air dry sections overnight prior to each staining run.
2. Place slides in a humidified chamber and incubate with a 1:5 dilution of normal rabbit serum (approx. 50 μl/well) for 10 min. This step is designed to block background staining.
3. Rinse off blocking serum in three PBS baths (200 ml each); leave the slides in the first bath for a few seconds, and in the other two baths for a total washing time of 15 min.
4. Take one slide at a time and remove the excess PBS from around each well. It is important to remove as much fluid as possible so as not to further dilute the antibody. However, it is equally important to ensure the section does not dry out completely as this can increase background staining. Add 50 μl of optimally diluted antibody to each well.

5. Incubate slides in a humidified chamber for 2 h.
6. Repeat the washing stage, as in step 3 above (using fresh PBS).
7. Add 1:100 diluted, biotinylated second antibody as for the first layer (i.e. 50 μl/well, one slide at a time, and removing excess PBS), paying particular attention to the species that the primary antibody was raised in.
9. Wash slides as in step 3 (using fresh PBS).
10. Add the final layer reagent (peroxidase- or FITC-conjugated streptavidin) exactly as for the second layer
11. Incubate the slides in a humidified chamber for 35 min.
12. Wash the slides in fresh PBS as before (step 3).
13. If using peroxidase, develop slides using a suitable chromogen.[b] To allow later quantification of staining, care must be taken to ensure that any variation in staining intensity is due to levels of protein in the sample and not due to varying development times. Make up 225 ml of AEC reagent (see part B) in a bath and submerge the racked slides for exactly 5 min.
14. Stop the reaction by transferring the racks to a bath of PBS.
15. Finally, counterstain the developed slides in Mayer's Haemalum (30 sec to 5 min depending on the strength of colour required). Alternatively, if slides are to be quantified by image analysis, mount directly in aqueous mountant (AEC is alcohol soluble) without counterstaining (the blue background complicates image analysis).
16. If both optical density measurements and photography of sections are necessary, first mount slides without counterstaining and then, when analysed, remove the coverslips by an overnight soak in PBS or water, counterstain, and remount for photography.

[a] Inclusion of protein to stabilize the antibody once diluted may take the form of bovine serum albumin or fetal calf serum. We routinely use a 'magimix' combination of 400 ml Dulbecco's MEM basal medium, 10% fetal calf serum containing 0.4 ml of 20% sodium azide, but care must be exercised as serum may contain the growth factor under analysis, or a soluble receptor or binding protein for it.
[b] If using FITC, counterstain slides in propidium iodide solution, mount in anti-fade, put nail varnish around the coverslip to prevent movement, and view using an ultraviolet microscope.

3. Semi-quantification of immunohistochemistry by computer image analysis

3.1 Introduction

To establish protein expression profiles from very limited volumes of tissue an '*in situ*' method which maintains tissue morphology is ideal. Whilst a certain amount of information is available by histometric scoring of immunohistochemistry, this is a very subjective approach. The human interpretation system for visual information is designed to be modified as new information becomes available and is therefore relatively plastic. Recognition of features within an image is far less a problem than applying consistent interpretation criteria. The resolution of visual measurement of intensity is very coarse. Image analysis systems can rapidly make accurate and consistent measurements of intensity. Provided staining variability and lighting instability are controlled, image analysis data should be reproducible.

Optical density (OD), a measurement of the extent to which an object prevents light from passing through it, is expressed by the equation:

$$OD = -\log_{10}(\text{transmitted light/incident light}).$$

Information must be taken from two images to make optical density measurements. As long as neither the intensity of illumination nor the configuration of the image forming optics has changed between capturing the two images, simply dividing the sample image by the image of the incident illumination will result in an image in which the only information is for each point the unlogged optical density of the image at that point.

3.2 Average OD measurement

Various computer image analysis systems are available on the market. In choosing an appropriate system ease of programming and ready access to a trouble-shooter at the source company are vital. Our experience has been with the Seescan System (Seescan plc., Cambridge), and this is used as an example. For average OD measurement of immunohistochemistry, five readings are taken from three 5 μm cryostat sections staggered throughout each tissue block. If these do not prove consistent, more readings may be required.

3.2.1 Some technical considerations

The image of a stained slide is encoded numerically by the computer. The image is split into an array of points (512 × 512), and the intensity of the image at each point measured and assigned a number. The convention is to assign numbers in the scale from 0 (black), linearly through shades of grey to 255 (white). For the measurement of optical density, greyshade digitization must be carefully controlled. The response of the camera to light must be linear. Ideally, a titration of Seescan OD versus nominal optical density should be performed before tissue analysis (see *Figure 2*). Such image analysis systems have been designed to self-calibrate thousands of times per second to ensure that optical black is set to greyshade zero with an accuracy of less than one greyshade.

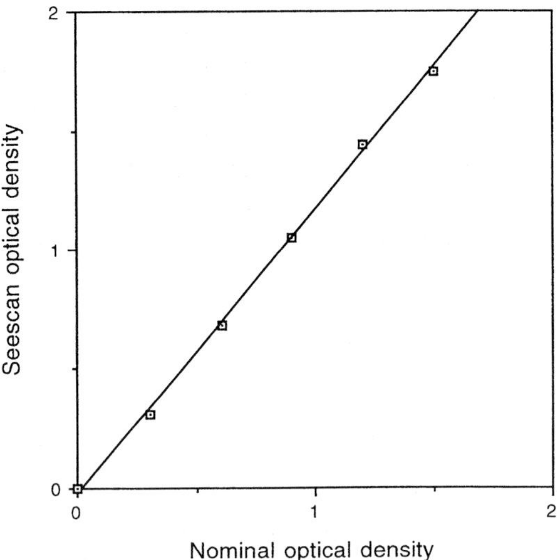

Figure 2. Seescan OD reading versus nominal OD. Titration of measured optical density (OD) against that indicated by a standardized grey-scale filter demonstrates a linear relationship.

3.2.2 Stability of the light source

Because the two images of incident and transmitted light are captured at different points in time, it is critical that the illumination comes from a stabilized light source.

3.2.3 Stray ambient lighting and glare

Any information in the image which is not directly that due to the incident illumination, such as that caused by overhead lighting entering the microscope via the eyepieces, adds an offset to the intensity of the image sensed by the camera. This is minimized by working in a darkened room.

4. Interpretation of immunohistochemistry results

Historically, gene expression has been examined at the protein level. The quantity and distribution of the RNA that encodes that protein is at least as important in relating the source of a factor to its site of action and function, and receives increasing interest. Immunohistochemistry, however, remains a rapid and straightforward method to assay protein level and distribution using a minimum of tissue. It avoids the problems of growth factor activation or inactivation by protein extraction procedures, but does require the use of well-characterized antibodies. Semi-quantification of immunohistochemistry is a refinement which, when well controlled, increases sensitivity and the retrieval of information from precious tissue (*Figure 3*). As the term 'semi-quantification' implies it is seldom possible to provide an absolute measure of

(a)

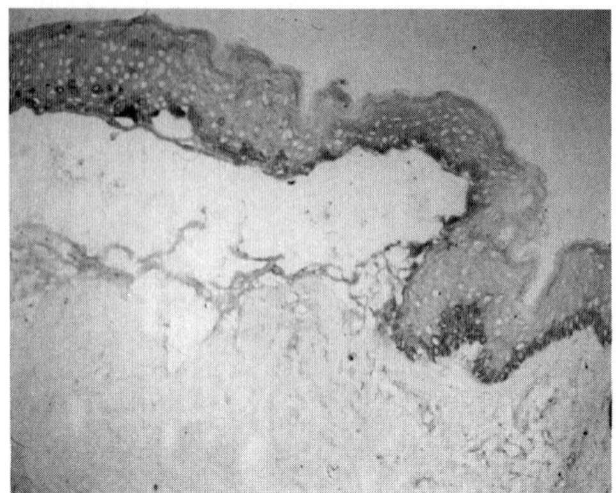

(b)

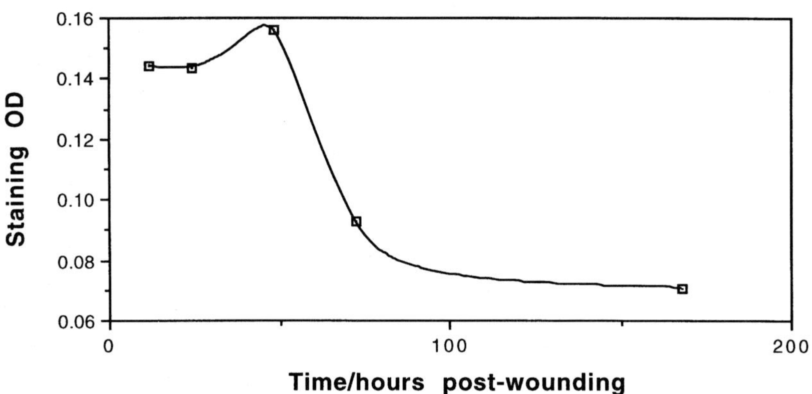

Figure 3. Profile of IL-6 expression using the Seescan system. (a) Immunohistochemistry of a 24 h post-wounding blister section using the antibody AB-206-NA. (b) Basal keratinocyte staining has been quantified over the week following superficial cutaneous suction blister injury in a single individual.

protein level with this technique. The mathematical relationship (linear/logarithmic) between staining intensity and antigen level can, however, be ascertained from pre-absorption. The lowest detectable level of antigen can be estimated by suspending serial dilutions of the antigen in collagen gels, snap-freezing and sectioning, staining with the appropriate antibody, and semi-quantifying by image analysis.

5. Semi-quantification of mRNA levels using non-competitive RT-PCR

5.1 Introduction

Absolute quantification of mRNA levels is similarly frought with difficulties, and the strength of RT-PCR methodologies lie in defining relative changes or differences. Where small amounts of total RNA are available, Northern blotting is insufficient, and a method based on PCR technology required. RT-PCR is a sensitive method of indirect detection of specific mRNAs which, once established in a laboratory is quick and straightforward to perform. It may be considered in four steps:

(a) Extraction of total RNA (or total mRNAs).

(b) Reverse transcription of all extracted RNAs to cDNAs.

(c) Amplification of specific cDNAs by the polymerase chain reaction using synthesized oligonucleotide primers.

(d) Detection of amplified DNA.

Before any test tissue can be processed, however, a number of control titrations are necessary. The simplest system possible is the place to start, although it may not prove adequate for tissue processing at the end of the day.

The first control usually considered is an internal control for the amount of RNA extracted from each specimen. In some, but not all, situations it is safe to assume that housekeeping transcripts, such as those from the β-actin gene, remain at constant levels in test cells under varying conditions. The level of expression of the transcript under investigation can then be measured as a percentage of this control. In practice this would entail amplifying cDNA from each specimen with primers for both the test transcript and housekeeping transcript and comparing, in a simple assay, the optical density of the DNA products as bands in an ethidium bromide agarose gel. Inclusion of the primer pair for the transcript under investigation and the pair for actin in the same PCR tube to control for intertube variation, may result in unwanted primer competition (9). Titration of the RT-PCR products of varying dilutions of extracted RNA by such a simple assay will eventually uncover two problems. First, the PCR ampifications will be seen to rapidly plateau since straightforward ethidium bromide gel electrophoresis is not sensitive enough to demonstrate the linear relationship between starting RNA and DNA product at low starting RNA concentrations. Secondly, the slope of the titration curves for test and housekeeping transcript PCR product band densities may not be the same. This suggests that the efficiency of each reaction is different, and that difference may cause unacceptable error if not corrected for. The first problem can be overcome by using autoradiography to detect radiolabelled RT-PCR products. The second is more difficult to resolve. At this stage it may be appropriate to move on to a more sophisticated, and at the end of the day, more rewarding competitive RT-PCR system which will compensate for some of the inadequacies of the system just described.

If using β-actin, or some other housekeeping gene, as an internal control in a non-competitive system, titrations of starting total RNA concentration versus PCR product band optical density are required for both actin and the species under investigation. Subconfluent 'activated' keratinocyte cultures provide a ready source of growth factor RNA for such titrations, although in some instances a cell line may be more useful (e.g. interleukin-6 and the SiHa line) (10). Such titrations will demonstrate the efficiencies of each primer pair relative to the β-actin pair. They also demonstrate the concentration of starting total RNA at which a plateau is reached for a given detection system. If the slope of log OD versus log RNA concentration is of similar gradient for a test primer pair and actin primer pair, and the quantity of test sample in the PCR can be titrated to give a product whose optical density falls on the linear portion of the titration curve, then such a system may be useful for semi-quantification.

5.2 Extraction of total RNA from small tissue specimens

There are various methods for extracting RNA suitable for RT-PCR from cells or tissue. mRNA rather than total RNA can be extracted using the Dynal magnetic bead system (Dynabeads mRNA DIRECT kit). A number of extraction kits are also available, for example the Promega RNAgents Total RNA Isolation System (Z5110). Kits tend to be relatively expensive, but facilitate the introduction of the method with well controlled reagents. The method below uses a guanidinium isothiocyanate-based reagent, RNAzol B, and provides a simple but effective extraction, suitable for cells or tissue if a kit is not being used.

Protocol 4. Extraction of total RNA

Equipment and reagents

- Culture flask (T75, base area 75 cm^2) of subconfluent cells
- Small tissue biopsy, e.g. 6 mm punch
- RNAzol B (Biogenesis CS-105)
- Chloroform
- Isopropanol
- Absolute ethanol
- 0.05% EDTA (w/v) in water
- Diethyl pyrocarbonate (DEPC)-treated water (11)

A. *Extraction of total RNA from subconfluent cultured keratinocytes*

1. Wash subconfluent keratinocyte cultures in EDTA (once) and PBS (three times) in a laminar flow cabinet.
2. To each T75 flask add 5 ml of RNAzol B and pipette across the flask base until all the cells are broken down to a translucent 'slime' (approx. 3 min).

3. Transfer the cell lysate to a 50 ml Falcon tube, and add 0.1 of the volume (i.e. 500 μl) of chloroform.
4. Vortex mix briefly and centrifuge the tube at 1750 *g* for 10 min.
5. Gently pipette the upper aqueous phase into a 14 ml Falcon tube and add an equal volume (approx. 2.5 ml) of isopropanol.
6. Leave overnight at –20 °C.
7. Centrifuge the tube in a microcentrifuge at 7000 *g* for 20 min, and wash the pellet of RNA once in absolute alcohol. If the pellet is dislodged during washing, spin for a further 5 min.
8. Dry the pellet and resuspend the RNA in DEPC-treated water for immediate use, keep at –20 °C for use within the following 48 h, or precipitate with ethanol for longer storage.
9. Electrophorese RNA through an agarose gel prior to transcription to assess its quality (11).

B. *Extraction of total RNA from small tissue biopsies*

1. Take small tissue biopsies (e.g. punch biopsy material) straight from the –70 °C freezer ready for step 2.
2. Place tissue in a 1.5 ml microcentrifuge tube with 1 ml of RNAzol and homogenize with a tissue homogenizer (Anachem).
3. Add 150 μl of chloroform.
4. Vortex mix and centrifuge the tube in a microcentrifuge at 7000 *g* for 2 min.
5. Pipette the upper aqueous phase into another microcentrifuge tube and add an equal volume of isopropanol.
6. Leave overnight at –20 °C.
7. Centrifuge the tube in a microcentrifuge at 7000 *g* for 5 min, and wash the pellet of RNA once in absolute ethanol. If the pellet is dislodged during washing, spin for a further 2 min.
8. Continue as in part A, steps 8 and 9.

5.3 Reverse transcription of mRNA to cDNA

The use of a reliable reverse transcription kit is simple and ensures reproducible results. The enzyme reverse transcriptase requires a primer site from which to commence transcription. In this example single-stranded cDNA is generated from total RNA using oligo(dT) primers which bind to the poly(A) tail of mRNAs.

Protocol 5. Reverse transcription of mRNA to cDNA

Equipment and reagents

- Thermal cycler (e.g. Hybaid Omnigene), heating block, or water-bath
- Total RNA from *Protocol 4*
- First-Strand cDNA Synthesis kit (Pharmacia Biotech 27-96-01—containing Bulk First-Strand cDNA Reaction mix, DTT solution, and *Not*I-$(dT)_{18}$ primer)

Method

1. Place the aqueous RNA sample in a microcentrifuge tube and make up to a total volume of 20 μl with RNase-free water.
2. Heat the tube and contents to 65 °C for 10 min using the manual control of a Hybaid Omnigene PCR block, a heating block, or a water-bath, and then chill on ice.
3. Gently pipette the Bulk First-Strand cDNA Reaction Mix to form a uniform suspension.
4. Add 11 μl of Bulk First-Strand cDNA Reaction mix to a sterile microcentrifuge tube.
5. To this add 1 μl of DTT solution and 1 μl of 1/25 *Not*I-$(dT)_{18}$ primer.
6. Add the heat denatured RNA and pipette to mix.
7. Incubate at 37 °C for 1 h in the PCR block, heating block, or water-bath.

5.4 PCR amplification of mRNA-derived cDNA with specific oligonucleotide primers

Protocol 5 provides cDNA derived from the extracted total RNA. Primers designed from the sequence of the gene to be quantified are used to amplify a specific template. Primer sequences may be taken from the published literature, or designed 'in-house' with or without the aid of computer soft-ware (e.g. *OLIGO*™ produced for PC or Mac by National Biosciences Incorporated, and distributed in Europe by MedProbe AS, Norway). Each should be around 20 base pairs in length, with around 40–60% GC content, and amplify a template of 200–400 base pairs. Ideally, primers should span an intron to limit the possibility of detecting amplified genomic DNA. PCR with the designed primers must yield a product of the predicted size when compared to a DNA ladder (e.g. φX174 RF DNA/*Hae*III fragments, Gibco BRL, 15611–015) on gel electrophoresis. Examples of primer pairs used for amplification of mRNA encoding growth factors are shown in *Table 2*, and in refs 12–14.

The PCR yield will depend on the annealing temperature and cycle number employed. The lower melting temperature of two primers is an appropriate annealing temperature for that pair. In a non-competitive system the number of cycles should be the lowest that produces reasonably dense bands with the

Table 2. An example of the PCR programme used with primers for growth factors[a]

cDNA species	Forward primer[b]	Reverse primer[b]	Ref
TGFα	5′-ATGGTCCCCTCGGCTGGACAG-3′	5′-TCAGACCACTGTTTCTGAGTGGCA-3′	12
TGFβ1	5′-GCCCTGGACACCAACTATTGC-3′	5′-GCTGCACTTGCAGGAGCGCAC-3′	12
β-actin	5′-GTTTGAGACCTTCAACACCCC-3′	5′-GTGGCCATCTCCTGCTCGAAGTC-3′	13

PCR step[c]	Temperature (°C)	Time (min)/cycles
Denaturing	94	5/1
Denaturing	94	1/30
Annealing	56	1/30
Extension	72	1/30
Extension	72	5/1

[a] The annealing temperature and cycle number should be optimized for each primer pair. When a large number of samples are being amplified with a large number of different primer pairs, a single programme may be used.
[b] Primer pairs used to amplify selected growth factors.
[c] PCR programme for use with the primers shown.

detection system used (often 30–35 cycles). Over-amplification increases the chance of reaching the plateau phase where quantification becomes impossible.

Protocol 6. PCR amplification, and resolution of reverse transcribed RNA

Equipment and reagents

- Thermal cycler (e.g. Hybaid Omnigene)
- cDNA from *Protocol 5*
- Oligonucleotide primers prepared at 25 μg/ml
- dNTPs: 10 mM stocks (Advanced Biotechnologies, AB-0241) mixed in equal volumes to give a 2.5 mM stock of each
- Reaction buffer (Advanced Biotechnologies, AB-0289)
- DEPC-treated water (11)
- 25 mM magnesium chloride solution (Advanced Biotechnologies, supplied with polymerase)
- *Taq* polymerase ('Red Hot DNA polymerase', Advanced Biotechnologies, AB-0406)
- ϕX174 RF DNA/*Hae*III fragments (Gibco BRL, 15611–015)
- DNA sample buffer (11)

Method

1. Use 50 μl reactions in 0.5 ml microcentrifuge tubes throughout. The reaction includes 4 μl mixed dNTPs, 5 μl reaction buffer, 2 μl magnesium chloride solution, 0.2 μl *Taq* polymerase, and generally 2 μl of each primer pair.
2. Prepare on ice a master mix containing the required volumes of dNTPs, reaction buffer, *Taq* polymerase, and magnesium chloride to make up sufficient for all the reactions in the planned run.
3. Prepare master mixes of primer pairs, i.e. combine sufficient of each pair of primers for the planned number of reactions. If only one

Protocol 6. *Continued*

template is being amplified, the primers can be included in the master mix in step 2.

4. Add to each sterile microcentrifuge tube DEPC-treated water, at a volume calculated to achieve, in combination with the cDNA, a reaction volume of 50 μl.
5. Add cDNA as calculated, the volume being dictated by prior titration of the primers with test cell-derived cDNA as described in Section 5.4 and *Figure 4*.
6. Add 12 μl of master mix, followed by 4 μl of the appropriate mixed primer pairs.
7. Gently pipette the reaction components to mix.
8. Add two drops of mineral oil to each tube, and centrifuge by pulsing at 7000 *g* in a microcentrifuge.
9. Transfer the tubes to a PCR block and run the PCR (see *Table 2* for examples of primers and conditions).
10. Add DNA sample buffer and run 10 μl of the PCR products out on 1.5% agarose gel (11) along with the actin controls and ϕX174/*Hae*III marker fragments.

Any volume of test RNA will have different quantities of specific mRNAs, with, for example, lower levels of a short-lived growth factor mRNA than the control, housekeeping mRNA species. A greater volume of total RNA will therefore be required in a PCR to reach the detection range of the former. Also the efficiency of amplification with two different sets of primer pairs may be dissimilar, one pair reaching the plateau at a greater rate. *Figure 4* is designed to illustrate the amplification of transforming growth factor-α (TGFα) mRNA using β-actin as a control. Titration of serial dilutions of total RNA from test tissue or cells against the optical density of bands of their respective RT-PCR products in agarose gels defines the range of detection of each primer pair. Through a trial and error process, it became apparent that ten times the volume of cDNA was required in a PCR reaction using the TGFα primers compared with the control primers to bring the PCR product bands into the control range. Within the detection range, the efficiency of amplification of the actin primers is higher than that for the TGFα primers.

5.5 Detection of RT-PCR product and quantification

Detection of the RT-PCR product bands in the agarose gel can be achieved using a variety of different techniques including:

(a) Ethidium bromide staining of the gels.
(b) Southern blot analysis using non-radioactive or radioactive probes.
(c) Direct labelling of bands in gels using short radioactive probes.

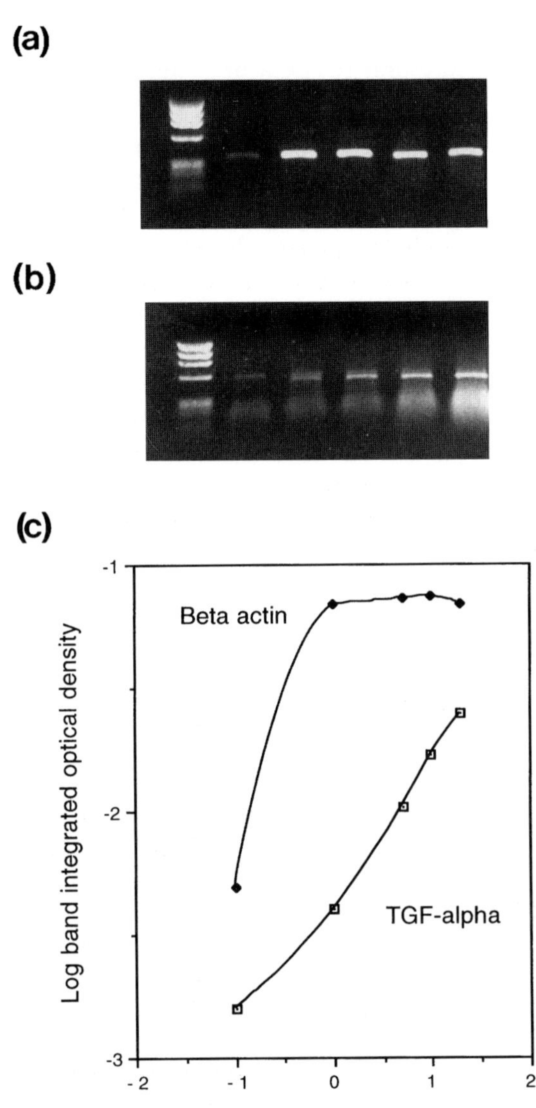

Figure 4. TGFα versus β-actin RT-PCR titrations: serial dilutions of starting total RNA have been reverse transcribed and amplified by PCR using both (a) TGFα and (b) β-actin primers. (c) The slope of the $\log_{10}$ [OD] vs $\log_{10}$ [concentration] for each is not the same due to the different efficiencies of the primer pairs under the same conditions.

Quantification depends on the method used for detection but will usually employ an image capture unit (camera or scanner hooked up to a PC) and image analysis software. Various methods for quantification include:

(a) Cutting out radioactively labelled bands directly from gels or from Southern blots of gels and determining their radioactivity by Cerenkov counting in a beta radiation counter.
(b) Autoradiography of Southern blots (where radioactive or chemiluminescent probes are used), followed by scanning the autoradiogram into an image file on a PC for quantification using appropriate image analysis software.
(c) Scanning the Southern blot directly into an image file (where non-radioactive probes are detected colorimetrically).
(d) Exciting fluorescence of ethidium stained bands with UV light followed by camera-capture of the gel image in a file for quantification.

Whatever the technique employed, it is necessary in each case to:

(a) Measure the integrated optical density or radioactivity for any bands of the appropriate size.
(b) Express all test band measurements as a percentage of the corresponding actin measurement, provided neither are on or above plateau as judged by preliminary titrations.
(c) If plateau has been reached, process all specimens once more with a smaller starting volume of extracted RNA.

6. Competitive RT-PCR quantification of specific mRNA levels

6.1 Introduction

The principal of a competitive RT-PCR system is to provide a mutant PCR template of known concentration, which amplifies with the same efficiency as the test template, and is easily distinguishable at the detection stage. Both templates compete for amplification, and the ratio of mutant to test template remains constant throughout amplification, allowing quantification outside the linear portion of the log (OD) versus log (starting RNA concentration) curve. This allows the test template to be amplified to a greater extent, so that the poor sensitivity of ethidium bromide gel electrophoresis becomes less of a problem.

6.2 Mutant templates

Various methods have been devised to generate mutant templates and to distinguish them from test templates at detection:

(a) The mutant competitor can be engineered slightly smaller or larger than the test template (15, 16). *Protocol 7* describes a method for generating a shortened mutant using internal primers.

(b) A unique restriction site may be added or removed from the mutant (17).

(c) A mutant differing from the test by a single nucleotide can be distinguished by a PCR–temperature gradient gel electrophoresis system (18).

(d) A heterologous DNA fragment of different size may be generated as a mimic with only the primer binding sites homologous to the test cDNA (19).

(e) Other strategies aim to control for variations in the efficiency of the reverse transcription step by generating mRNA standards (20).

6.2.1 Design of internal primers to generate shortened mutant templates

Based on available sequences and the pair of external primers in use to generate a specific template, internal composite primers are designed. These will allow two fragments to be amplified by PCR from the original template, which will, in a further PCR, provide the shortened mutant template (*Figure 5*). Each fragment is generated in a separate PCR reaction, the first with the forward external primer and the reverse internal primer, and the second with the reverse external primer and forward internal primer. The internal composite primer for the first reaction is designed at the beginning of the sequence to be dropped, but with a tail sequence complimentary to a 5′ to 3′ sequence at some distance downstream. The reverse is true for the second reaction. When these two are combined in a further PCR using the both external primers alone, a shortened template is generated. The internal primer sequences should have appropriate GC content, and enable loss of a sequence of sufficient size that original and mutant templates can be distinguished on electrophoresis. The amount of mutant generated must be enough for reliable spectrophotometer quantification, and this may necessitate cloning into a plasmid vector.

Protocol 7. Generation of shortened mutant template using internal oligonucleotide primers

Equipment and reagents

- PCR block and PCR reagents
- Internal and external composite primers (see *Figure 5*)
- Low melting point agarose (Flowgen, 50102)
- Phenol
- Chloroform
- Image analysis system
- φX174 RF DNA/*Hae*III fragments (Gibco BRL, 15611–015)

Method (see Figures 5 and 6)

1. Generate an 'original' template from cDNA (derived either from reverse transcribed RNA or cDNA cloned into a plasmid vector) using both external primers (see *Protocols 5* and *6*).[a]

Protocol 7. *Continued*

2. Using approx. 100 ng of original template in a 50 μl PCR reaction generate the first fragment with the forward external and reverse composite internal primers (see *Protocol 6*).
3. Using approx. 100 ng of original template in a 50 μl PCR reaction generate the second fragment with the reverse external and forward composite internal primers (see *Protocol 6*).
4. Run both fragments on a 1.5% low melting point agarose gel along with ϕX174/*Hae*III marker fragments, and retrieve the cDNA by phenol and chloroform extractions (11).
5. In a further PCR reaction, amplify 50 ng of each extracted fragment in combination, using both external primers (see *Protocol 6*).
6. Once again run the product on a low melting point gel and extract it.
7. At this point, there must be enough mutant to reliably quantify by spectrophotometry. If not, repeat the process, reamplifying the product or cloning the product into a vector for amplification.

[a] Ideally the product should be of an amount roughly quantifiable by spectrophotometry.

6.3 Use of mutant templates in tissue analysis

Protocol 8 shows how the amount of mutant template required to act as control for test RNA expression can be estimated and used to determine the concentration of test RNA in tissue analyses. An excellent discussion of this method of quantification and an example using interleukin-1β may be found in ref. 19.

Protocol 8. Analysis of RNA expression in tissues by competitive RT-PCR[a]

Equipment and reagents

- Thermal cycler and PCR reagents (see *Protocol 6*)
- Mutant template of known concentration
- Test cDNA

Method

1. Prepare a series of tenfold dilutions of the mutant template in five to ten steps starting with a concentration of 1 amol/μl.
2. Amplify the serially diluted mutant template using PCR with a constant quantity of test cDNA (see *Protocols 5* and *6*).[b]
3. Based on the relative amounts of the products (see Section 5.5), set-up

a 'fine-tuning' twofold serial dilution of mutant, starting with a ratio of mutant template:test cDNA of 10:1.

4. Repeat step 2.
5. Plot $\log_{10}$ (test band OD/mutant OD) against $\log_{10}$ concentration of mutant for each dilution series. Where $\log_{10}$ (test band OD/mutant OD) is equal to zero, equal masses of test and mutant template are present, and the quantity of test template can be derived from the corresponding point on the $\times$ axis.[c]
6. Calculation of the quantity of test mRNA assuming 100% efficiency of transcription will give a minimum number of test RNA molecules.

[a] In this instance the method for quantitation uses optical densities of ethidium stained bands for quantification of PCR products. In principle the same calculations could be applied to quantities derived by any of the techniques listed in Section 5.5.
[b] This volume will be broadly based on the amount of total RNA extracted, which can only be approximated by spectrophotometry in small extracts of tissues, but which should be around 2 μg.
[c] The different lengths of original and mutant templates will have to taken into consideration when determining the actual molar concentrations.

7. Interpretation of RT-PCR amplification analyses of gene expression

Great care must be taken to avoid over-interpetation of RT-PCR results, even those from a competitive system. Such assays are far more useful in the description of relative change or differences than absolute levels. As has been described, several levels of sophistication (and expense) can be used in these assays. Resource becomes an issue particularly at the detection level. The development of competitive systems has been driven in part by a lack of sensitivity in detection, since they allow amplification and quantification of PCR product outside the linear portion of the product OD versus starting RNA concentration curve. Competitive systems are, however, laborious and will soon be superceded by real-time analysis systems.

The real-time progress of a PCR reaction can now be monitored using a fluorogenic probe system and appropriate hardware (e.g. the ABI PRISM 7700 Sequence Detection System, Applied Biosystems, Perkin Elmer). An oligonucleotide fluorogenic probe with both a reporter and quencher dye attached anneals between forward and reverse primers. With each cycle reporter fluorescence results as the probe is cleaved by the 5′ nuclease activity of DNA polymerase. Reactions are characterized, not by the the end-point product of a number of cycles, but by the fractional cycle number at which fluorescence passes a threshold level. Such information on the reaction kinetics during the course of a PCR should revolutionize the approach to quantification.

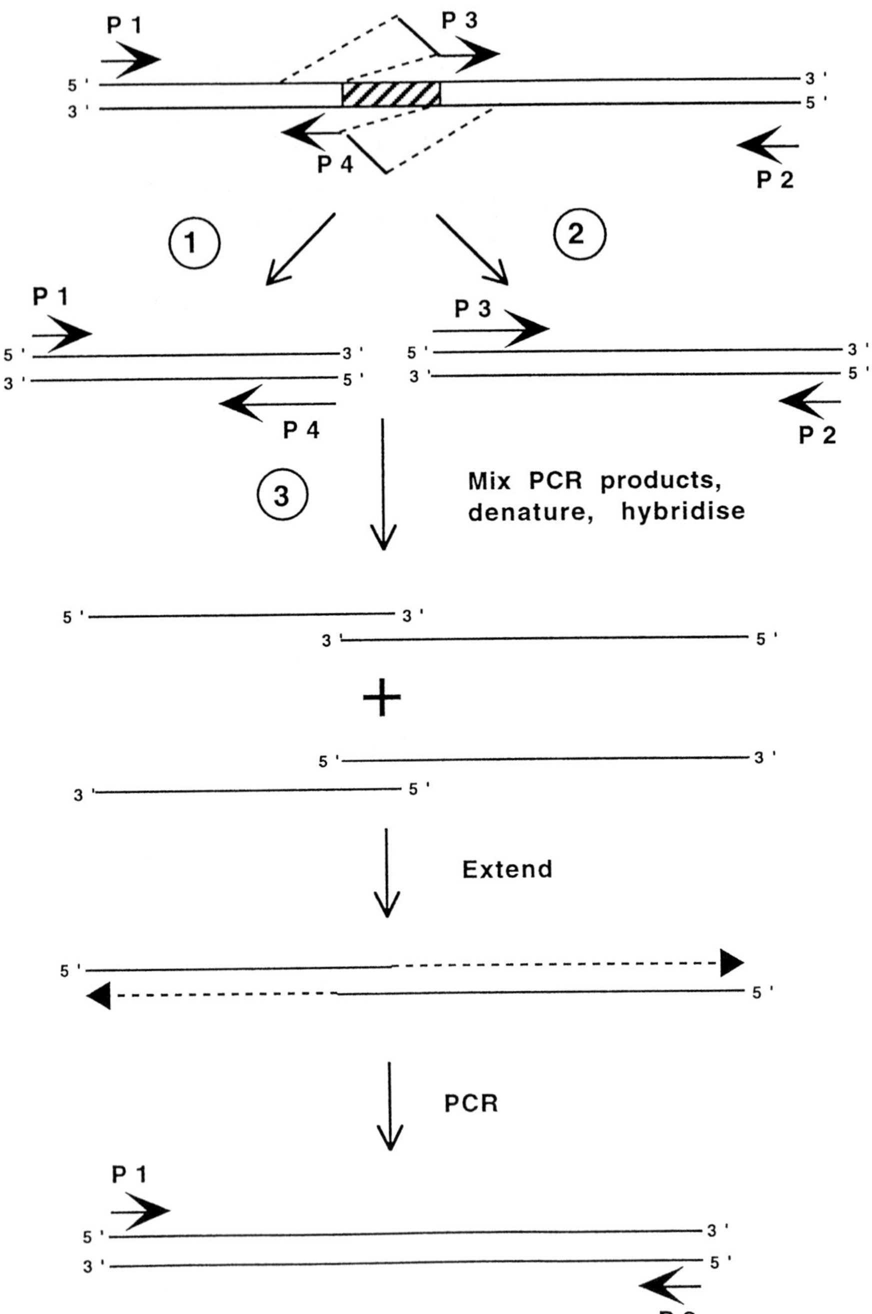

Figure 5. Diagrammatic representation of the generation of a shortened mutant template using internal oligonucleotide primers.

(a)

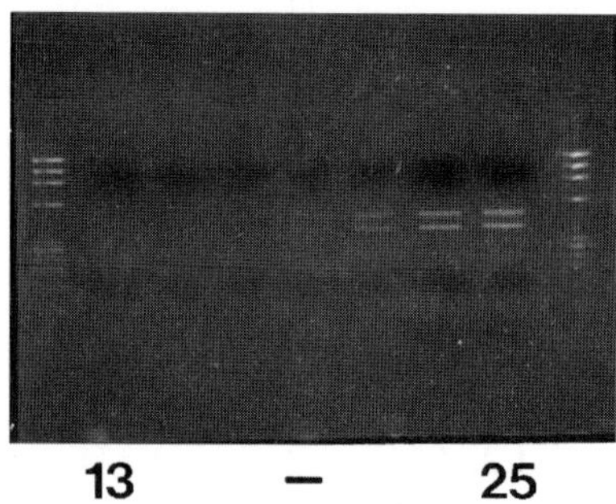

(b)

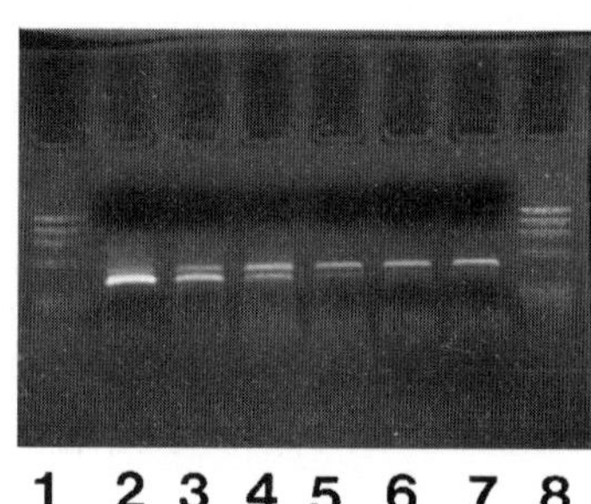

Figure 6. PCR products on ethidium bromide stained gels demonstrating (a) the similar kinetics of amplification for control and mutant TGFα templates, (b) the point at which control and mutant templates are present at equivalent concentrations. In (a), 0.01 amol of control and mutant template were co-amplified and 10 μl aliquots drawn at alternate cycles from 13–25. In (b), fivefold serial dilutions of mutant (0.025 amol to 8×10^{-6} amol) were co-amplified with 0.001 amol of control template. Equivalent masses of control and mutant products would occur at a ratio between those represented in lanes 3 and 4.

References

1. Capper, S. J. (1993). In *Growth factors: a practical approach* (ed. I. A. McKay and I. M. Leigh), p. 181. IRL Press, Oxford.
2. Kane, C. J. M., Hebda, P. A., Mansbridge, J. N., and Hanawalt, P. C. (1991). *J. Cell. Physiol.*, **148**, 157.
3. Chen, W. H., Morriss-Kay, G. M., and Copp, A. J. (1995). *Development*, **121**, 681.
4. O'Driscoll, L., Daly, C., Saleh, M., and Clynes, M. (1993). *Cytotechnology*, **12**, 289.
5. Barcellos-Hoff, M. H., Ehrhart, E. J., Manu, K., Jirtle, R., Flanders, K., and Tsang, M. L-S. (1995). *Am. J Pathol.*, **147**, 1228.
6. Linge, C. and Green, M. R. (1993). In *Growth factors: a practical approach* (ed. I. A. McKay and I. M. Leigh), p. 201. IRL Press, Oxford.

7. McKay, I. A. and Leigh, I. M. (1995). *Clin. Dermatol.*, **13**, 105.
8. Polak, J. M. and Van Noorden, S. (ed.) (1986). *Immunocytochemistry, modern methods and applications*, p. 36. Wright, Bristol.
9. Siebert, P. D. and Larrick, J. W. (1992). *Nature*, **359**, 557.
10. Bryan, D., Sexton, C. J., Williams, D., Leigh, I. M., and McKay, I. A. (1995). *Cell Growth Differ.*, **6**, 1245.
11. Sambrook, I., Fritsch, E. F., and Maniatis, T. (ed.) (1989). *Molecular cloning: a laboratory manual*, 2nd edn. Cold Spring Harbor Laboratory Press, NY.
12. Nickoloff, B. J. and Naidu, Y. (1994). *J. Am. Acad. Dermatol.*, **30**, 535.
13. Carroll, J. M., Albers, K. M., Garlick, J. A., Harrington, R., and Taichman, L. B. (1993). *Proc. Natl. Acad. Sci. USA*, **90**, 10270.
14. Van Zoelen, E. J. J., Delaey, B., Van der Burg, B., and Huylebroeck, D. (1993). In *Growth factors: a practical approach* (ed. I. A. McKay and I. M. Leigh), p. 13. IRL Press, Oxford.
15. Compton, C., Tong, Y., Trookman, N., Zhao, H., and Roy, D. (1994). *J. Invest. Dermatol.*, **103**, 127.
16. Gilliland, G., Perrin, S., Blanchard, K., and Bunn, H.F. (1990). *Proc. Natl. Acad. Sci. USA*, **87**, 2725.
17. Becker-Andre, M. and Hahlbrook, K. (1989). *Nucleic Acids Res.*, **17**, 9437.
18. Kang, J., Kuhn, J. E., Schafer, P., Immelmann, A., and Henco, K. (1994). In *PCR 2: a practical approach* (ed. M. J. McPherson, B. D. Hames, and G. R. Taylor), p. 119. IRL Press, Oxford.
19. Siebert, P. D. and Kellogg, D. E. (1994). In *PCR 2: a practical approach* (ed. M. J. McPherson, B. D. Hames, and G. R. Taylor), p. 135. IRL Press, Oxford.
20. Heuvel, J. P. V., Tyson, F. L., and Bell, D. A. (1993). *Biotechniques*, **14**, 395.

8

Defining growth factor function through tissue-specific expression of dominant-negative receptor mutants

SABINE WERNER

1. Introduction

Elucidating the function of growth factors and their receptors during development and in the adult organism is one of the most challenging tasks in growth factor biology. A few years ago, only descriptive expression data were available which suggested multiple roles of growth factors in development. However, the development of transgenic mouse technologies has provided new insights into the function of growth factors and their receptors *in vivo*. These technologies allow gain-of-function experiments (over-expression of ligands or receptors) as well as loss-of-function experiments (gene knock-outs by homologous recombination in embryonic stem cells). Knock-out studies have shed light on the function of many growth factors in embryonic development. While these studies can reveal the earliest essential function of a particular growth factor or its receptor, the specific role of these genes at later stages of development may be obscured due to the early embryonic lethality. In addition, the loss of one member of a gene family may be compensated for by overlapping expression of another, related gene (1). Such a compensation could be particularly important for growth factors which are members of a large family of related ligands, e.g. the fibroblast growth factors (FGFs) and the epidermal growth factor (EGF) receptor ligands. This hypothesis is supported by the very 'mild' phenotypes of FGF-5, FGF-7, and transforming growth factor-α (TGFα) knock-out mice (2–5). These animals reveal only a defect in the hair, although FGF-5, FGF-7, and TGFα are widely expressed in the embryo and—in the case of FGF-7 and TGFα—also in the adult animal (6–8).

Given the above mentioned limitations of knock-out experiments it would be extremely helpful to have a strategy which allows:

(a) Inhibition of growth factor function in a tissue-specific manner.

(b) Blocking the action of all ligands of a particular receptor.

These goals can be achieved by tissue-specific expression of dominant-negative receptor mutants in transgenic mice.

1.1 The concept of dominant-negative growth factor receptors

It has long been known that the action of tyrosine kinase receptors depends on dimerization which is a prerequisite for receptor transphosphorylation on tyrosine residues (9). The phosphorylated receptors subsequently bind and activate several signal transduction molecules which induce appropriate cellular responses. Dominant-negative receptor mutants are characterized by the lack of a functional tyrosine kinase (or serine/threonine kinase) domain (10, 11). This loss-of-function can be achieved by complete truncation of the kinase domain (*Figure 1*) but also by mutating the ATP binding site. Upon ligand binding, the mutant receptors form non-functional heterodimers with the full-length wild-type receptors, thereby blocking signal transduction (10–12) (*Figure 1*). Most importantly, the dominant-negative action is specific for each growth factor receptor; for example, a truncated FGF receptor only blocks signal transduction through FGF receptors but not through the epidermal growth factor receptor or platelet-derived growth factor receptor. Dominant-negative receptors are therefore useful to block selectively the action of a specific growth factor receptor *in vitro* and *in vivo*. The only prerequisite for the dominant-negative effect of the truncated receptor is that both the mutant and the wild-type receptor bind to the same ligand (13, 14). For example, a truncated FGFR1 will not only block signalling through wild-type FGFR1, but also through wild-type FGFR2 or FGFR3, provided that the truncated and the wild-type receptor bind to the same type of FGF (13, 14). Thus a truncated FGF receptor should inhibit the response of all FGF receptors which bind the same FGF.

In order to act as a dominant-negative mutant, the kinase-deficient receptor mutant must be expressed at much higher levels than the endogenous wild-type receptor. From our experience, a 20–50-fold excess of the truncated compared to the endogenous receptor is essential to block signal transduction *in vivo*. Thus it is important to choose a strong promoter for the expression of dominant-negative receptor mutants. Furthermore, the promoter should be tissue-specific in order to avoid side-effects in other organs. Most importantly, the promoter should allow expression of the truncated receptor in the same cells which express the wild-type receptor. Therefore, a detailed knowledge of the expression pattern of the receptor is essential. In addition, an extensive expression analysis of related receptors which bind to the same ligands as well as of the ligands themselves should be performed in order to correctly interpret the observed phenotypic abnormalities. In each of the experiments where expression of the appropriate receptor was directed to the appropriate tissue at the right time, a specific phenotype was obtained.

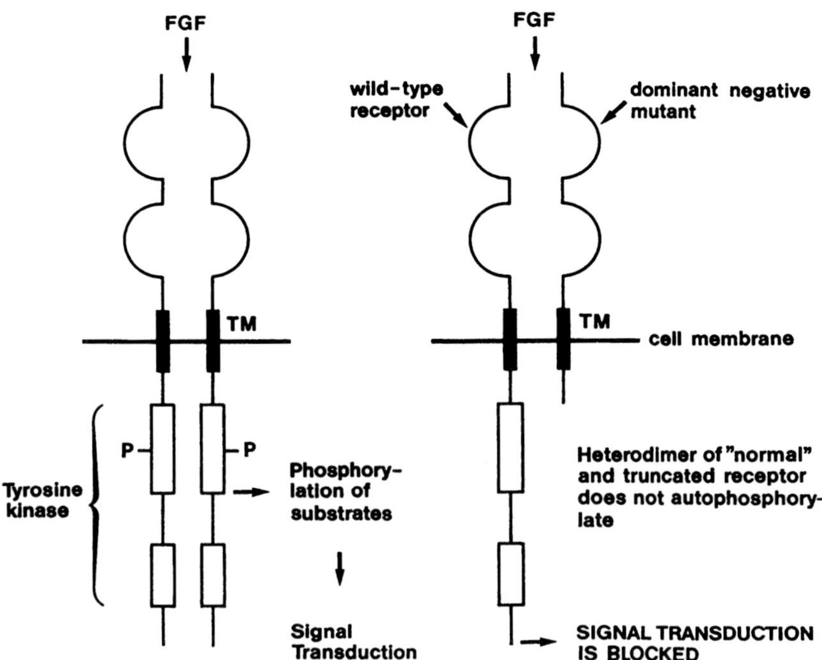

Figure 1. The mechanism of action of dominant-negative FGF receptor mutants. The effect of ligand binding to endogenous receptors and dominant-negative receptors is shown schematically. The transmembrane domain (TM) is indicated as a black box. Tyrosine kinase domains 1 and 2 are indicated as open boxes. P indicates phosphorylated tyrosine residues. Left-hand side: binding of FGF to its receptor induces receptor dimerization which is followed by transphosphorylation of the receptor on tyrosine residues. Receptor phosphorylation results in the recruitment of intracellular signalling proteins which bind to specific phosphotyrosine residues. Some of the signalling molecules become phosphorylated by the receptor kinase. Recruited signalling molecules are then activated and induce appropriate cellular responses. Right-hand side: overexpression of truncated receptors which lack a functional kinase domain favours the formation of non-functional heterodimers which fail to become phosphorylated. These unphosphorylated heterodimers fail to associate with intracellular signalling molecules and thus cannot activate signalling pathways or induce cellular responses.

1.2 Examples of efficient use of dominant-negative growth factor receptors in transgenic mice

The first successful *in vivo* use of a dominant-negative receptor was reported by Amaya *et al.* (15). In this study, a dominant-negative FGF receptor mutant was globally expressed in early *Xenopus* embryos, where it prevented formation of posterior mesoderm. Subsequently, several other truncated receptors were expressed in the *Xenopus* embryo. Most remarkably, this approach was also successful with dominant-negative serine/threonine kinase receptors (16, 17), demonstrating that this strategy can also be applied for this type of receptor.

Given these promising results in *Xenopus*, we decided to test the potency of dominant-negative receptors to block growth factor function in a tissue-specific manner in transgenic mice. Due to its accessibility and the availability of strong and highly tissue-specific keratin promoters, we chose the skin as a model system to test this strategy. Keratin promoters do not only allow a skin-specific expression but in addition make it possible to target transgenes selectively to the proliferating basal cells or to the differentiated suprabasal cells, respectively (18, 19). Using a basal keratin 14 promoter, we were able to demonstrate an important role of keratinocyte growth factor and its receptor in skin morphogenesis and wound re-epithelialization (20), and targeted expression of a dominant-negative FGFR1 revealed a novel role of FGF in keratinocyte organization and differentiation (21). A similar approach was used for the EGF receptor. Thus expression of a dominant-negative EGF receptor under the control of a basal keratin 5 promoter elicited striking alterations in hair follicle development and skin structure (22).

Expression of dominant-negative receptors has also been used successfully to study growth factor function in other tissues. For example, expression of a dominant-negative keratinocyte growth factor receptor (FGFR2-IIIb) in the developing lung epithelium under the control of the lung surfactant protein C promoter (23) blocked lung branching morphogenesis and alveolar differentiation (24). In the eye, the lens-specific expression of a truncated FGF receptor disturbed lens development (25), and retinal degeneration was observed after expression of a dominant-negative FGF receptor under the control of a rhodopsin promoter (26). This approach is therefore suitable for blocking growth factor function in a wide variety of tissues and organs. In this chapter I will focus on the characterization of mice which express truncated receptors in the skin, but the same methods can also be used for the characterization of other tissues.

2. Design and synthesis of a gene encoding a dominant-negative receptor mutant

Our experience with the use of dominant-negative receptors has resulted in the development of a few guidelines for the application of the approach:

(a) As mentioned above, a prior knowledge of the biology and in particular the expression pattern of the receptor is essential.

(b) The promoter has to be carefully chosen. It should direct the transgene to the appropriate cells at the right developmental stage. Furthermore, the promoter has to be very strong, since at least a 20-fold over-expression is required to completely inhibit receptor function. Therefore, it is essential to use a well-characterized promoter which has been shown to be strong and tissue-specific. A wide variety of tissue-specific promoters is now upstream from it. The completeness of the restriction digest should be determined by gel electrophoretic analysis.

(c) The template DNA is subsequently extracted twice with phenol and once with chloroform, ethanol precipitated, dissolved in diethyl pyrocarbonate-treated water (DEPC–H_2O) at a concentration of 1 mg/ml, and stored at –20 °C.

(d) The linearized plasmid is transcribed in the presence of ^{32}P-labelled rNTPs with the appropriate bacteriophage DNA-dependent RNA polymerase to produce an antisense RNA that extends from the initiation site of the promoter to the site where the DNA has been digested. Optimal results are obtained with probes of 150–350 nucleotides. For the detection of mRNAs encoding truncated receptors it is helpful to use a probe which discriminates between the endogenous and the truncated receptor. Thus templates which include the truncation site are useful for this purpose (*Figure 2*).

(e) For reasons of economy, the radioactive UTP is added at a limited concentration. This might lead to shorter transcripts due to premature termination at U residues. These smaller transcripts will give rise to smaller protected fragments, resulting in a smear on the final gel. Therefore, it is essential to purify the riboprobe using preparative polyacrylamide gel electrophoresis (PAGE).

(f) An excess of the radiolabelled purified antisense RNA is hybridized in solution with the RNA being tested, so that all complementary sequences form ^{32}P-labelled RNA:RNA hybrids. After unhybridized material has been removed by digestion with RNase, the radiolabelled RNA:RNA hybrid is then detected and quantitated by PAGE under denaturing conditions, followed by autoradiography.

(g) For all protection assay experiments it is important to include appropriate controls:

   (i) tRNA (50 μg) is used as a negative control. Bands which are obtained by hybridization of the probe with tRNA result from incomplete RNase digestion and/or self-hybridization of the probe.

   (ii) If available, an appropriate positive control should be used, e.g. unlabelled sense riboprobe or RNA from a tissue where the gene of interest has already been shown to be expressed.

   (iii) A sample of the radiolabelled riboprobe (approx. 1000 c.p.m.) is loaded onto the analytical gel to compare the size of the protected fragments with the size of the original riboprobe. Since the probe includes polylinker sequences of variable length, the protected fragments should be shorter than the original probe. Bands which are identical in size with the riboprobe are therefore a result of incomplete digestion of the probe.

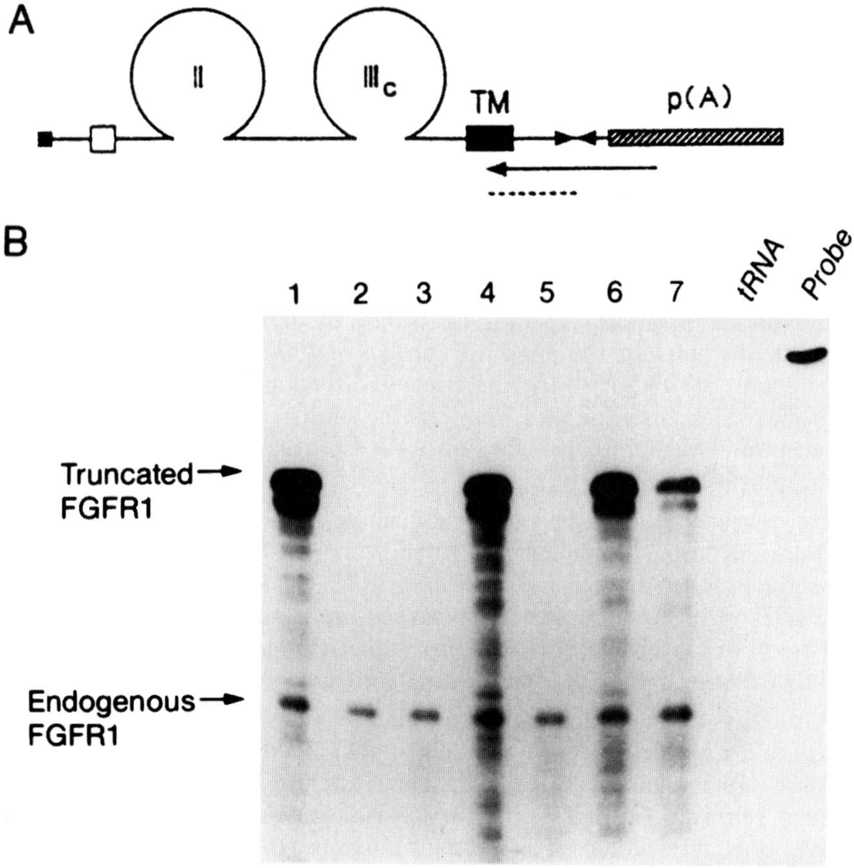

Figure 2. Detection of truncated and endogenous FGF receptor mRNAs by RNase protection assay. (A) Diagram of the transgene mRNA. Functional elements include the ATG translation initiation codon (small black box on the left), the highly acidic region of FGFR1 (open box), immunoglobulin-like domains II and IIIc (II and IIIc), the transmembrane region (TM), the stop codon (double arrow), and the human growth hormone poly(A) (dashed box). To discriminate between the endogenous and the truncated receptor we used the probe which is indicated with an arrow. It includes part of the transmembrane domain, the truncation site, and part of the human growth hormone poly(A). Transcripts encoding the truncated receptor give rise to the full-length protected fragment (upper band in B). Transcripts encoding the endogenous receptor give rise to a shorter protected fragment (lower band in B; indicated with a dotted line in A). (B) 50 μg total cellular RNA from mouse tail skin was analysed. The protected fragments obtained by mRNA encoding the endogenous and the truncated receptor are indicated. 1000 c.p.m. of the hybridization probe were loaded in the lane labelled probe and used as a size marker. 50 μg tRNA were used as a negative control. The numbers 1–7 indicate transgenic founder mice No. 1–7. (Adapted from ref. 21 with permission from the *EMBO Journal*.)

available which allow expression of a dominant-negative receptor in many different organs.

(c) An effective dominant-negative receptor should be available. Ideally, a truncated receptor should be used which has already been tested *in vivo*. If such a transgene is not available, the appropriate receptor cDNA should be truncated 3′ from the sequence encoding the transmembrane domain. From our experience, at least 15–20 amino acids of the intracellular domain should be retained in the truncated receptor in order to allow correct membrane anchoring. In any case, the transgenes should be extensively tested in cultured cells or in *Xenopus* oocytes prior to using them in transgenic mice. This is particularly important if the transgene includes a carboxy terminal tag. In our experience, short tags did not disturb the dominant-negative effect but a tag which was longer than 20 amino acids resulted in a partial loss of the dominant-negative effect (S. Werner, unpublished data).

(d) An appropriate vector should be constructed. The latter should include the promoter, the truncated receptor cDNA followed by a stop codon, and a polyadenylation site. If cDNAs are used, the transgene should also contain an intron. We have successfully used the rabbit β-globin intron for these purposes. This sequence is also suitable for subsequent dot-blot or Southern blot analyis (see below). Unique restriction sites should be present at both ends of the transgene to allow release of the insert from vector sequences. The purified insert can then be used for the generation of transgenic mice as described by Hogan *et al.* (27).

3. Identification of transgenic mice

Several different methods can be used to identify the transgenic animals, including the polymerase chain reaction (PCR), dot-blot analysis, and Southern blot analysis. All these methods require the isolation of chromosomal DNA which is usually extracted from the tips of the tails (*Protocol 1*). Tails of mice three weeks of age are routinely used, although mice at any age can be tested.

Protocol 1. Preparation of genomic DNA from a mouse tail

Equipment and reagents

- Shaking water-bath at 55 °C
- Proteinase K (Boehringer Mannheim)
- Tail blot solution: 50 mM Tris–HCl pH 8.0, 100 mM EDTA (ethylenediamine tetraacetic acid), 100 mM NaCl, 1% SDS

Method

1. Anaesthetize a three-week-old mouse according to local procedures.[a]
2. Cut 0.5–1 cm from the tip of its tail with a scalpel.

Protocol 1. *Continued*

3. Cut the tip of the tail into five to eight pieces and transfer to a labelled 1.5 ml microcentrifuge tube.
4. Add 700 μl tail blot solution to the tube.
5. Add 35 μl proteinase K solution (10 mg/ml).
6. Incubate at 55 °C overnight in a shaking water-bath.
7. Extract twice with 500 μl phenol, twice with 500 μl phenol:chloroform (1:1), and twice with 500 μl chloroform.[b]
8. Precipitate DNA from the aqueous phase by adding 800 μl isopropanol.
9. Place the tubes on a rocking platform at room temperature until a discrete precipitate forms.
10. Remove liquid, leaving DNA precipitate behind, and wash the DNA precipitate first with 70% ethanol and subsequently with 100% ethanol.
11. Remove ethanol completely with a Pipetman and air dry pellet.
12. Dissolve DNA in 50–100 μl sterile distilled water and store at –20 °C.

[a] All animal experiments should be performed according to governmental guidelines.
[b] Vigorous shaking and vortexing should be avoided, since this causes shearing of the DNA.

The quickest and easiest way to analyse the DNA is by PCR. However, this method is subject to artefacts, such as contamination of mouse tail DNA samples with plasmid DNA, leading to false positive results. Furthermore, PCR is often less efficient with genomic DNAs, leading to false negative results. Therefore, we always identify potential founder transgenic mice by Southern blot analysis. If bands of the predicted size are obtained, the mouse is likely to be transgenic, since contaminating plasmid DNA will most often produce bands that differ in size from the expected bands. Furthermore, a Southern blot provides information on the integrity and copy number of the inserted DNA sequences and on the number of integration loci. Once a transgenic line is established, we routinely test the progenitors by dot-blot analysis (*Protocol 2*). Best results are obtained with a probe which does not cross-hybridize with chromosomal mouse DNA. The rabbit β-globin intron which is usually included in our transgene constructs (see above) is very suitable for this purpose. Dot-blot analysis allows a rapid screening of multiple samples and gives more reliable results than PCR analysis. Hybridization of the blot can be performed with ^{32}P-labelled probes but also with non-radioactive hybridization probes.

Protocol 2. Identification of transgenic mice by dot-blot analysis

Equipment and reagents

- Shaking water-bath at 65°C
- Hybond N^+ nylon membrane (Amersham)
- Hybridization mix: 1 g dextran sulfate in 10 ml 7% SDS/0.25 M sodium phosphate pH 7.2, 1 mg sonicated salmon sperm DNA,[a] and 1–2 × 10^7 c.p.m. of a ^{32}P-labelled DNA probe[a]
- Sonicated salmon sperm DNA (Sigma)
- Pre-hybridization mix: 7% SDS, 0.25 M sodium phosphate pH 7.2, 100 μg/ml sonicated salmon sperm DNA[a]
- 20 × SSC: 3 M NaCl, 0.3 M trisodium citrate dihydrate pH 7.4

Method

1. Mix 5 μl of genomic DNA (from mouse tails) and 5 μl 0.8 N NaOH in a 1.5 ml microcentrifuge tube.
2. Draw 1.5 cm square grids on a Hybond N^+ membrane and apply 3.3 μl of the DNA/NaOH mix into the middle of each square. Repeat twice until the complete 10 μl DNA/NaOH mix is applied to the same spot on the membrane.
3. Neutralize the filter by placing in 20 × SSC for 20 min on a rocking platform.
4. Wash the filter twice for 20 min using 2 × SSC.
5. Air dry the filter.
6. Pre-hybridize for 30–120 min at 65°C with pre-hybridization solution in a sealed plastic bag.
7. Remove the pre-hybridization solution and replace with hybridization solution.
8. Hybridize in a sealed plastic bag at 65°C overnight with continuous shaking in a shaking water-bath.
9. Wash the filter with 0.2 × SSC/0.1% SDS for 5 min at room temperature, then twice for 30 min at 65°C.
10. Air dry the filter and expose it to film for 1–24 h.

[a] Pre-heat salmon sperm DNA and radiolabelled DNA probe for 10 min to 95°C before adding to the rest of the mix.

4. Analysis of transgene expression

4.1 Advantages and disadvantages of various RNA detection methods

Methods for detecting the presence of specific mRNAs in cells or tissues are mostly based on hybridization with a suitable complementary probe sequence.

Usually, the RNA is first immobilized on a membrane, either directly, for dot- or slot blot-analysis, or after separation in denaturing agarose gels, as for Northern blot analysis. To detect the mRNA, the membrane is subsequently incubated with a radioactively labelled probe of sufficient length. Dot-blot hybridization has the advantage of being very fast and easily quantifiable, but requires strict control of hybridization and washing conditions in order to avoid non-specific signals. Northern blotting is more time-consuming, but has the bonus of yielding significant information on the length and complexity of the detected RNA species. However, filter hybridization assays generally require relatively large amounts of high quality RNA (2–5 μg of polyadenylated RNA for most transcripts).

Alternatively, the presence of specific RNAs may be detected by the reverse transcription PCR (RT-PCR) technology which is described in detail in Chapter 7. This is a very sensitive method but, as mentioned above, false positive results may be obtained. A second disadvantage of this technique lies in the difficulty of its quantitation. This is due to the inherent dependency of RT-PCR efficiency on a number of reaction conditions. Therefore, reliable quantitation can only be carried out by inclusion at different concentrations of an internal control template that contains the same primer hybridization sites as the target sequence.

4.2 RNase protection assay

4.2.1 Principle of the method

A powerful alternative to Northern blotting or RT-PCR is the mapping of RNA with antisense transcripts using ribonuclease (RNase) and radiolabelled RNA probes (RNase protection assay, see *Protocol 3*). Although this method does not allow determination of transcript length, it is highly sensitive and as little as 0.1 pg of mRNA can be detected. Therefore, RNase protection assays can be carried out with less than 10 μg of total cellular RNA. Furthermore, the method is very specific. Thus every single mismatch between a template and a particular RNA can be detected, since the RNA will be cleaved by the RNases at all sites where Watson–Crick base pairing does not occur. This method is therefore suitable for the mapping of transcription start sites and of exon/intron boundaries.

The strategy for mapping mRNA with radiolabelled RNA probes generated *in vitro* is as follows:

(a) A segment of DNA containing all or part of the gene of interest is inserted into a polycloning site immediately downstream from a bacteriophage T3, T7, or SP6 promoter, in an orientation producing antisense RNA. Many vectors, e.g. the 'Bluescript' vector (Stratagene), are suitable for this purpose.

(b) The recombinant plasmid is digested with a restriction enzyme that cleaves at a convenient site within the gene or at a site in the plasmid upstream from it. The completeness of the restriction digest should be determined by gel electrophoretic analysis.

(c) The template DNA is subsequently extracted twice with phenol and once with chloroform, ethanol precipitated, dissolved in diethyl pyrocarbonate-treated water (DEPC–H_2O) at a concentration of 1 mg/ml, and stored at –20 °C.

(d) The linearized plasmid is transcribed in the presence of ^{32}P-labelled rNTPs with the appropriate bacteriophage DNA-dependent RNA polymerase to produce an antisense RNA that extends from the initiation site of the promoter to the site where the DNA has been digested. Optimal results are obtained with probes of 150–350 nucleotides. For the detection of mRNAs encoding truncated receptors it is helpful to use a probe which discriminates between the endogenous and the truncated receptor. Thus templates which include the truncation site are useful for this purpose (*Figure 2*).

(e) For reasons of economy, the radioactive UTP is added at a limited concentration. This might lead to shorter transcripts due to premature termination at U residues. These smaller transcripts will give rise to smaller protected fragments, resulting in a smear on the final gel. Therefore, it is essential to purify the riboprobe using preparative polyacrylamide gel electrophoresis (PAGE).

(f) An excess of the radiolabelled purified antisense RNA is hybridized in solution with the RNA being tested, so that all complementary sequences form ^{32}P-labelled RNA:RNA hybrids. After unhybridized material has been removed by digestion with RNase, the radiolabelled RNA:RNA hybrid is then detected and quantitated by PAGE under denaturing conditions, followed by autoradiography.

(g) For all protection assay experiments it is important to include appropriate controls:

   (i) tRNA (50 μg) is used as a negative control. Bands which are obtained by hybridization of the probe with tRNA result from incomplete RNase digestion and/or self-hybridization of the probe.

   (ii) If available, an appropriate positive control should be used, e.g. unlabelled sense riboprobe or RNA from a tissue where the gene of interest has already been shown to be expressed.

   (iii) A sample of the radiolabelled riboprobe (approx. 1000 c.p.m.) is loaded onto the analytical gel to compare the size of the protected fragments with the size of the original riboprobe. Since the probe includes polylinker sequences of variable length, the protected fragments should be shorter than the original probe. Bands which are identical in size with the riboprobe are therefore a result of incomplete digestion of the probe.

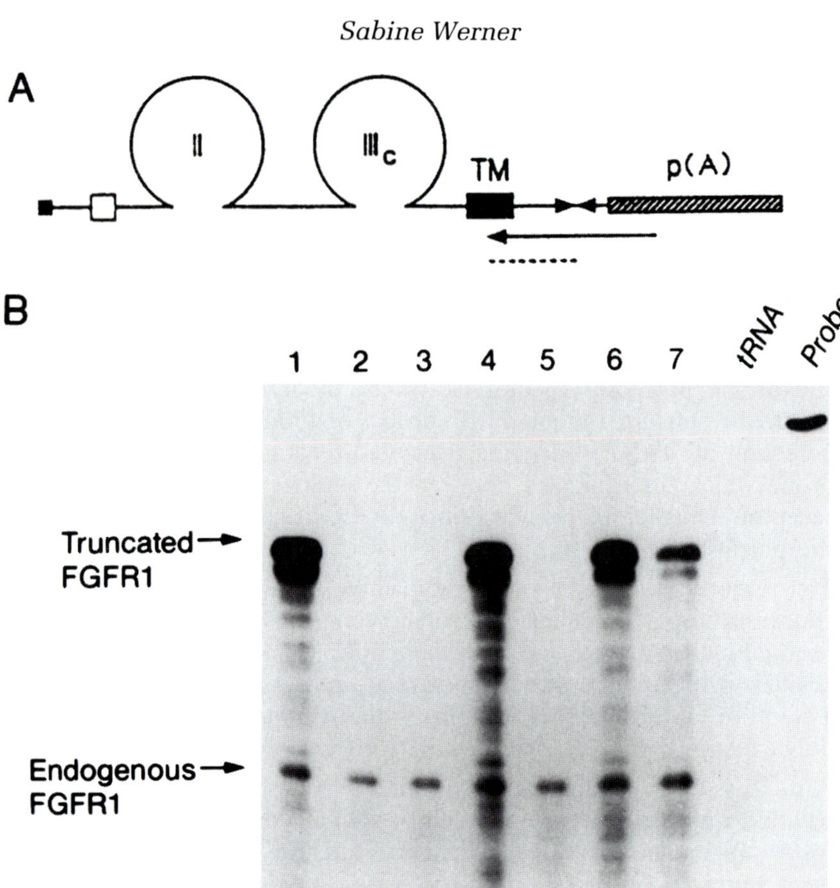

Figure 2. Detection of truncated and endogenous FGF receptor mRNAs by RNase protection assay. (A) Diagram of the transgene mRNA. Functional elements include the ATG translation initiation codon (small black box on the left), the highly acidic region of FGFR1 (open box), immunoglobulin-like domains II and IIIc (II and IIIc), the transmembrane region (TM), the stop codon (double arrow), and the human growth hormone poly(A) (dashed box). To discriminate between the endogenous and the truncated receptor we used the probe which is indicated with an arrow. It includes part of the transmembrane domain, the truncation site, and part of the human growth hormone poly(A). Transcripts encoding the truncated receptor give rise to the full-length protected fragment (upper band in B). Transcripts encoding the endogenous receptor give rise to a shorter protected fragment (lower band in B; indicated with a dotted line in A). (B) 50 μg total cellular RNA from mouse tail skin was analysed. The protected fragments obtained by mRNA encoding the endogenous and the truncated receptor are indicated. 1000 c.p.m. of the hybridization probe were loaded in the lane labelled probe and used as a size marker. 50 μg tRNA were used as a negative control. The numbers 1–7 indicate transgenic founder mice No. 1–7. (Adapted from ref. 21 with permission from the *EMBO Journal*.)

Protocol 3. RNase protection assay

Equipment and reagents

- Water-baths at 37°C and 42°C
- 'Saran wrap'
- 0.2 M dithiothreitol (Boehringer Mannheim) (store at −20°C)
- RNasin (40 U/μl) (Promega) (store at −20°C)
- T3, T7, or SP6 RNA polymerase (20–40 U/μl) (Boehringer Mannheim) (store at −20°C)
- [^{32}P]UTP (800 Ci/mmol) (Amersham) (store at −20°C)
- 5 × transcription buffer: 200 mM Tris–HCl pH 7.5, 30 mM $MgCl_2$, 10 mM spermidine (store at −20°C)
- 10 × NTP mix: 5 mM ATP, CTP, GTP (Boehringer Mannheim) (store at −20°C)
- RNase-free tRNA (Boehringer Mannheim) (store at −20°C)
- FLB80: 80% deionized formamide, 1 × TBE, 1 mM EDTA, 0.05% bromophenol blue, 0.05% xylene cyanol (store at −20°C)
- FAB: 80% deionized formamide, 400 mM NaCl, 40 mM Pipes (piperazine-*N,N'-bis*(2-ethanesulfonic acid)) pH 6.4, 1 mM EDTA
- RNase digestion buffer: 300 mM sodium acetate pH 7.0, 10 mM Tris–HCl pH 7.5, 5 mM EDTA (store at 4°C)
- RNase A (10 μg/μl) (Boehringer Mannheim) (store at −20°C)
- RNase T1 (100 U/μl) (Boehringer Mannheim) (store at 4°C)
- RNase mix for one sample: 297 μl RNase digestion buffer, 1 μl RNase A, 2 μl RNase T1, freshly prepared
- 10 mg/ml proteinase K (Boehringer Mannheim) (store at −20°C)
- tRNA from *E. coli* (10 μg/μl in DEPC–H_2O) (Boehringer Mannheim)
- 1 × TBE: 89 mM Tris, 89 mM boric acid, 10 mM EDTA
- Ultra-pure urea (Boehringer Mannheim)
- 40% acrylamide/*bis*acrylamide (29:1) (store at 4°C)
- *N,N,N',N'*-tetramethylethylenediamine (TEMED)
- 10% (w/v) ammonium persulfate (APS) (store at 4°C)

A. *Synthesis of ^{32}P-labelled riboprobe*

1. To a sterile 1.5 ml microcentrifuge tube, add the following reagents in the specified order:
   - 3.2 μl DEPC–H_2O (28)
   - 2 μl 10 × NTP mix
   - 4 μl 5 × transcription buffer (use the same buffer for all polymerases)
   - 1 μl 0.2 M dithiothreitol
   - 0.8 μl RNasin
   - 1 μl polymerase
   - 7 μl [^{32}P]UTP
   - 1 μl linearized DNA template (pre-warmed to 65°C for 5 min)
2. Incubate for 60 min at 37°C.
3. Add 80 μl DEPC–H_2O, extract with 100 μl phenol:chloroform (1:1, v/v) pH 7.5, then with 100 μl chloroform.
4. Precipitate the DNA template and the riboprobe by adding 40 μl 7.5 M ammonium acetate, 1.5 μl tRNA (10 μg/μl), and 350 μl ethanol.
5. Incubate at −70°C for at least 30 min.
6. Centrifuge in a microcentrifuge at full speed for 15 min at 4°C.

Protocol 3. *Continued*

7. Dissolve the pellet in 30 μl FLB80 and measure the radioactivity of a 1 μl sample in a beta counter. There should be more than 10^6 c.p.m. in the sample.

B. *Purification of the riboprobe by PAGE*

1. Dissolve 192 g urea in 332 ml distilled water (final volume), add 50 ml 40% acrylamide/*bis*acrylamide solution, and filter through a folded filter (Schleicher & Schuell).
2. Assemble a vertical gel apparatus using 14 × 16 cm glass plates and gel spacers of 3 mm thickness.
3. Make a denaturing polyacrylamide gel by mixing together the following: 76 ml acrylamide/urea solution (see step 1), 3.2 ml 25 × TBE, 24 μl TEMED, and 500 μl 10% APS.
4. Immediately pour the solution into the gel mould, insert the comb (ten teeth), and leave to set for at least 2 h.
5. Assemble the gel apparatus, placing 1 × TBE buffer in both reservoirs.
6. Using a syringe, wash out the wells of the gel with 1 × TBE buffer.
7. Load the sample and run the gel at 250 V until the bromophenol blue has run out of the gel (2–3 h).
8. Cover the gel with Saran wrap.
9. Expose the gel to film for 60 sec, and label the film and the gel during this period with a waterproof pen. The label on the film and on the gel should allow the subsequent identification of the gel area which was exposed to film and therefore the location of the full-length riboprobe in the gel.
10. Cut out a strip of the gel corresponding to the migration position of the full-length transcript (3 mm × 1 cm). This piece of gel should be cut into pieces of 3 mm × 2 mm.[a]
11. Extract the probe from gel fragments with 300 μl 0.1 × TBE, 0.2% SDS under vigorous shaking for approx. 1–2 h at room temperature.
12. Spin for 1–5 min at 13 000 r.p.m. in a microcentrifuge, take off supernatant, and measure the radioactivity in a 1 μl sample.[b]
13. Use for hybridization on the same day.[c]

C. *Hybridization*

1. Precipitate 10^5 c.p.m. of the freshly purified riboprobe in a 1.5 ml microcentrifuge tube together with 10–20 μg of the RNA that is to be analysed[d] by adding 0.1 vol. 3 M sodium acetate pH 5.2 and 2.5 vol. ethanol.

2. Incubate at –70 °C for at least 30 min.
3. Centrifuge in a microcentrifuge at full speed for 15 min at 4 °C.
4. Air dry the precipitate for 5 min (do not dry in a desiccator).
5. Dissolve the precipitate *well* in 30 μl FAB by vigorous pipetting.
6. Heat the solution to 85 °C for 10 min in a heating block to denature the probe and target RNAs.
7. *Quickly* transfer the tube to a 42 °C water-bath and leave overnight to allow hybridization.

D. *RNase treatment and analysis of the protected fragments*

1. Remove the samples from the water-bath, place the tubes on ice, and add 300 μl freshly prepared RNase mix.
2. Incubate for 60 min at 30 °C in a water-bath.
3. Place the samples on ice, add 6.6 μl 10% SDS, and 4.4 μl proteinase K (10 μg/μl).
4. Incubate for 15 min at 42 °C.
5. Extract once with 400 μl phenol:chloroform (1:1, v/v) and once with 400 μl chloroform.
6. Precipitate the RNA with 880 μl ethanol and 1.5 μl tRNA (10 μg/μl).
7. Incubate at –70 °C for at least 30 min.
8. Centrifuge in a microcentrifuge at full speed for 15 min at 4 °C.
9. Dissolve the precipitate in 30 μl FLB80.
10. Heat to 95 °C for 5 min.
11. Load on 5% acrylamide/8 M urea gels (1.5 mm thickness; for preparation of gel see part B).
12. Run gel at 250 V until bromophenol blue runs out of the gel.
13. Dry gel for 2 h at 80 °C and expose to film for one to seven days at –80 °C.

[a] Do not cut smaller pieces, since it will be impossible to separate the acrylamide particles from the eluate.
[b] The probe can be used for hybridization if 1 μl contains more than 10^4 c.p.m. The total volume of the probe used for hybridization should not exceed 10 μl, since the SDS in the probe solution may interfere with hybridization.
[c] Do not store the radiolabelled probe for more than 48 h, since it will undergo radiolysis.
[d] The cellular RNA to be used should not contain SDS.

4.2.2 Interpretation of results and problem solving

Some common problems that occur in RNase protection assays are listed below.

(a) Lack of specific signals:
  (i) The gene is not expressed or only expressed at extremely low levels.
  (ii) The protected fragments are partially degraded by the RNase due to mutations in the probe.

(b) Lack of specific signals combined with smear:
   (i) There has been incomplete hybridization due to poor resuspension of cellular RNA and probe.
   (ii) There has been incomplete hybridization due to residual ethanol in the hybridization mix.
   (iii) There has been incomplete hybridization due to inappropriate hybridization conditions.
   (iv) The FAB solution is too old (FAB can be stored at –20 °C for up to three months).
   (v) The RNase concentration used was too high.
   (vi) Partial degradation of the probe had occurred prior to hybridization with the sample RNA.

(c) Bands which are identical in size with the probe:
   (i) This may be due to incomplete digestion of the probe caused by poor resuspension of cellular RNA and probe.

4.3 *In situ* detection of transgene mRNA and protein

As a next step, expression of the transgene should be localized. If no specific antibody against the protein is available, *in situ* hybridizations should be performed to detect the mRNA. The protocol for *in situ* hybridization has been described in detail by Akhurst (28). Since the transgene is often expressed at much higher levels than the endogenous gene, the cDNA encoding the truncated receptor can be used as a probe. After a short exposure time, the transgene, but not the endogenous gene, will be detected. This can easily be controlled by the use of sections from a non-transgenic mouse. Alternatively, the poly(A) tract which is specific for the transgene can be used as a probe. However, the AT-rich sequences present in the poly(A) tract frequently cause higher background compared to other sequences.

If a specific antibody is available, the truncated receptor can be detected by immunohistochemistry (see Section 5.2). However, in most cases the truncated receptor cannot be distinguished from the endogenous receptor. This problem can be solved by addition of an epitope tag to the amino or carboxy terminus of the transgene. In this case, the transgene product can be detected by immunohistochemistry using a specific antibody directed against the tag, but see Section 2 regarding tag length.

5. Characterization of the skin of transgenic mice

5.1 Histological analysis

The characterization of the skin of transgenic mice should be started with a detailed histological analysis (*Protocol 4*). For this purpose, newborn and adult back skin as well as tail skin should be tested.

Protocol 4. Histological analysis of mouse skin

Equipment and reagents

- Scissors and scalpel
- HV filters 0.45 μm (Millipore)
- PBS: 8 g NaCl, 0.2 g KCl, 1.15 g $Na_2HPO_4.2H_2O$, 0.2 g KH_2PO_4, adjust volume to 1 litre with distilled water
- 4% paraformaldehyde in PBS[a]
- Paraffin
- Paraffin-embedding system
- Microtome
- Scott's water (Sigma): 20 g/litre $MgSO_4.7H_2O$, 2 g/litre $NaHCO_3$
- Haematoxylin Gill No. 2 (Sigma)
- 1% eosin (Sigma) in 90% ethanol
- Organic mounting medium S/P™ Accu Mount 60™ (Baxter)

A. *Preparation of sections of mouse skin*

1. Cut hair on the back with fine scissors.
2. Separate the tail skin from the underlying bone: make a cut through the skin of the tail and remove the bone with your fingers.
3. Cut out a piece of skin (0.5 × 0.5 cm), including epidermis, dermis, and muscle layer.
4. Stretch skin on a Millipore HV filter (dermis side down) and trim edges with sharp razor or scalpel.
5. Fix skin overnight at 4 °C in 4% paraformaldehyde.
6. Embed in paraffin as described in ref. 29, positioning sample along one edge. The skin is still on the filter at this stage.
7. Cut 6 μm sections (filter does not disturb sectioning).

B. *Haematoxylin/eosin staining of skin sections*

1. Dewax sections twice for 5 min in xylene.
2. Rehydrate sections by passing the slides through a series of filtered alcohols: twice for 2 min 100% ethanol, 30 sec 95% ethanol, 30 sec 60% ethanol.
3. Wash slides quickly with distilled water.
4. Stain sections for 45 sec with haematoxylin.
5. Wash the slides bearing sections three times for 10 sec with distilled water.
6. Incubate for 30 sec in Scott's water.
7. Wash for 10 sec in distilled water.
8. Incubate for 10 sec in 70% ethanol.
9. Stain for 10 sec in 1% eosin in 90% ethanol.
10. Dehydrate sections: incubate twice for 10 sec in 80% ethanol, twice

Protocol 4. *Continued*

for 10 sec in 95% ethanol, twice for 10 sec in 100% ethanol, and twice for 10 min in xylene.

11. Mount coverslips with organic mounting medium.

[a] 4% paraformaldehyde should be used within one week after preparation. Heat PBS to 70 °C on a hot plate in a fume-hood, and add the paraformaldehyde with continuous stirring. Once the solution clears, filter the solution through a folded filter (Schleicher & Schuell). Immediately cool the solution on ice to prevent the breakdown of the paraformaldehyde. Since paraformaldehyde is harmful by inhalation and contact, precautions should be taken during preparation and use.

5.2 Analysis of differentiation-specific proteins in the epidermis

As a next step, the effect of the truncated receptor on keratinocyte differentiation can be determined. For this purpose, several differentiation markers are available. The most commonly used marker proteins are the differentiation-specific keratins K5, K14, K1, K10, K6, and K16. In normal epidermis, keratins 14 and 5 are expressed at high levels in the basal layer and only at low levels in the suprabasal layers (see *Figure 3A* for keratin 14). When basal cells become committed to terminal differentiation and move to the suprabasal layers, they synthesize a new pair of keratins, K1 and K10, which are the major differentiation products of mature epidermis (see *Figure 3B* for keratin 10). K6 and K16 are restricted to the hair follicles in normal skin and are transiently expressed in the proliferating epidermis of wounds. However, these keratins are aberrantly expressed in the suprabasal epidermal layers in hyperplastic, neoplastic, and psoriatic skin (for review see ref. 30). Thus, the expression pattern of various keratins can be used to assess the effects of the transgene on keratinocyte differentiation. A large variety of antibodies are available against different keratins, and most of these antibodies are now commercially available. These antibodies can be used for immunohistochemistry, with the following controls:

- immunostaining without primary antibody
- use of different concentrations of the primary antibody
- immunostaining with pre-immune serum (if available)
- pre-incubation of primary antibody with immunization peptide or protein
- if a new antibody is used, staining of serial sections with a well-characterized antibody

5.2.1 Immunohistochemistry: choice of fixation and tissue preparation methods

Most histological studies are carried out on paraformaldehyde-fixed, paraffin-embedded tissue samples. This preparation of samples gives the best

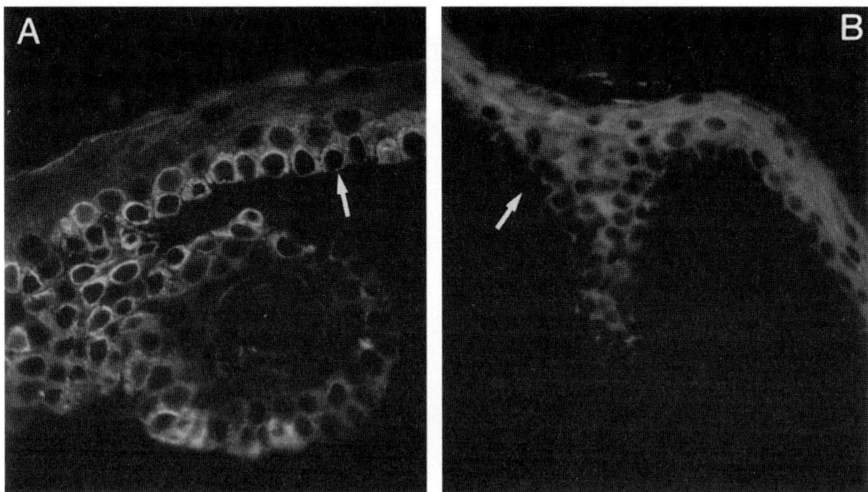

Figure 3. Immunohistochemical detection of differentiation-specific keratins. Mouse tail skin was fixed in Carnoy's fixative and paraffin-embedded. 6 μm sections were deparaffinized and incubated with an antibody against the basal keratin 14 (A) or the suprabasal keratin 10 (B) and a FITC-conjugated secondary antibody. Basal keratinocytes are indicated with an arrow. Note the strong expression of keratins 14 and 10 in basal and suprabasal keratinocytes, respectively.

preservation of cell structure and morphology. However, the fixation and embedding procedures are harsh and many antigens are not well preserved. Thus, not all antibodies can be used with paraformaldehyde-fixed and paraffin-embedded tissue samples. The choice of a different fixation method can often reduce these problems. For example fixation in 70% ethanol or in Carnoy's fixative and subsequent paraffin embedding gives good results with most antibodies against differentiation-specific keratins. However, several antibodies cannot be used with paraffin sections but only with frozen sections. This is the gentlest method for the preparation of samples and gives good preservation of cell structure and antigens. Its principal disadvantages are that the specimens must be stored frozen and a special microtome, known as a cryostat, is required.

Several detection methods can be used, which are based on the use of enzyme labelled reagents (alkaline phosphatase, horse-radish peroxidase), gold labelled reagents, or fluorochrome labelled reagents. Due to the high resolution and to the possibility of double staining we prefer the use of fluorochrome labelled reagents (*Protocol 5*, *Figure 3A* and *3B*).

Protocol 5. Immunofluorescence staining of skin sections with antibodies against differentiation-specific keratins

Equipment and reagents

- Tissue freezing medium (Leica)
- Tissue-embedding moulds: Peel-A-Way (Polysciences)
- Cryostat
- Carnoy's fixative: 60% ethanol, 30% chloroform, 10% glacial acetic acid
- Paraffin
- Paraffin-embedding system
- Microtome
- PBS (see *Protocol 4*)
- 3% bovine serum albumin (Fraction V, Sigma) in PBS
- Primary antibody
- Secondary antibody, conjugated to a fluorescent dye, e.g. anti-rabbit IgG, conjugated to fluorescein isothiocyanate (FITC) (Boehringer Mannheim)
- Haematoxylin Gill No. 2 (Sigma)
- Water-based mounting medium (e.g. Faramount, Dakopatts)
- Fluorescence microscope

A. *Preparation of frozen sections*

1. Cut out a piece of skin as described in Section 5.1.
2. Stretch skin on a Millipore filter (dermis side down).
3. Fill embedding moulds with tissue freezing medium, insert the skin sample, and freeze skin sample on dry ice.
4. Cut 6 μm frozen sections with a cryostat and store them at –70 °C.
5. Remove slides from the freezer and dry them at room temperature.
6. Fix slides with cold acetone (pre-cooled to –20 °C) for 10 min at room temperature.
7. Incubate sections for 5 min in PBS.
8. Stain sections as described below (part C).

B. *Preparation of paraffin sections*

1. Cut out a piece of skin as described in Section 5.1.
2. Stretch skin on Millipore filter (dermis side down) and trim edges.
3. Fix overnight at 4 °C in Carnoy's fixative.
4. Transfer into absolute ethanol until embedding in paraffin.
5. Embed in paraffin, positioning sample along one edge of the embedding mould.
6. Cut 6 μm sections using a microtome.
7. Dewax sections twice for 5 min in xylene.
8. Rehydrate slides by passing through a series of filtered alcohols: twice for 2 min 100% ethanol, 30 sec 95% ethanol, 30 sec 80% ethanol.
9. Wash slides quickly with distilled water.

10. Incubate in PBS for 10 min.
11. Stain sections as described below.

C. *Immunofluorescence staining of skin sections*

1. Block non-specific binding sites by a 30 min incubation in PBS/3% BSA.
2. Wash slides three times for 10 min with PBS.
3. Incubate with primary antibody diluted in PBS/3% BSA overnight at 4°C or for 2 h at room temperature. The dilution of the primary antibody is usually specified by the supplier. Otherwise the optimal dilution has to be tested using serial dilution of the antibody. A 1:250 dilution of a rabbit antiserum is appropriate for most purposes.
4. Wash slides three times for 10 min with PBS.
5. Incubate with secondary antibody conjugated to a fluorescent dye for 1 h at room temperature. The dilution of the secondary antibody is specified by the manufacturer.
6. Wash slides three times for 10 min with PBS.
7. Transfer slides into distilled water.
8. Mount with water-based mounting medium (e.g. Faramount, Dakopatts).

5.3 Labelling of proliferating cells with 5-bromo-2′-deoxyuridine (BrdU)

BrdU is a thymidine analogue and is specifically incorporated into DNA during DNA synthesis. It can therefore be used to label proliferating cells. Cells which have incorporated this modified nucleotide can be identified by immunostaining with a BrdU-specific antibody. For immunohistochemical analysis of BrdU labelled cells, the tissue is either frozen in tissue freezing medium or embedded in paraffin. We prefer the use of paraffin sections due to the better preservation of histological features (*Protocol 6*, *Figure 4*).

Protocol 6. BrdU treatment of mice and immunostaining of tissues with anti-BrdU antibodies

Equipment and reagents

- Microtome
- Millipore filters HV 0.45 μm (Millipore)
- Fluorescence microscope
- Paraffin
- Paraffin-embedding system
- 5-bromo-2′-deoxyuridine (Sigma)
- FITC-conjugated monoclonal antibody to BrdU (Boehringer Mannheim)

Protocol 6. *Continued*

Method

1. Dissolve 100 mg BrdU in 6 ml 0.9% NaCl solution, incubate at 37 °C for at least 10 min until the BrdU is completely dissolved.
2. Inject 300 μl BrdU solution intraperitoneally into a 20 g mouse (5 mg BrdU/20 g mouse).
3. Sacrifice mouse 2 h later.
4. Cut the skin into pieces of 0.5 × 0.5 cm.
5. Stretch skin on Millipore filter and trim edges with sharp razor or scalpel.
6. Fix tissue in 70% ethanol overnight at 4 °C.
7. Transfer skin to 100% ethanol until embedding in paraffin.
8. Embed skin in paraffin, positioning sample along one edge of the embedding mould.
9. Cut 6 μm sections.
10. Dewax sections twice for 10 min in xylene.
11. Rehydrate sections by passing through a series of filtered ethanols: 30 sec each in 100%, 95%, and 80% ethanol.
12. Soak in distilled water, followed by PBS for 5 min each.
13. Denature DNA by incubating the slides in 2 M HCl for 30 min.
14. Wash slides once with PBS, twice with 70% ethanol in 0.1 M Tris–HCl pH 7.6, and once with 70% ethanol in H_2O for 10 min each.
15. Soak in distilled water for 10 sec.
16. Wash twice in PBS for 10 min each.
17. Add 50–70 μl of 1:3 diluted FITC-coupled anti-BrdU antibody to each section.
18. Incubate at room temperature in a humidified chamber for 4 h, protecting slides from light.
19. Wash with PBS, changing the solution three times over a 30 min period.
20. Soak in distilled water and mount with a water-based mounting medium (e.g. Faramount, Dakopatts).

Note: if frozen sections are used, they should be fixed for 10 min in ice-cold acetone and subsequently treated as described in *Protocol 6*, steps 12–20.

The techniques described above are sufficient for an initial characterization of a skin phenotype. In subsequent steps, electron microscopy can be performed to further characterize the morphological features of the tissue. In

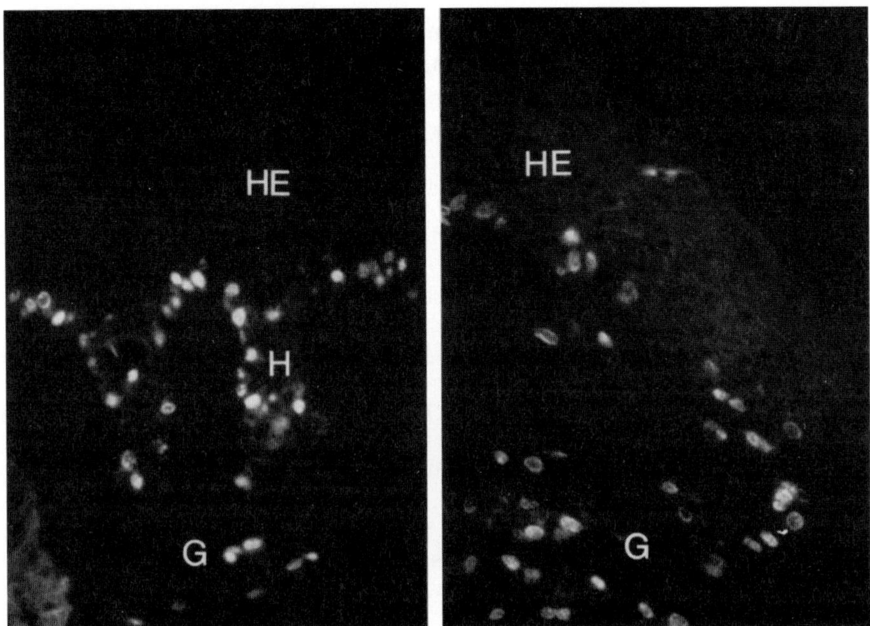

Figure 4. Detection of proliferating cells at the wound edge of a 5d mouse full-thickness wound. A full-thickness excisional wound was generated on the back of a mouse. Five days later the animal was injected intraperitoneally with BrdU and later was sacrificed. The complete wound was isolated, bisected, fixed in 70% ethanol, and paraffin-embedded. 6 μm sections were stained with an FITC-conjugated antibody against BrdU. Stained nuclei are present in the hyperproliferative epidermis at the wound edge (HE), in the hair follicles (H), and in the underlying granulation tissue (G).

addition, hair growth experiments and wound healing studies can be performed to analyse the effect of the truncated receptor on hair growth and tissue repair. Finally, the targeted cells may be cultured and analysed for their responsiveness to the ligand of the truncated receptor in order to prove the dominant-negative effect.

Acknowledgements

The first transgenic mice which express dominant-negative receptors were generated in the laboratory of Dr L. T. Williams, University of California, San Francisco. I would like to thank Dr Williams for his support and his invaluable contribution to this work. I also thank my co-workers who have helped set-up the techniques described above in the laboratory. Many thanks to all the people who have given advice, particularly Dr Wendy Weinberg, NIH Bethesda, and Dr Kevin Peters, Duke University, Durham. Work in the author's laboratory is funded by the Max-Planck-Gesellschaft, the Deutsche

Forschungsgemeinschaft, the Bundesministerium für Bildung und Forschung, the Human Frontier Science Program, and the Hermann-und-Lilly Schilling Stiftung.

References

1. Schneider, J.Q., Gu, W., Zhu, L., Mahdavi, V., and Nadal-Ginard, B. (1994). *Science*, **264**, 1467.
2. Hebert, J.M., Rosenquist, T., Götz, J., and Martin, G.R. (1994). *Cell*, **78**, 1017.
3. Guo, L., Degenstein, L., and Fuchs, E. (1996). *Genes Dev.*, **10**, 165.
4. Mann, G.B., Fowler, K.J., Gabriel, A., Nice, E.C., Williams, R.L., and Dunn, A.R. (1993). *Cell*, **73**, 249.
5. Luetteke, N.C., Qiu, T.H., Peiffer, R.L., Oliver, P., Smithies, O., and Lee, D.C. (1993). *Cell*, **73**, 263.
6. Haub, O. and Goldfarb, M. (1991). *Development*, **112**, 397.
7. Mason, Y.J., Fuller-Pace, F., Smith, R., and Dickson, C. (1994). *Mech. Dev.*, **45**, 15.
8. Derynck, R. (1992). *Adv. Cancer Res.*, **58**, 27.
9. Fantl, W., Johnson, D.E., and Williams, L.T. (1993). *Annu. Rev. Biochem.*, **62**, 453.
10. Honegger, A.M., Schmidt, A., Ullrich, A., and Schlessinger, J. (1990). *Mol. Cell. Biol.*, **10**, 4035.
11. Ueno, H., Colbert, H., Escobedo, J.A., and Williams, L.T. (1991). *Science*, **252**, 844.
12. Kashles, O., Yarden, Y., Ullrich, A., and Schlessinger, J. (1991). *Mol. Cell. Biol.*, **11**, 1454.
13. Ueno, H., Gunn, M., Dell, K., Tseng, A., and Williams, L.T. (1992). *J. Biol. Chem.*, **267**, 1470.
14. Ueno, H., Escobedo, J.A., and Williams, L.T. (1993). *J. Biol. Chem.*, **268**, 22814.
15. Amaya, E., Musci, T., and Kirschner, M. (1991). *Cell*, **66**, 257.
16. Hemmati-Brivanlou, A. and Melton, D.A. (1994). *Cell*, **77**, 273.
17. Suzuki, A., Thies, R.S., Yamaij, N., Song, J.J., Wozney, J.M., Murakami, K., *et al.* (1994). *Proc. Natl. Acad. Sci. USA*, **91**, 10255.
18. Vassar, R., Rosenberg, M., Ross, S., Tyner, A., and Fuchs, E. (1989). *Proc. Natl. Acad. Sci. USA*, **86**, 1563.
19. Bailleul, B., Surani, M.A., White, S., Barton, S.C., Brown, K., Blessing, M., *et al.* (1990). *Cell*, **62**, 697.
20. Werner, S., Smola, H., Liao, X., Longaker, M.T., Krieg, T., Hofschneider, P.H., *et al.* (1994). *Science*, **266**, 819.
21. Werner, S., Weinberg, W., Liao, X., Blessing, M., Peters, K.G., Yuspa, S., *et al.* (1993). *EMBO J.*, **12**, 2635.
22. Murillas, R., Larcher, F., Conti, C.J., Santos, M., Ullrich, A., and Jorcano, J.L. (1995). *EMBO J.*, **14**, 5216.
23. Korfhagen, T.R., Glasser, S.W., Wert, S.E., Bruno, M.D., Daugherty, C.C., McNeish, J., *et al.* (1990). *Proc. Natl. Acad. Sci. USA*, **87**, 6122.
24. Peters, K.G., Werner, S., Liao, X., Wert, S., Whitsett, J., and Williams, L.T. (1994). *EMBO J.*, **13**, 3296.
25. Robinson, M.L., MacMillan-Crow, L.A., Thompson, J.A., and Overbeek, P.A. (1995). *Development*, **121**, 3959.

26. Campochiaro, P.A., Chang, M., Ohsato, M., Vinores, S.A., Nie, Z., Hjelmeland, L., *et al.* (1996). *J. Neurosci.*, **16**, 1679.
27. Hogan, B., Beddington, R., Costantini, F., and Lacy, E. (1994). *Manipulating the mouse embryo: a laboratory manual*, 2nd edn. Cold Spring Harbor Laboratory Press, NY.
28. Sambrook, J., Fritsch, E.F., and Maniatis, T. (ed.) (1989). *Molecular cloning: a laboratory manual*, 2nd edn. Cold Spring Harbor Laboratory, Cold Spring Harbor, NY.
29. Akhurst, R. (1993). In *Growth factors: a practical approach* (ed. I. McKay and I. Leigh), p. 109. Oxford University Press.
30. Fuchs, E. (1993). *J. Cell Sci.*, **Suppl. 17**, 197.

9

Growth factor–toxin chimeras and their applications

PAMELA A. DAVOL, A. RAYMOND FRACKELTON, JR., and PAUL CALABRESI

1. Introduction

Clinically, the most successful drugs in use today are those which were designed to target specific metabolic pathways. Not surprisingly, growth factor receptors expressed on the surfaces of cells have been recognized as potentially important therapeutic targets. The specific affinity of growth factor ligands for their receptors combined with the observation that many growth factor receptors are rarely expressed at high levels on normal, non-proliferating cells have made growth factor ligands attractive molecules for the delivery of toxins for the treatment of neoplastic, autoimmune, cardiovascular, and proliferative eye diseases and transplant rejection (see reviews in refs 1–3).

Linking a toxic protein derived from bacteria or plants, which has the ability to inhibit protein synthesis in eukaryotic cells, to a growth factor protein can be accomplished chemically through a disulfide bond or through genetic engineering of a recombinant fusion protein. The resulting hybrid molecule binds with the affinity of the growth factor ligand for its cell surface receptor, internalizes by receptor-mediated endocytosis, and once released from the endosome, the toxin catalytically inhibits protein synthesis causing cell death. Mechanistically, plant toxins such as ricin and saporin enzymatically remove a single nucleotide base from specific sequences of ribosomal RNA thereby rendering the ribosomes unable to bind elongation factor II. In contrast, bacterial toxins such as diphtheria toxin and *Pseudomonas aeruginosa* exotoxin enzymatically catalyse ADP ribosylation of elongation factor II. Complicating their therapeutic use, many of these naturally occurring toxins have in addition to a toxic domain (A chain) their own cell binding domain (B chain) which binds to glycoproteins and glycolipids on the cell surface and results in non-growth factor-specific killing of cells. Toxins which possess a cell binding domain and which will be used in the construction of growth factor–toxin chimera must be modified, therefore, either biochemically or through recombinant engineering to remove the B chain to ensure that the hybrid molecule will exhibit growth factor receptor specificity. Examples of modified toxins include ricin A-chain toxin and PE40, a mutant form of *Pseudomonas* exotoxin which lacks the cell binding domain.

Much of the early work in the field of growth factor–toxin chimeras utilized chemical conjugates of epidermal growth factor (EGF) with diphtheria toxin (DT), ricin A-chain toxin (RTA), or *Pseudomonas* exotoxin (PE). More recently, growth factors and cytokines as diverse as transforming growth factor-α (TGFα), interleukin-2 (IL-2), IL-4, IL-6, insulin-like growth factor-1 (IGF-1), and fibroblast growth factor (FGF) have been combined with PE40 and saporin (SAP) (*Table 1*). All have demonstrated cytotoxicity to cell lines expressing the respective target receptors whereas cells devoid of target receptors are unaffected by exposure to the growth factor–toxin chimera.

In this chapter, procedures for evaluating activity and specificity, clinical applications, and potential problems related to the use of biologically active growth factor–toxin chimeras will be discussed.

2. Detecting chimeric function

After constructing and purifying a growth factor–toxin chimera (see refs 1–3), it is important to confirm that the hybrid molecule is active both in terms of the enzymatic ability of the toxin moiety to inhibit protein synthesis and the growth factor moiety's ability to bind to its respective receptor.

2.1 Evaluating cytotoxicity as a function of toxin activity

In order for the toxin moiety to facilitate cell death, once internalized by receptor-mediated endocytosis, the toxin must translocate across an intracellular membrane to gain access to the cell cytoplasm and the protein synthesis machinery. To date, the most effective and sensitive method for estimating translocation of the toxin moiety is by cell killing assays (see *Protocol 1*; typical results are presented in *Figure 1*).

Protocol 1. Cell killing assay

Equipment and reagents

- Coulter particle counter (Coulter Electronics, Inc.)
- SK-MEL-5 human melanoma cells (American Type Culture Collection)
- Minimum essential medium (Life Technologies, Inc.)
- Fetal calf serum (FCS) (Life Technologies, Inc.)
- Recombinant basic fibroblast growth factor–saporin (45 kDa), basic fibroblast growth factor (18 kDa), saporin (30 kDa)
- Trypsin/EDTA (Life Technologies, Inc.)

Method

1. Seed SK-MEL-5 cells (10^4) into 48-well tissue culture plates and incubate overnight in minimum essential medium supplemented with 10% FCS at 37 °C in a humidified atmosphere of 95% air/5% CO_2.
2. The following day, remove medium and add growth factor–toxin chimera, growth factor protein, toxin protein or a mixture of unconjugated growth factor and toxin diluted to various, equimolar concentrations (0.001 nM to 100 nM) with fresh supplemented medium, to each well in triplicate. Treat cells for 72 h at 37 °C.
3. After treatment, suspend cells by treating them with trypsin/EDTA.
4. Count the number of cells from each well using a Coulter particle counter.
5. Determine cell kill by comparing the cell counts from treated wells to medium treated controls.

If the toxin chimera is cytotoxic to cells, then it is empirically concluded that the toxin moiety is capable of translocation. If it is not cytotoxic, however, other explanations are also possible. For instance, it has been observed that linking ligands to toxins can reduce activity of the toxin molecule (4, 5).

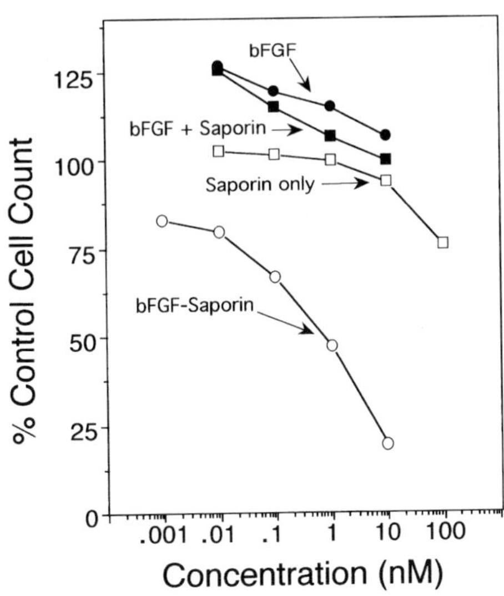

Figure 1. Cell killing assay. Cytotoxic effect of the growth factor–toxin chimera comprised of basic fibroblast growth factor and the ribosomal toxin, saporin (bFGF–SAP) on human melanoma cells (SK-MEL-5) after 72 hours exposure according to *Protocol 1*. Negative controls include growth factor alone (bFGF), toxin alone (saporin only), and growth factor admixed with saporin (bFGF + saporin).

Quantitative analysis of the ribosome inactivating activity of chimeras composed of plant toxins can be performed *in vitro* using rabbit reticulocyte lysates to measure cell-free protein synthesis (see *Protocol 2*).

Protocol 2. *In vitro* assay for measuring ribosome inactivating activity of chimeras composed of plant toxins (4)

Equipment and reagents

- 96-well filtration plates (Millititer HA, Millipore)
- Millipore vacuum holder and vacuum source (Millipore)
- Phosphate-buffered saline (PBS) (Life Technologies, Inc.)
- Growth factor–toxin chimera and free toxin
- Nuclease-treated rabbit reticulocyte lysate (Promega)
- Reaction mixture: 0.5 μg Brome mosaic virus RNA, 1 μl amino acid mixture without leucine (1 mM), 5 μCi [^{3}H]leucine, 3 μl water

Method

1. Reduce the growth factor–toxin chimera by treating with 50 mM dithiothreitol (DTT) for 1 h at 37 °C.
2. Dilute the samples, free toxin, and growth factor–toxin chimera in PBS, and add on ice 5 μl of sample to 35 μl of nuclease-treated rabbit reticulocyte lysate and 10 μl of reaction mixture.
3. Incubate for 1 h in a 30 °C water-bath.
4. Stop the reaction by transferring the tubes to ice.
5. Add 5 μl of the assay mixture in triplicate to 75 μl of 1 M sodium hydroxide, 2.5% hydrogen peroxide in the well of 96-well filtration plate.
6. When the red colour bleaches from the samples, add 300 μl of ice-cold 25% trichloroacetic acid (TCA) to each well and place the plate on ice for 30 min.
7. Place plate on a Millipore vacuum holder attached to a vacuum source and vacuum liquid through.
8. Wash the wells three times with 300 μl of ice-cold 8% TCA.
9. After drying, punch the filter paper circles out of the 96-well plate and measure radioactivity using a scintillation counter.
10. Determine protein synthesis inhibiting activity by comparing [^{3}H]leucine incorporation in samples containing free toxin or growth factor–toxin chimera to control samples.

For chimera composed of bacterial toxins, an *in vitro* assay which measures the transfer of adenosine diphosphate ribose, derived from NAD, to a crude preparation of elongation factor 2 can be performed (see *Protocol 3*).

Protocol 3. *In vitro* assay for measuring ADP ribosylating activity of chimeras composed of plant toxins (6)

Reagents

- Growth factor–toxin chimera and free toxin
- [^{14}C]NAD, uniformly labelled in the adenosine moiety (New England Nuclear), serves as the ADP ribose donor
- Rabbit reticulocyte lysate
- Buffer: 40 mM DTT, 1 mM EDTA, 50 mM Tris–HCl pH 8.2

Method

1. On ice, combine rabbit reticulocyte lysate containing 10 μmoles of elongation factor 2 prepared as described (6), various amounts of the growth factor–toxin chimera or for comparison, the free toxin (typically bracketing 1 μg), 40 pmoles of [^{14}C]NAD (about 60 nCi), and buffer to a final volume of 500 μl.
2. Transfer the reaction mixtures to 37°C and incubate for 30 min.
3. Add 0.5 ml of ice-cold 10% trichloroacetic acid (TCA) to terminate the reaction and to precipitate elongation factor 2. Place the tubes on ice for 15 min.
4. Collect the TCA precipitate, either by vacuum filtration using Whatman GF/C filters and washing four times with 5 ml of 5% TCA, or by centrifuging the tubes at 3000 *g* for 10 min at 4°C and washing the pellet with 1 ml of cold 6% TCA.
5. Determine the ^{14}C radioactivity in the collected precipitates using a scintillation counter. The amount of radioactivity is a measure of the amount of ADP ribosylating activity of the growth factor–toxin chimera.

2.2 Evaluating cytotoxicity as a function of receptor specificity

Once it has been established that the growth factor–toxin chimera is indeed toxic to cells, the next step is to ascertain that cytotoxicity of the hybrid molecule is dependent upon interaction of its growth factor moiety with the respective cell surface receptors. Functional activity of the chimera's growth factor moiety can be qualitatively determined by cross-linking iodinated growth factor–toxin chimera to the cell surface receptor (see *Protocol 4*).

When cross-linking chemically conjugated growth factor–toxin chimeras to cell receptors, it should be noted that the procedure may cleave the disulfide bond linking the growth factor with the toxin. As a result, the autoradiograph will show a high molecular weight band consistent in size with the cross-linked receptor and growth factor, as well as a lower molecular weight band consistent in size with the toxin protein.

Protocol 4. Cross-linking of growth factor–toxin chimeras to cell receptors

Equipment and reagents

- 6-well tissue culture trays (Corning, Cat. No. 25810)
- Vertical slab gel apparatus (Model SE 400, Hoefer)
- Kodak X-AR film (Eastman Kodak Co.)
- Kodak X-Omat film processor (Model M35A, Eastman Kodak Co.)
- Lightening Plus intensifying screens (Dupont)
- Na[^{125}I] (Dupont New England Nuclear)
- Iodobeads iodination reagent (Pierce Chemical Co.)[a]
- Growth factor–toxin chimera
- Cells (receptor bearing)
- Binding buffer: unsupplemented medium, 25 mM Hepes pH 7.5, 0.1% bovine serum albumin
- Growth factor protein
- PBS (Life Technologies, Inc.)
- Disuccinimidyl suberate (DSS): prepare 100 mM stock in DMSO (Boehringer Mannheim)
- Wash buffer: 50 mM Tris–HCl, 100 mM glycine pH 7.4
- Lysing buffer: 150 mM sodium chloride, 20 mM Tris–HCl pH 8.0, 1 mM magnesium chloride, 0.1 mM zinc chloride, 0.5% NP-40, 1 mU/ml aprotinin, 1 μg/ml leupeptin, 2 mM phenylmethylsulfonyl fluoride
- 2 × SDS buffer: 0.125 M Tris, 20% glycerol, 4 mM dithiothreitol, 4.6% SDS pH 6.8
- β-mercaptoethanol (BioRad Laboratories)
- SDS–polyacrylamide gel (7% SDS–PAGE)

Method

1. Label the growth factor–toxin chimera with Na[^{125}I][a] (1 mCi/ml reaction volume) using Iodobeads according to the manufacturer's instructions.
2. For cross-linking studies, seed cells (2×10^5/well) into 6-well tissue culture trays and incubate overnight at 37 °C.
3. Remove medium, wash cells three times with binding buffer, then add 1 ml of binding buffer to each well. Incubate at 37 °C for 1 h.
4. Pre-cool cells by washing them twice with 4 °C binding buffer.
5. Add fresh 4 °C binding buffer containing 1×10^6 c.p.m. [^{125}I]growth factor–toxin chimera with or without an excess of unlabelled growth factor protein (5 μg) to each well, and incubate for 2 h at 4 °C.
6. Wash cells twice with PBS.
7. Cross-link growth factor–toxin chimera to receptors by treating cells with 0.15 mM DSS diluted in PBS for 15 min at room temperature.
8. Wash once with wash buffer.
9. Extract cells by adding a small volume (100 μl) of lysing buffer.
10. Remove cellular debris by centrifuging at 8000 *g* for 15 min at 4 °C.
11. Add β-mercaptoethanol (1:50 dilution) to the 2 × SDS buffer, then add an equal volume of 2 × SDS buffer to each sample, and boil for 6 min.
12. Resolve proteins by 7% SDS–PAGE.

13. Vacuum dry gel and expose to Kodak XAR-5 pre-sensitized film with Lightening Plus intensifying screens at –70 °C for up to one week to visualize radiolabelled proteins.

[a] Although many growth factors and toxins can be labelled satisfactorily with Iodobeads®, some may require radioiodination with lactoperoxidase or the Bolton and Hunter reagent (see ref. 7).

Alternatively, a receptor binding assay (see *Protocol 5*) followed by Scatchard analysis (8) provides a quantitative method for measuring binding affinity of the growth factor–toxin chimera compared to its free growth factor form.

Protocol 5. Receptor binding assay

Equipment and reagents

- 6-well tissue culture trays (Corning, Cat. No. 25810)
- Na[^{125}I] (Dupont New England Nuclear)
- Iodobeads iodination reagent (Pierce Chemical Co.)[a]
- Cells (receptor bearing)
- Unconjugated growth factor protein
- Binding buffer: serum-free medium, 0.15% gelatin, 25 mM Hepes pH 7.4
- PBS (Life Technologies, Inc.)
- Solution A: 2 M NaCl in 20 mM Hepes pH 7.4
- Solution B: 2 M NaCl in 20 mM sodium acetate pH 4.0

Method

1. Label the growth factor–toxin chimera with Na[^{125}I] (1 mCi/ml reaction volume) using Iodobeads according to the manufacturer's instructions. Following labelling, determine the amount of incorporated radioactivity in a known concentration of growth factor–toxin chimera by counting in a gamma scintillation counter.
2. For receptor binding studies, seed cells (2×10^5/well) into 6-well tissue culture trays and incubate overnight at 37 °C.
3. Wash cells with PBS and incubate in receptor binding medium for 1 h at 37 °C.
4. Replace medium with receptor binding medium (4 °C) containing [^{125}I]growth factor–toxin chimera (5–50 ng/ml), with or without a 500-fold excess of unlabelled growth factor protein, and incubate for 2–4 h at 4 °C on a platform rocker.
5. Wash cells once with cold PBS.
6. For bFGF chimeras:
   (a) Collect and save [^{125}I]bFGF–toxin bound to extracellular matrix (low affinity receptors) by washing twice with solution A.
   (b) Collect and save [^{125}I]growth factor–toxin bound to high affinity receptors by washing twice with solution B.

Protocol 5. *Continued*

7. For most other growth factor chimeras:
   (a) Wash cells three more times with PBS.
   (b) Solublize and save [^{125}I]growth factor–toxin bound to receptors by extracting the cell monolayer with 1% Triton X-100 containing 0.1% BSA or 0.15% gelatin.
8. For all growth factor chimeras determine the amount of bound [^{125}I]growth factor–toxin chimera in samples using a gamma scintillation counter.
9. Perform Scatchard analysis to calculate the dissociation constant, K_d, for growth factor chimera binding with the growth factor receptor (detailed in ref. 8).

[a] As mentioned in *Protocol 4*, although many growth factors and toxins can be labelled satisfactorily with Iodobeads®, some may require radioiodination with lactoperoxidase or the Bolton and Hunter reagent (see ref. 7).

Additionally, some growth factor receptors contain tyrosine kinase domains which are activated in response to ligand binding. The ability of the growth factor moiety of the toxin chimera to stimulate tyrosine phosphorylation of the receptor or a second messenger adaptor protein such as Shc would provide further evidence that the hybrid molecule is interacting with its respective receptor (see *Protocol 6*).

Protocol 6. Detecting tyrosine kinase activation

Equipment and reagents

- Cell scrapers (Costar, Cat. No. 3010)
- Vertical slab gel apparatus (Model SE 400, Hoefer)
- Nitrocellulose filters (0.45 μm pore size, Hybond, Amersham)
- Transfer apparatus (Model TE50 or TE50X, Hoefer)
- Rotating cylinder (The Navigator, BioComp)
- Kapak pouches (Kapak Corp.)
- Kodak X-AR film (Eastman Kodak Co.)
- Kodak X-Omat film processor (Model M35A, Eastman Kodak Co.)
- Cells (receptor bearing)
- Growth factor–toxin chimera, growth factor protein, toxin protein
- Extraction buffer: 1% Triton X-100, 10 mM Tris, 5 mM EDTA, 50 mM NaCl, 30 mM sodium pyrophosphate, 50 mM sodium fluoride, 100 μM sodium orthovanadate pH 7.6—add phenylmethylsulfonyl fluoride (final concentration 1 mM) just prior to use
- β-mercaptoethanol (BioRad Laboratories)
- 1G2 anti-phosphotyrosine antibody (Oncogene Science)
- Elution buffer: 1 mM hapten phenyl phosphate prepared in extraction buffer
- 2 × SDS buffer: 0.125 M Tris, 20% glycerol, 4 mM dithiothreitol, 4.6% SDS pH 6.8
- SDS–polyacrylamide gel (7.5%)
- Transfer buffer: 0.2 M glycine, 0.01 M Tris, 0.1% (w/v) SDS, 10% (v/v) methanol pH 7.4
- Fixing solution (optional): 7% acetic acid, 40% methanol, 3% glycerol
- TBST: 0.01 M Tris, 0.15 NaCl, 0.1% (w/v) Tween 20 pH 7.6
- Blocking buffer: 0.1% (w/v) bovine serum albumin, 0.1% (w/v) ovalbumin in TBST
- 4G10 monoclonal phosphotyrosine antibody (UBI)
- Sheep anti-mouse immunoglobulin and horse-radish peroxidase (Amersham)
- Enhanced Chemiluminescent Kit (ECL, Amersham)

Method

1. Serum starve cells (10^6) by incubating them in medium supplemented with only 0.5% serum overnight.
2. The following day, add equimolar concentrations (approx. 1 nM) of either growth factor–toxin chimera, growth factor protein, toxin protein, or the combination of unconjugated growth factor protein and toxin protein to the cells, and incubate at 37 °C for 10–30 min.
3. Place cells on ice and extract by adding 1 ml of ice-cold extraction buffer to each plate. Scrape cells from well and collect into 1.5 ml microcentrifuge tubes. Vortex several times.
4. Centrifuge at 8000 *g* for 15 min at 4 °C to remove debris.
5. Collect supernatant and incubate with monoclonal 1G2 anti-phosphotyrosine antibody linked to Sepharose beads for 1 h, rotating microcentrifuge tubes end-over-end at 4 °C.
6. Wash immunocomplexes three times with extraction buffer before eluting the phosphotyrosine proteins from the antibody with approximately 40 μl of elution buffer.
7. Add β-mercaptoethanol (1:50 dilution) to the 2 × SDS buffer, then add an equal volume of 2 × SDS buffer to each sample, and boil for 6 min.
8. Resolve proteins by 7.5% SDS–polyacrylamide gel electrophoresis.
9. Transfer proteins from the gel onto a nitrocellulose filter by electrophoresis at 1.5 amp for 1 h in transfer buffer using a cooled transfer apparatus.
10. Fix protein to the filter by thoroughly air drying overnight or by immersing filter for 10 min in fixing solution.
11. Saturate sites on the nitrocellulose filter that bind proteins non-specifically by incubating the filter in blocking buffer for 1–3 h at 37 °C.
12. Incubate filter with 10–100 ng/ml of 4G10 monoclonal phosphotyrosine antibody in TBST for 2 h at room temperature in a rotating cylinder.
13. Wash filter with 5 ml of TBST five times for 5 min each, then incubate with a 1/5000 diluted conjugate of sheep anti-mouse immunoglobulin and horse-radish peroxidase in TBST for 1 h at room temperature.
14. Wash filter again as in step 13 before exposing to ECL reagents according to manufacturer's directions.
15. Place filter in a Kapak pouch, expose to Kodak XAR-5 pre-sensitized film for 1–60 sec, and develop film in a processor to visualize immunoreactive proteins.

A true test for cytotoxic specificity, however, is accomplished in a cell killing assay by treating cells that display the receptors with equimolar concentrations of either the growth factor–toxin chimera or free, unconjugated toxin, growth factor, or their combination (see *Protocol 1*). The growth factor–toxin chimera, when compared to the uncoupled moieties, should demonstrate a dramatic increase in cell kill. In contrast, cells that lack the specific receptors should be insensitive to the growth factor–toxin chimera. As a further test for specificity, the cells are treated with the hybrid molecule in the presence of excess, unconjugated ligand which will compete with the growth–factor toxin chimera for receptor binding or in the presence of neutralizing antibodies to ligand which will block the growth factor from binding to its receptor. If cytotoxicity of the growth factor–toxin chimera is inhibited under these conditions, this constitutes strong evidence that the hybrid molecule is acting through the respective growth factor receptor.

2.3 Quantitating cytotoxic activity on various cell lines

The number of target receptors expressed on the surface of the cells, the nature of the extracellular matrix which may trap the chimeric toxin, and the ability of cells to produce endogenous growth factor which may compete for receptor binding are factors which vary from cell line to cell line and which influence cell sensitivity to the growth factor–toxin chimera. For this reason, sensitivity of various cell lines to the hybrid molecule must be determined. Ways to determine sensitivity include testing the amount of the growth

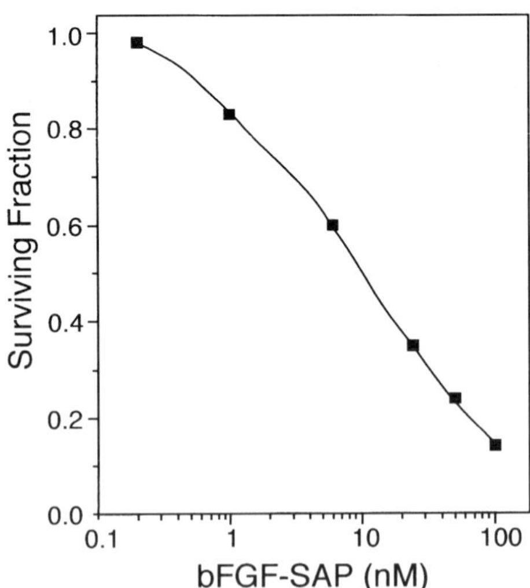

Figure 2. Cell survival assay. Survival of human prostatic carcinoma cells (DU-145) treated for one hour with growth factor–toxin chimera (bFGF–saporin), (or as a control, saporin alone according to *Protocol 7*).

factor–toxin required to inhibit 50% of cell proliferation (IC_{50}) in a cell survival assay (see *Protocol 7* and *Figure 2*) or inhibit 50% of protein synthesis (EC_{50}) (see *Protocol 8*) as compared to untreated controls.

Protocol 7. Cell survival assay

Equipment and reagents

- 6-well tissue culture plates (Corning, Cat. No. 25810)
- Coulter particle counter (Coulter Electronics, Inc.)
- DU 145 human prostatic carcinoma cells (American Type Culture Collection)
- Minimum essential medium (Life Technologies, Inc.)
- Trypsin /EDTA (Life Technologies, Inc.)
- Fetal calf serum (FCS) (Life Technologies, Inc.)
- Recombinant basic fibroblast growth factor–saporin (bFGF–SAP; 45 kDa), basic fibroblast growth factor (bFGF; 18 kDa), saporin (30 kDa)
- Optional: bFGF neutralizing antibody (Upstate Biotechnology, Inc.)

Method

1. Seed cells at a concentration of 2×10^4 cells/cm^2 in 6-well tissue culture plates. Incubate overnight in minimum essential medium supplemented with 10% FCS at 37 °C in a humidified atmosphere of 95% air/5% CO_2.
2. The following day, wash cells with serum-free medium and add bFGF–SAP (1–150 nM) with or without bFGF neutralizing antibody (10 mg/litre) or 100-fold excess bFGF, or equimolar concentrations of bFGF alone, saporin alone, or a mixture of unconjugated bFGF and saporin diluted with serum-free medium to wells. Treat cells for 1 h at 37 °C.
3. After treatment, wash cells three times with fresh serum-supplemented medium, and suspend cells by treating them with trypsin/EDTA.
4. Centrifuge, wash, and reseed cells from each treatment group into new plates at 3×10^3 cells/cm^2 in duplicate.
5. After 7–14 days, determine cell survival by suspending cells with trypsin/EDTA and counting the number of cells from each plate using a Coulter particle counter.

Protocol 8. Inhibition of protein synthesis

Reagents

- SK-MEL-5 human melanoma cells (American Type Culture Collection)
- Minimum essential medium (Life Technologies, Inc.)
- FCS (Life Technologies, Inc.)
- [^{35}S]Trans Label (ICN Radiochemicals, Inc.)
- Recombinant basic fibroblast growth factor–saporin (bFGF–SAP; 45 kDa)
- Methionine-free medium (ICN Radiochemicals, Inc.)
- PBS (Life Technologies, Inc.)

Protocol 8. *Continued*

Method

1. Seed cells (5×10^4/well) into 48-well tissue culture plates in medium supplemented with 10% FCS, and maintain overnight at 37 °C in a humidified atmosphere of 95% air/5% CO_2.
2. The following day, remove medium, wash cells twice with serum-free medium, and add growth factor–toxin chimera, diluted in serum-free medium to various concentrations (1–150 nM), to each well. Add serum-free medium to control wells. Incubate 1 h.
3. Following treatment, remove medium, wash cells three times with fresh serum-supplemented medium, then add fresh serum-supplemented medium, and incubate for 24 h.
4. Remove medium and add cysteine and methionine-free medium to each well. Incubate at 37 °C for 30 min.
5. Pulse cells with 100 μCi Trans Label added to each well. Incubate at 37 °C for 30 min.
6. Wash cells once with ice-cold PBS (0.5 ml/well).
7. Wash cells twice with ice-cold 10% trichloroacetic acid (TCA) (0.5 ml/well).
8. Wash cells with ice-cold methanol (0.5 ml/well).
9. Add 0.5 M NaOH (0.5 ml/well) and incubate for 30 min at 37 °C.
10. Neutralize the samples from each well with 10% TCA (150 μl/well) before adding to 4 ml scintillation fluid.
11. Determine protein synthesis in each sample by counting the incorporation of [^{35}S]methionine into cellular protein using a liquid scintillation counter.

Although inhibitory concentrations of growth factor–toxin chimeras do vary from cell line to cell line, and even though translocation of the toxin from the endosome into the cytoplasm is very inefficient, normally only nanomolar concentrations are required to produce 50% inhibition in receptor bearing cells.

2.4 Assessing antitumour effects in animals

A human cell line which is sensitive to the growth factor–toxin chimera in culture can be evaluated for specific targeting when xenografted into athymic mice. The use of mice as predictors for drug efficacy and toxicity is acceptable both in terms of cost efficiency and therapeutic comparisons with humans: the use of more expensive, larger animal species has not demonstrated any safety advantages. For drug developmental purposes, athymic mice are implanted

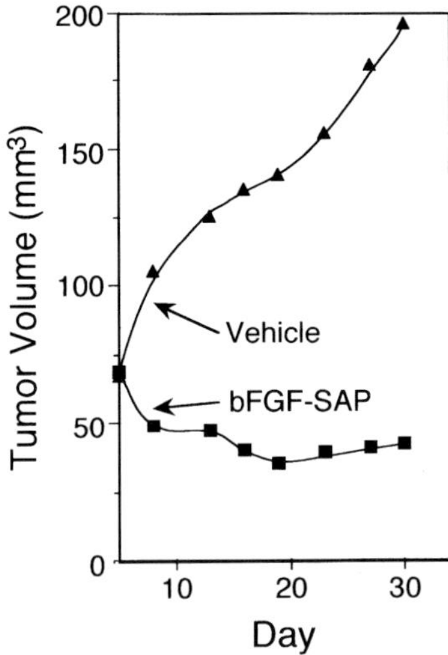

Figure 3. Tumour growth delay assay. Athymic mice bearing SK-MEL-5 human melanoma tumours were treated five days after implantation with a bolus i.v. dose of growth factor–toxin chimera (bFGF–SAP) or vehicle (vehicle), according to *Protocol 9*.

with a suspension of human tumour cells, then after the tumour develops, they are treated with various concentrations of the growth factor–toxin chimera to generate a dose–response curve. For this type of study, the tumour growth delay assay is used in which tumour cells are implanted subcutaneously in a conspicuous area to allow for easy monitoring of tumour growth (see *Protocol 9* and *Figure 3*).

Protocol 9. Tumour growth delay study

Equipment and reagents

- 1 ml syringes
- 26 gauge needles
- Calipers
- Tumorigenic cell line or stock tumours growing in mice
- Mice (conventional strains for murine tumours; athymic nude or SCID strains for human xenografts)[a]
- Trypan blue
- Serum-free medium

Method

1. Passage tumour cells either from tumours previously grown in mice or from cells grown in culture. Under the former conditions, excise tumours from stock animals, gently homogenize in serum-free medium, and filter the solution to create a single cell suspension. Cells grown in culture should be in exponential growth at the time they are harvested with trypsin/EDTA. Wash cells twice before suspending them in serum-free medium.

Protocol 9. *Continued*

2. Count cells by Trypan blue exclusion, then dilute cells so that the appropriate number of cells for implantation will be delivered in 0.1 ml of serum-free medium (cell concentration necessary to form tumours will vary from cell line to cell line, however, usually 1–2 $\times$ 10^6 cells/mouse for murine models and 0.5–1 $\times$ 10^7 cells/mouse for human xenograft models is sufficient).
3. Inject 0.1 ml of tumour cell suspension subcutaneously into one rear flank of each animal using a 26 gauge needle attached to a 1 ml syringe.
4. Initiate intravenous, intraperitoneal, or local subcutaneous treatments with growth factor–toxin chimera, equimolar concentrations of unconjugated growth factor or toxin moiety, or vehicle when tumours reach approx. 50 mm^3.
5. Perform external caliper measurements of tumours beginning the first day of treatment and twice weekly thereafter.
6. Calculate tumour volume using the formula:

   volume = [(minimum measurement)^2 (maximum measurement)] $\div$ 2.

[a] All animal work must comply with government regulations.

One disadvantage of the tumour growth delay assay is the length of the time required to generate results: typically 30–60 days from the time of tumour cell implantation. An alternative and less costly method is the tumour excision assay in which mice are implanted with the tumour cells and treated with the growth factor–toxin chimera, as before. However, within a short interval following drug treatment (usually within 24–48 h) the animals are sacrificed and the tumours excised, processed into single cell suspensions, seeded into tissue culture, and then assessed for viability compared to cells of tumours excised from vehicle treated animals (see *Protocol 10*).

Protocol 10. Tumour excision assay

Equipment and reagents

- 50 ml centrifuge tubes (Falcon 2098)
- 60 mm or 100 mm tissue culture dishes (Falcon 3002 or 3003)
- DNase I (Sigma)
- Tumorigenic cell suspension (10^6 to 10^7/0.1 ml)
- Collagenase IV (Sigma)
- Trypan blue

Filter sterilize all reagents through 0.2 μm membranes and handle all materials and samples aseptically when performing the following procedures.

Method

1. Prepare and implant tumour cells as described in *Protocol 9*. For this procedure, tumour cells may be implanted on both rear flanks. When

tumours reach approx. 50 mm^3, treat animals with various doses of growth factor–toxin chimera, equimolar doses of unconjugated growth factor or toxin moiety, or vehicle by intravenous, intraperitoneal, or local subcutaneous injection.

2. 24–48 h following treatment, kill the mice, and soak them in 70% ethanol.
3. Excise leg tumours under sterile conditions in a laminar flow-hood, pool four tumours from each treatment group to make approx. 1 g of tumour tissue, and mince tumours into a fine brie using crossed scalpels motion. Transfer each sample to a clean 50 ml centrifuge tube.
4. Wash tumour samples with unsupplemented medium, allow cells and tissue to settle, then decant off the wash medium.
5. Resuspend tumour samples in 20 ml of medium containing DNase I (0.1 mg/ml) and collagenase IV (450 U/ml), and incubate at 37 °C for 20 min in a shaking water-bath.
6. Centrifuge samples, remove and discard supernatant, then resuspend as above in enzyme-containing medium, and incubate at 37 °C for an additional 20 min in a shaking water-bath.
7. Thoroughly mix each cell suspension, then filter through two layers of sterile gauze into a clean centrifuge tube.
8. Centrifuge the single cell suspensions, remove and discard the supernatant, and add fresh medium. Wash samples two more times by this method, then resuspend cells in serum-supplemented medium.
9. Remove an aliquot from each sample and count cells by Trypan blue exclusion.
10. Seed cells at three different concentrations in duplicate tissue culture plates for a cell survival assay.

This method is more of a direct measurement of antitumour activity as a function of tumour cell targeting *in vivo*, whereas antitumour activity resulting in the former assay may be the result of an alternative or combination of events related to growth factor–toxin chimera targeting. For example, tumour growth and development is dependent upon the formation of new vasculature, a process known as angiogenesis. Vascular endothelial cells involved in angiogenesis selectively express growth factor receptors and, therefore, may be targeted by some growth factor–toxin chimeras. As a result, antitumour activity demonstrated by a growth factor–toxin chimera *in vivo* may be a primary effect of targeting and inhibiting tumour vasculature which leads to secondary tumour cell death. However, though the tumour excision assay is useful for discriminating *in vivo* targeting of tumour cells by the growth factor–toxin chimera, it will not replace tumour growth delay studies necessary for determining efficacy of multiple treatments or animal studies to determine lethal dose (LD) and to assess tissue and organ toxicities.

3. Clinical applications of growth factor–toxin chimeras

Growth factors and their receptors can cause cells to proliferate, to differentiate, and to remain viable. Since many disease processes are dependent upon these functions, they are likely targets for therapy with growth factor–toxin chimeras. Though there has been extensive pre-clinical testing of growth factor–toxin chimeras, to date, only a few phase I and II clinical trials have been conducted.

3.1 Cancer therapy

A role for growth factor receptors in cell malignancy and invasiveness has been suggested for a number of neoplasms. In comparison to normal, non-proliferating cells which may transiently express low levels of growth factor receptors, many tumour cell lines express high levels of high affinity growth factor receptors and proliferate in response to paracrine or autocrine growth factor ligands. In many instances, inhibiting growth factor production in these cells inhibits tumorgenicity, further supporting a relationship between growth factors and malignant transformation. It follows, therefore, that if growth factors and their receptors are necessary for the progression of neoplastic disease, then using a growth factor ligand to deliver a cytotoxic agent to these cells may offer a viable approach for targeting cancer cells.

Based upon this rationale, pre-clinical studies assessing the efficacy of growth factor–toxin chimeras against a broad range of malignant cells and human tumour xenografts has spanned more than a decade of research. Despite promising results obtained in pre-clinical testing, in comparison to antibody–toxin chimeras, few clinical trials utilizing growth factor–toxin chimeras have been conducted. To date, the most clinically explored growth factor–toxin chimera has been the recombinant fusion toxin DAB_{486}IL-2 in which the native receptor binding domain of diphtheria toxin has been replaced with human interleukin-2 (9). The IL-2 receptor is constitutively expressed on cells associated with haematologic malignancies such as lymphomas and leukaemias. Early clinical trials with DAB_{486}IL-2 as an anticancer therapeutic agent began in the late 1980s and included patients with histologically confirmed IL-2 receptor expressing Hodgkin's disease, non-Hodgkin's lymphoma, cutaneous T cell lymphoma, and chronic lymphocytic leukaemia whose disease had been refractory to standard or other experimental therapy. In the first phase I trial, the purpose was to determine the maximal tolerated dose (MTD) while evaluating drug efficacy (10). Intravenous dosing of patients was based on pre-clinical studies and began at a dose that was 1.5% of the non-toxic dose in monkeys when administered over 14 daily doses. Results from this study established the MTD for a daily intravenous bolus of DAB_{486}IL-2 as 0.1 mg/kg per day for ten doses. Furthermore, DAB_{486}IL-2 treatment produced tumour reduction and, in some patients, disease remission: important and encouraging results in terms of assessing the feasibility of targeting such neoplastic diseases with a growth factor–toxin chimera. Subsequent clinical trials (11–14) assessing DAB_{486}IL-2 safety and efficacy as a function of treatment regimen, including administration of multiple doses, continuous infusion, or dose escalation, have continued to demonstrate objective responses in some patients. Now in phase II trials (14, 15), DAB_{486}IL-2 is only the first of many such growth factor–toxin chimeras which are undergoing pre-clinical evaluation for targeting malignancies of various origins (see *Table 1*).

Another growth factor–toxin chimera to undergo clinical evaluation was TP–40, a chimera composed of transforming growth factor-α (TGFα) fused to *Pseudomonas* exotoxin (16). The TGFα growth factor has an affinity for epidermal growth factor (EGF) receptors which are present on many epithelial cells and which play a role in the development and progression of various malignancies. In a phase I trial (3), patients with superficial bladder cancer, which tested positive for EGF receptors and had failed to respond to conventional therapeutics, were treated with escalating doses of TP–40 administered intravesically. Though therapeutic response was limited to patients with only carcinoma *in situ*, there was resolution of lesions in 70% of this group and one complete remission suggesting therapeutic potential for TP–40.

3.2 Other applications

In addition to having a role in the malignant transformation of cells, growth factor receptors are involved in the development of new vasculature (angiogenesis), wound healing, hyperproliferative diseases such as psoriasis, and autoimmune diseases. In regard to the last, it has been found that activated T lymphocytes, which appear to be involved in autoimmune diseases, express IL-2 receptors thereby offering a potential target for treatment with a growth factor–toxin chimera. An initial phase I dose escalation study of DAB_{486}IL-2 in patients with refractory rheumatoid arthritis indicated significant improvement in half of the patients (17). In a subsequent phase II, double blind, placebo controlled trial (18), objective responses were demonstrated by 18% of patients with refractory rheumatoid arthritis treated with DAB_{486}IL-2 compared to no response in patients receiving the placebo. Clinical response improved to greater than 30% when patients received more than one treatment cycle with DAB_{486}IL-2.

Table 1. Growth factor–toxin chimeras undergoing pre-clinical evaluation[a]

Growth factor	Toxin	Target	Reference
IL-2	PE40	Ocular tissue, T cells	33, 34
	PE(PE66(4)GLU	B cells	35
	Shigella	T lymphoblasts	36
	Ricin A	T cells	37
IL-4	PE4E	Lymphocytes, leukaemia, lymphoma, colon, breast, stomach, liver, adrenals, and prostate carcinomas, melanoma and epidermoid carcinoma	38–41
	PE38QQR	Haematopoetic and epithelial tumours	38
IL-6	PE40	B lymphocytes and myeloma	42
IL-13	PE A(PE38QQR)	Human renal cell carcinomas	43, 44
EGF	PE(delta 34–220)	Epidermoid carcinoma	45
	Ricin A	Hepatocytes	46, 47
	Diphtheria	Hepatocytes	47–49
FGF-1 (aFGF)	PE40 and PE4EKDEL	Human aortic smooth muscle cells, breast, hepatocellular, and epidermoid carcinomas	50
FGF-2 (bFGF)	PE40 and PE4EKDEL	Human aortic endothelial cells, human aortic smooth muscle cells, breast, hepatocellular, and epidermoid carcinomas	50
	Saporin	Corneal and aortic arch endothelial cells, fibroblasts, melanoma, neuroblastoma, ovarian teratocarcinoma, prostatic carcinoma, fibrosarcoma, and Kaposi's sarcoma	31, 51–59
TGFα	PE40	Bladder, breast, colon, prostate, and lung carcinoma	60–62
VEGF	Ricin A and diphtheria	Umbilical vein and microvascular endothelial cells	63
	PE38(PEG-TCP)	Epidermoid carcinoma	26
IGF-1	PE	Hepatocellular carcinoma	64

[a] Abbreviations: IL, interleukin; EGF, epidermal growth factor; aFGF, acidic fibroblast growth factor; bFGF, basic fibroblast growth factor; TGFα, transforming growth factor-α; VEGF, vascular endothelial growth factor; IGF, insulin-like growth factor; PE, *Pseudomonas* exotoxin.

4. Potential problems in using biologically active chimeras

The success of a growth factor–toxin chimera to target and kill specific cells in a patient is dependent upon both intrinsic factors which include the activity of the toxin moiety and the affinity of the growth factor moiety for target receptors, as well as extrinsic factors such pharmacokinetics, immunogenicity, and resistance. Pre-clinical studies have been invaluable for predicting limitations of growth factor–toxin chimeras and exploring methods to improve efficacy through prevention or circumvention of associated problems.

4.1 Pitfalls associated with the toxin moiety

In comparison to their free form, toxin moieties of some growth factor–toxin chimeras, such as those containing saporin or ricin A, often have reduced ribosome inactivating protein activity when examined in a cell-free system (4, 5). However, because these growth factor–toxin chimeras are cytotoxic to whole cells bearing the target receptors, it has been suggested that in order to be activated, such toxins may depend upon lysosomal proteolytic cleavage following cellular internalization to separate the toxin from the growth factor. Studies comparing cytotoxicity of chemically conjugated or recombinant forms of the same growth factor–toxin chimera also suggest that some cell lines are more efficient at cleaving the disulfide linkage of the chemical conjugate than they are at separating the toxin from the growth factor in a recombinant molecule, thereby making the conjugated molecule more potent in terms of cytotoxicity.

Methods to enhance toxin efficiency have been the focus of many laboratories involved in the development of growth factor–toxin chimeras as therapeutic agents. Fishman *et al.* inserted the DNA sequence for lysosomal alkaline phosphatase within the chimeric gene encoding the recombinant growth factor–toxin IL2–PE40 (19). Incorporation of this LAP signal which targets the lysosomal compartment within cells increased cytotoxic activity of IL2–PE40 by two- to threefold.

Target cells treated with lysosomotropic agents which increase lysosomal pH and inactivate lysosomal proteases are more effectively killed by growth factor–toxin chimeras with ricin A or *Pseudomonas* exotoxin moieties (20). This suggests that these toxins may be susceptible to proteolytic degradation. This proteolytic lability has been suggested to be the cause for the relative ineffectiveness of ricin A-chain immunotoxins in targeting T cells (21), lymphomas (22), and myeloid leukaemias (23). Griffin *et al.*, however, have reported very dramatic potentiation of ricin A-chain immunotoxins when monensin is administered in liposomes (24). Curiously, potentiation was achieved with amounts of monensin far less than necessary to increase lysosomal pH. In contrast, immunotoxins composed with the saporin toxin are resistant to proteolytic degradation and as a result, are approximately 100-fold more cytotoxic than ricin A immunoconjugates (22) while failing to be potentiated by lysosomotropic agents.

4.2 Adverse effects associated with growth factor–toxin chimeras

Growth factor receptors are generally expressed at low levels in normal tissues. This, combined with pharmacokinetic observations that many growth factor–toxin chimeras are rapidly cleared from circulation offers, in many instances, a narrow window for therapeutic efficacy. Dose-limiting toxicity presents a major concern in terms of clinical applications with the liver being the predominant site for toxicity.

In the clinical trial with TP–40 (3), liver toxicity had been anticipated, given the fact that liver cells express EGF receptors, and was circumvented by administering TP–40 locally into the bladder. When administered in this manner, no evidence of local or systemic toxicity was noted in patients. However, local administration of a growth factor–toxin chimera is not always applicable, particularly when treating patients with systemic malignancies or diseases such as those participating in clinical trials with DAB_{486}IL-2.

In phase I clinical trials with DAB_{486}IL-2, potential concern was that administration of DAB_{486}IL-2 might result in immune dysfunction since it is cytotoxic to activated lymphocytes. However, no evidence of this was found in patients. Instead, hypersensitivity-like effects and reversible hepatic transaminase elevations were the determining factors for the maximal tolerated dose in the systemic treatment of patients with haematologic malignancies and immunologic diseases. The former reaction in these patients correlated with the presence of antibodies to the diphtheria toxin and represented the most severe adverse effects. In fact, antibody formation in patients as a result of immunologic reaction to the toxin component of the growth factor–toxin chimera represents a major challenge in terms of the effectiveness of these agents. For example, one method to circumvent non-specific tissue toxicity is to administer the chimeric agent in multiple treatment cycles at low or escalating doses. Antibody development in patients following initial treatment, however, presents limitations to follow-up treatments in terms of neutralizing growth factor–toxin chimera activity as well as producing hypersensitivity symptoms in patients. Pre-clinical studies conducted in the adjuvant arthritis rat model in which the animals were pre-immunized with DT antibodies prior to treatment with DAB_{486}IL-2, however, demonstrated that despite development of antibodies to the diphtheria moiety, therapeutic efficacy of DAB_{486}IL-2 was not altered (25). Similarly, it was found that although patients in clinical trials receiving DAB_{486}IL-2 developed or had pre-existing positive antibody titres for DT, the presence of antibodies did not preclude some patients from responding to treatment (10, 14, 18). Possible explanations for these observations are that circulating antibodies bind to the diphtheria component of the chimeric molecule and, therefore, do not impede the IL-2 moiety from binding to its receptor, and that although antibodies bind to the diphtheria moiety, they do not neutralize toxin activity.

Methods to reduce immunogenicity of other growth factor–toxin chimeras have been explored in pre-clinical studies. For example, Wang *et al.* have chemically modified the growth factor–toxin chimera, TGFα–PE38, with monomethoxy-polyethylene glycol (mPEG) (26). Curiously, the modified chimera demonstrates reduced toxic and receptor binding activity compared to the unmodified chimera, yet exhibits marked antitumour activity when administered *in vivo* despite the observation that its unmodified form does not. The investigators suggest that because the modified form of the chimera is five to ten times less immunogenic and circulates five to ten times longer in blood, these characteristics compensate for its reduced activity and serve to broaden the therapeutic window.

Though immunogenicity does not appear to present a major concern with DAB_{486}IL-2, one limitation is its short monophasic half-life of approximately five minutes (14). Bacha *et al.* have examined reconstruction of the growth factor–toxin chimera as a means to address this problem (27). In their investigations they found that removing the second disulfide loop of the diphtheria toxin resulted in a growth factor–toxin chimera, DAB_{389}IL-2, that in pre-clinical studies demonstrated increased binding to the IL-2 receptor, increased cytotoxicity to target cells, and a slower clearance time compared to DAB_{486}IL-2.

4.3 Mechanisms of resistance

One attractive feature of growth factor–toxin chimeras is that multidrug-resistant cell lines have not demonstrated cross-resistance to these agents. This is of particular importance from a clinical standpoint since patient tumours which develop resistance to one chemotherapeutic agent are often resistant to other agents. However, as with conventional chemotherapeutic agents, genetic instability of neoplastic cells and strong selective pressures exerted by growth factor–toxin chimeras combine to establish favourable conditions for the emergence of variant cells which may be resistant to treatment with growth factor–toxin chimeras, as well. In fact, in clinical trials evaluating TGFα–PE and DAB_{486}IL-2, not all patient malignancies responded to treatment despite the fact that they expressed the appropriate receptors. Besides variable chimera immunogenicity, pharmacokinetics, tissue distribution, and penetration of tumours, all of which may alter efficacy, conceptually, cells could become resistant to toxin conjugates through, for example:

(a) Reduction in the number or affinity of cell surface targets.
(b) Failure to internalize the growth factor–toxin chimera.
(c) Alteration of the internal compartmentalization of the toxin conjugates so that they are rapidly destroyed in lysosomes.

Because resistance is often the main cause of treatment failure, several laboratories have developed growth factor–toxin chimera-resistant cell lines to investigate mechanisms responsible for emergence of resistance as well as devise strategies for its prevention or circumvention. Resistance to ricin A-chain immunotoxins in somatic cell mutants has been linked to deficient endocytosis (28). Similarly, resistance to ricin A-chain growth factor–toxin chimeras in other cell lines have been attributed to altered transport and increased lysosomal degradation of the toxin (29). Reduced sensitivity to

EGF–PE growth factor–toxin chimera in mutagenized human epidermoid carcinoma cells may also be related to enhanced lysosomal degradation, however, in addition, these cells have a decrease in the number of EGF receptors though the binding constant remains unchanged (30).

In our own laboratory we have developed and subcloned resistant cell lines through step by step exposure to escalated concentrations of a growth factor–toxin chimera composed of basic fibroblast growth factor linked to saporin (bFGF–SAP) (see *Protocol 11*) (31).

Protocol 11. Development and cloning of resistant cells

Equipment and reagents

- 96-well tissue culture plates (Costar, Cat. No. 3595)
- 60 mm tissue culture dishes (Falcon 3002) or T25 tissue culture flasks (Falcon 3082)
- SK-MEL-5 human melanoma cells (American Type Culture Collection)
- FCS (Life Technologies, Inc.)
- Minimum essential medium (Life Technologies, Inc.)
- Recombinant basic fibroblast growth factor–saporin (bFGF–SAP; 45 kDa)
- PBS (Life Technologies, Inc.)
- Trypsin /EDTA (Life Technologies, Inc.)

A. *Development of resistant cell populations*

1. Treat semi-confluent cells, growing in minimum essential medium supplemented with 10% FCS, overnight with a concentration of bFGF–SAP sufficient to kill 90% of cells (10 nM).
2. After exposure, wash the cells three times with PBS.
3. Add fresh serum-supplemented medium and return the cells to 37 °C.
4. Each time the cells reach 75% confluence, expose them to growth factor–toxin chimera at a concentration increased 15–20% from the previous treatment.
5. The maximal tolerated concentration is defined as the drug concentration beyond which no cells survive. Surviving cells are cloned.

B. *Cloning of resistant cells*

1. Suspend adherent cells by treating with trypsin/EDTA; sediment cells at 400 *g* for 10 min, wash with fresh medium, and resediment cells.
2. Resuspend cells in fresh medium, remove an aliquot of cell suspension, and count viable cells in Trypan blue.
3. Serially dilute cell suspension to a final concentration of 25 cells/ml and seed the first of three 96-well plates with 0.12 ml/well (= 3 cells/well).
4. Dilute the remainder of the cell suspension 1:3 with fresh medium, and seed the second of three 96-well plates with 0.12 ml/well (= 1 cell/well).

5. Dilute the remainder of the cell suspension 1:3.3 with fresh medium, and seed the third of three 96-well plates with 0.12 ml/well (= 0.3 cells/well).
6. Incubate plates for two weeks at 37 °C.
7. Choose clones from the dilution plate in which 30% or less of the wells are positive for growth.
8. Expand the selected clones into 60 mm dishes or T25 flasks.
9. Assay subcloned lines to determine if they have maintained resistance throughout the cloning procedure, and freeze aliquots from selected clones in liquid nitrogen.
10. Periodically assay resistant lines maintained in culture to evaluate stability of resistance.

In cell survival assays, resistant human melanoma cells developed in this manner require a sixfold increase in the concentration of bFGF–SAP required to produce a 50% growth inhibition (IC_{50}). In light of the narrow therapeutic window for many anticancer agents, even this modest alteration in cellular sensitivity to a particular agent threatens therapeutic success. Because no change in sensitivity to unconjugated saporin was noted between parental and resistant cells, subsequent investigation to characterize the mechanism of resistance in the subclone focused on alterations in the basic growth factor receptor. Receptor binding and cross-linking studies demonstrated greater than a threefold reduction in the number of bFGF receptors on resistant cells. Because melanoma cells produce bFGF and it has been observed that endogenous growth factors can act in an autocrine or paracrine way to downregulate receptors, we pre-treated parental and resistant cells with suramin, a polyanionic agent which generally blocks growth factor binding to receptors, and alternatively with neutralizing antibody specific for bFGF, and then assessed cell sensitivity by cell survival assay (see *Protocol 12*).

Protocol 12. Pre-treatment of cells with suramin or neutralizing antibodies in cell survival assays

Equipment and reagents

- 6-well tissue culture plates (Corning, Cat. No. 25810)
- SK-MEL-5 human melanoma cells (American Type Culture Collection)
- Minimum essential medium (Life Technologies, Inc.)
- FCS (Life Technologies, Inc.)
- Suramin (FBA Pharmaceuticals)
- Dialysed fetal calf serum (if necessary;[a] Life Technologies, Inc.)
- Recombinant basic fibroblast growth factor–saporin (bFGF–SAP; 45 kDa)
- Basic fibroblast growth factor neutralizing antibody (Upstate Biotechnology, Inc.)
- Trypsin/ EDTA (Life Technologies, Inc.)

Protocol 12. *Continued*

Method

1. Seed cells ($2 \times 10^4/cm^2$) into an appropriate number of 6-well tissue culture trays in fresh minimum essential medium supplemented with 10% FCS. Incubate overnight at 37 °C in a 95% air/5% CO_2 humidified atmosphere.
2. The following day, replace medium with fresh serum-free medium[a] with or without suramin (1 mM) or neutralizing antibodies (10 mg/litre).
3. 15–18 h later, remove medium and wash the cells three times with unsupplemented medium before adding fresh unsupplemented medium[b] with or without various concentrations (1–150 nM) of bFGF–SAP for 1 h.
4. Following treatment, the cells are treated with trypsin/EDTA and reseeded for assay of cell survival.

[a] Extended incubation in serum-free medium may be fatal to some cell lines. In such cases, using the lowest concentration of serum required to preserve cell viability will assist in synchronizing cells. If the growth factor moiety of your chimera is of small size and present in serum, it may be removed by dialysis or by absorbing with specific immobilized antibodies.
[b] The washing step to remove excess suramin or neutralizing antibody is necessary to prevent their inhibiting growth factor–toxin chimera binding as well.

Exposure of resistant cells to either suramin or bFGF neutralizing antibody restored sensitivity to bFGF–SAP in resistant cells. Interestingly, pre-treatment with suramin increased sensitivity of resistant cells to bFGF–SAP compared to parental cells due to their release of autocrine aFGF, another member of the FGF family which binds to bFGF receptors. Importantly, when suramin was administered to nude mice bearing resistant xenografts 48 hours prior to administration of bFGF–SAP, tumours became sensitive to the growth factor–toxin chimera compared to those which had not received pre-treatment with suramin. In other preliminary studies, treatment cycles of suramin were followed by low doses of bFGF–SAP and administered to athymic nude mice bearing non-resistant human melanoma xenografts (32). This caused a greater than 20-fold enhancement in antitumour activity compared to treatment with the growth factor–toxin chimera alone. Thus in addition to circumventing resistance, treatment with suramin may also prevent subpopulations of cells with increased autocrine activity from escaping targeting with growth factor–toxin chimera. Since suramin has been used in patients for decades for the treatment of certain parasitic infections and more currently is in phase III trials for its ability to block autocrine growth factors in prostatic cancer, the observation that suramin increases efficacy and widens the therapeutic window of bFGF–SAP suggests that it may play a role in future clinical modalities designed to enhance growth factor–toxin chimera antitumour activity.

5. Conclusions

This chapter has examined the rationale and approach for linking toxins to growth factors to produce hybrid molecules known as growth factor–toxin chimeras for therapeutic purposes. Pre-clinical and clinical studies have provided credible evidence that cells which are associated with disease processes and express growth factor receptors make definitive targets for the delivery of these growth factor-based cytotoxic compounds. Unlike conventional chemotherapeutic drugs which do not discriminate between normal and diseased cells, the specificity of growth factor–toxin chimeras may explain why those evaluated in clinical trials have thus far been well tolerated. Furthermore, clinical trials in which patients with disease refractory to conventional chemotherapeutics have responded to therapy with growth factor–toxin chimeras confirms pre-clinical findings that there is no cross-resistance between other antitumour agents and growth factor–toxin chimeras. Though many obstacles to increasing therapeutic response still remain, taken together, these initial findings support the usefulness of growth factor–toxin chimeras in clinical therapy.

References

1. Lappi, D.A. and Baird, A. (1991). *Prog. Growth Factor Res.*, **2**, 223.
2. Pastan, I., Chaudhary, V., and Fitzgerald, D. (1992). *Annu. Rev. Biochem.*, **61**, 331.
3. Pai, L.H. and Pastan, I. (1994). In *Important advances in oncology* (ed. V.T. DeVita, S. Hellman, and S.A. Rosenberg), p. 3. J.B. Lippincott Company, Philadelphia.
4. Lappi, D.A., Ying, W., Barthelemy, I., Martineau, D., Prieto, I., Benatti, L., *et al.* (1994). *J. Biol. Chem.*, **269**, 12552.
5. O'Hare, M., Brown, A.N., Hussain, K., Gebhardt, A., Watson, G., Roberts, L.M., *et al.* (1990). *FEBS Lett.*, **273**, 200.
6. Hwang, J., Fitzgerald, D.J., Adhya, S., and Pastan, I. (1987). *Cell*, **48**, 129.
7. Bolton, A.E. and Hunter, W.M. (1973). *J. Endocrinol.*, **55**, 529.
8. Linge, C. and Green, M.R. (1993). In *Growth factors: a practical approach* (ed. I. McKay and I. Leigh), p. 201. Oxford University Press, Oxford, UK.
9. Williams, D.P., Snyder, C.E., Strom, T.B., and Murphy, J.R. (1990). *J. Biol. Chem.*, **285**, 11885.
10. LeMaistre, C.F., Meneghetti, C., Rosenblum, M., Reuben, J., Parker, K., Shaw, J., *et al.* (1992). *Blood*, **79**, 2547.
11. Hesketh, P., Caguioa, P., Koh, H., Dewey, H., Facada, A., McCaffrey, R., *et al.* (1993). *J. Clin. Oncol.*, **11**, 1682.
12. Kuzel, T.M., Rosen, S.T., Gordon, L.I., Winter, J., Samuelson, E., Kaul, K., *et al.* (1993). *Leuk. Lymphoma*, **11**, 369.

13. Platanias, L.C., Ratain, M.J., O'Brien, S., Larson, R.A., Vardiman, J.W., Shaw, J.P., *et al.* (1994). *Leuk. Lymphoma*, **14**, 257.
14. Tepler, I., Schwartz, G., Parker, K., Charette, J., Kadin, M.E., Woodworth, T.G., *et al.* (1994). *Cancer*, **73**, 1276.
15. Foss, F.M., Borkowski, T.A., Gilliom, M., Stetler-Stevenson, M., Jaffe, E.S., Figg, W.D., *et al.* (1994). *Blood*, **84**, 1765.
16. Heimbrook, D.C., Stirdivant, S.M., Ahern, J.D., Balishin, N.L., Patrick, D.R., Edwards, G.M., *et al.* (1990). *Proc. Natl. Acad. Sci. USA*, **87**, 4697.
17. Sewell, K.L., Parker, K.C., Woodworth, T.G., Reuben, J., Swartz, W., and Trentham, D.E. (1993). *Arthritis Rheum.*, **36**, 1223.
18. Moreland, L.W., Sewell, K.L., Trentham, D.E., Bucy, R.P., Sullivan, W.F., Schrohenloher, R.E., *et al.* (1995). *Arthritis Rheum.*, **38**, 1177.
19. Fishman, A., Bar-Kana, Y., Steinberger, I., and Lorberboum-Galski, H. (1994). *Biochemistry*, **33**, 6235.
20. Pirker, R., FitzGerald, D.J.P., Williangham, M.C., and Pastan, I. (1988). *Cancer Res.*, **48**, 3919.
21. Casellas, P., Ravel, S., Bourrie, J.P., Derocq, J.M., Jansen, F.K., Laurent, G., *et al.* (1988). *Blood*, **72**, 1197.
22. Tazzari, P.L., Bolognesi, A., de Totero, D., Falini, B., Lemoli, R., Soria, M.R., *et al.* (1992). *Brit. J. Haematol.*, **81**, 203.
23. Engert, A., Brown, A., and Thorpe, P. (1991). *Leuk. Res.*, **15**, 1079.
24. Griffin, T., Rybak, M.E., Recht, L., Singh, M., Salimi, A., and Raso, V. (1993). *J. Natl. Cancer Inst.*, **85**, 292.
25. Bacha, P., Forte, S.E., Perper, S.J., Trentham, D.E., and Nichols, J.L. (1992). *Eur. J. Immunol.*, **22**, 1673.
26. Wang, Q.C., Pai, L.H., Debinski, W., FitzGerald, D.J., and Pastan, I. (1993). *Cancer Res.*, **53**, 4588.
27. Bacha, P.A., Forte, S.E., McCarthy, D.M., Estis, L., Yamada, G., and Nichols, J.C. (1991). *Intl. J. Cancer*, **49**, 96.
28. Goldmacher, V.S., Anderson, J., Schultz, M.L., Blattler, W.A., and Lambert, J.M. (1987). *J. Biol. Chem.*, **262**, 3205.
29. Shimizu, N., Shimizu, Y., and Miskimins, K.W. (1984). *Cell Struct. Funct.*, **9**, 203.
30. Lyall, R.M., Hwang, J., Cardarelli, C., FitzGerald, D., Akiyama, S.-I., Gottesman, M.M., *et al.* (1987). *Cancer Res.*, **47**, 2961.
31. Davol, P., Beitz, J.G., and Frackelton, A.R. Jr. (1995). *J. Invest. Dermatol.*, **104**, 916.
32. Davol, P. and Frackelton, A.R. Jr. (1995). *Proc. Am. Assoc. Cancer Res.*, **36**, 3817A.
33. BenEzra, D., Maftzir, G., Hochberg, E., Anteby, I., and Lorberboum-Galski, H. (1995). *Curr. Eye Res.*, **14**, 153.
34. Volk, H.D., Muller, S., Yarkoni, S., Diamantstein, T., and Lorberboum-Galski, H. (1994). *J. Immunol.*, **153**, 2497.
35. Steinberger, I., Brenner, T., and Lorberboum-Galski, H. (1996). *Cell. Immunol.*, **169**, 55.
36. Seregin, S.V., Siniakov, A.N., Pozdniakov, S.G., Kozlov, I.V., Sakhno, L.V., Leplina, O., *et al.* (1993). *Mol. Biol. (Mosk)*, **27**, 763.
37. Cook, J.P., Savage, P.M., Lord, J.M., and Roberts, L.M. (1993). *Bioconjug. Chem.*, **4**, 440.

38. Debinski, W., Puri, R.K., and Pastan, I. (1994). *Int. J. Cancer*, **58**, 744.
39. Puri, R.K., Mehrotra, P.T., Leland, P., Kreitman, R.J., Siegel, J.P., and Pastan, I. (1994). *J. Immunol.*, **152**, 3693.
40. Puri, R.K., Debinski, W., Obiri, N., Kreitman, R., and Pastan, I. (1994). *Cell. Immunol.*, **154**, 369.
41. Debinski, W., Puri, R.K., Kreitman, R.J., and Pastan, I. (1993). *J. Biol. Chem.*, **268**, 14065.
42. Siegall, C.B., Chaudhary, V.J., FitzGerald, D.J., and Pastan, I. (1988). *Proc. Natl. Acad. Sci. USA*, **85**, 9738.
43. Puri, R.K., Leland, P., Obiri, N.I., Husain, S.R., Kreitman, R.J., Haas, G.P., *et al.* (1996). *Blood*, **87**, 4333.
44. Debinski, W., Obiri, N.I., Pastan, I., and Puri, R.K. (1995). *J. Biol. Chem.*, **270**, 16775.
45. Liao, C.W., Hseu, T.H., and Hwang, J. (1995). *Appl. Microbiol. Biotechnol.*, **43**, 498.
46. Cawley, D.B. and Herschman, H.R. (1980). *Cell*, **22**, 563.
47. Simpson, D.L., Cawley, D.B., and Herschman, H.R. (1982). *Cell*, **29**, 469.
48. Shimizu, N., Miskimins, W.K., and Shimizu, Y. (1980). *FEBS Lett.*, **118**, 274.
49. Shaw, J.P., Akiyoshi, D.E., Arrigo, D.A., Rhoad, A.E., Sullivan, B., Thomas, J., *et al.* (1991). *J. Biol. Chem.*, **266**, 21118.
50. Gawlak, S.L., Pastan, I., and Siegall, C.B. (1993). *Bioconjug. Chem.*, **4**, 483.
51. Lappi, D.A., Martineau, D., and Baird, A. (1989). *Biochem. Biophys. Res. Commun.*, **160**, 917.
52. Lappi, D.A., Martineau, D.G., Nakamura, S., Buscaglia, M., and Baird, A. (1990). *J. Cell. Biochem.*, **14E** (Suppl.), 222.
53. Lappi, D.A., Maher, P.A., Martineau, D., and Baird, A. (1991). *J. Cell. Physiol.*, **147**, 17.
54. Beitz, J.G., Davol, P., Clark, J.W., Kato, J., Medina, M., Frackelton, A.R. Jr., *et al.* (1992). *Cancer Res.*, **52**, 227.
55. Davol, P., Beitz, J.G., Mohler, M., Ying, W., Cook, J., Lappi, D.A., *et al.* (1995). *Cancer*, **76**, 79.
56. Davol, P. and Frackelton, A.R. Jr. (1996). *J. Urol.*, **156**, 1174.
57. Ying, W., Martineau, D., Beitz, J., Lappi, D.A., and Baird, A. (1994). *Cancer*, **74**, 848.
58. Beattie, G.M., Lappi, D.A., Baird, A., and Hayek, A. (1990). *Diabetes*, **39**, 1002.
59. David, T., Tassin, J., Lappi, D.A., Baird, A., and Courtois, Y. (1992). *J. Cell. Physiol.*, **153**, 483.
60. Kameyama, S., Kawamata, H., Pastan, I., and Oyasu, R. (1994). *J. Cancer Res. Clin. Oncol.*, **120**, 507.
61. Kirk, J., Carmichael, J., Stratford, I.J., and Harris, A.L. (1994). *Br. J. Cancer*, **69**, 988.
62. Xu, Y.H., Peng, S.F., Jiang, W.L., Zhu, Y., Chandhary, V., FitzGerald, D., *et al.* (1993). *Shih Yen Sheng Wu Hsueh Pao*, **26**, 289.
63. Ramakrishnan, S., Mohanraj, D., Olson, T., Gupta, K., and Roy, S. (1995). *Proc. Am. Assoc. Cancer Res.*, **36**, 532A.
64. Prior, T.I., Helman, L.J., FitzGerald, D.J., and Pastan, I. (1991). *Cancer Res.*, **51**, 174.

10

Gene therapy applications of growth factors

TOR SVENSJÖ, FENG YAO, BOHDAN POMAHAC, and
ELOF ERIKSSON

1. Introduction to gene therapy applications of growth factors

Gene therapy (GTh) involves transfer of DNA to somatic cells with the purpose of getting those cells to express certain proteins or antisense RNA. Currently all protocols for GTh are experimental, but with the ultimate goal of curing patients. Remarkable progress has marked the field of GTh in the past decade. Since the first Recombinant DNA Advisory Committee (RAC) approved treatment that employed gene transfer (GTr) to study and treat human disease (1) the number of clinical trials involving GTh has increased exponentially, and as of March 1996 there were 145 approved human GTh protocols in the USA (2). Diseases that have been targeted include several congenital disorders (e.g. adenosine deaminase (ADA) deficiency, cystic fibrosis, and familial hypercholesterolaemia) and acquired immunodeficiency syndrome (AIDS), but more commonly the therapies are targeted at cancer (3).

Growth factors (GFs) belong to the group of regulators that include colony-stimulating factors, interferons, interleukins, lymphokines, and monokines, collectively known as cytokines (4). They are molecules that interact with specific cells to communicate information regarding the status of the cell of origin and they result in a specific biological response in target tissue. The GFs are capable of inducing or inhibiting proliferation, differentiation, and migration of cells. Conceptually, many GFs would therefore be attractive candidates for treatment of a variety of common disorders, including cancer, cardiovascular disease, and impaired wound healing. Some GFs or cytokines have already demonstrated useful clinical effects, including acceleration of wound repair by epidermal growth factor (EGF) (5) and modulation of cancer cell growth by inteferons and interleukins (6, 7). There are however circumstances that limit the clinical use of cytokines. Oral delivery is inefficient or impossible because of breakdown and limited absorption in the digestive tract. The half-life of cytokines in blood is generally short due to breakdown or excretion and systemic administration usually lacks targeted specificity. This is particularly important for GFs, which generally affect many organs of the body (4). Systemic adverse effects have, for instance, limited the use of some cytokines in cancer treatment (8). This necessitates more targeted effect on specific organs and systems than can be achieved by systemic infusion. Administration of GFs by GTr may provide target-specific delivery of GFs for the treatment of disease. This chapter will discuss the GTr techniques commonly used with mammalian cells *in vitro* and *in vivo*. It will also present some GTh applications and will describe in detail some of the methods used in our laboratory to achieve GF expression in skin.

1.1 Gene transfer methods

Methods for GTr to animals can be divided into *ex vivo* and *in vivo* techniques (see *Figure 1*). *Ex vivo* refers to genetic modulation of cells *in vitro* and subsequent transplantation into a host. The cells can be of any origin but most commonly, at least in human subjects, autologous or allogeneic cells are preferred. *In vivo* GTr denotes that the gene is delivered directly into the cells of the recipient. Transfer can be achieved with a variety of techniques, some of which can be used both for *ex vivo* and *in vivo* GTr. The following common methods will be discussed:

- viral techniques
- chemical techniques
- electroporation
- injection of DNA
- particle-mediated GTr

1.1.1 Viral vectors

Viruses are obvious as GTr vectors because of their ability to efficiently transfer nucleic acid into a host cell. For reasons of safety, viral vectors used for GTr must be replication-defective. These viral vectors carry a genome lacking essential functions in one or more of the genes required for autonomous viral replication (9). The intention is to limit viral infection to the targeted cells or animal and avoid the potential spread of replication-competent viruses in the environment. There are, however, two possible ways, recombination and complementation, by which a defective virus can overcome its defective state (reviewed in ref. 10). Recombination is the physical interaction of the viral genome with viral sequences harboured in the cell. This could, in the worst case, lead to a replication-competent virus with the ability to spread to non-targeted cells within the host or other hosts. In complementation, there is only a functional interaction at the level of cellular or viral gene products, i.e. there is no change to the genome of the defective

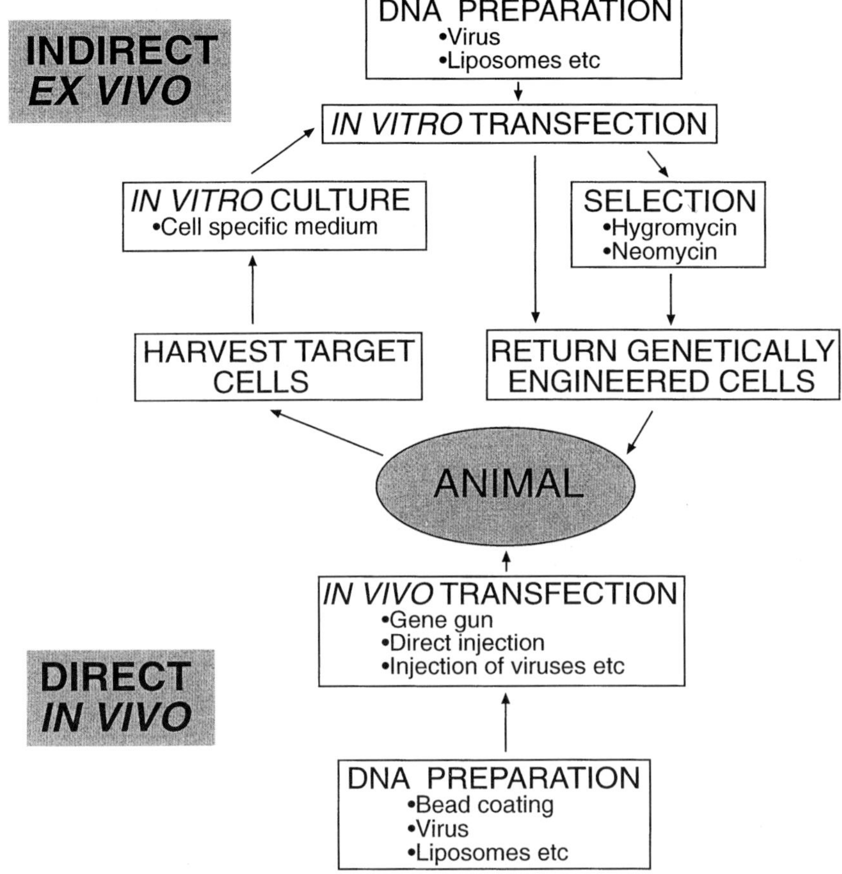

Figure 1. Schematic drawing exemplifying *ex vivo* and *in vivo* gene transfer techniques in gene therapy.

virus vector. Complementation may lead to spread of the defective virus but it will not become replication-competent.

A number of viruses have been used to introduce genes into cells. The most commonly used are retrovirus (11, 12), adenovirus (AV) (13), adeno-associated virus (AAV) (14, 15), vaccinia virus (16, 17), and herpes simplex virus (18). They have been constructed by replacing one or more viral genes with the DNA or RNA of interest and most often the viral genes, involved in infectivity, replication, and/or pathology, have been removed or disrupted, to prevent unwanted infection.

Retroviral GTr is attractive because it leads to a comparatively high incidence of stable integration of the transferred genes into the host chromosome, making them a permanent part of the host genome throughout the cell's lineage. Retroviruses also exhibit high transduction efficiencies *in vitro*, commonly more than 20% of replicating cells, and they have an amphotropic host range, i.e. they infect cells from different species, including human (19). In GTh trials, they have mainly been utilized in *ex vivo* GTr approaches (3), and there are problems that may limit extensive use of retroviruses in human GTh:

(a) The target cells must be dividing to integrate the provirus into the genome (20). This is of little concern in *ex vivo* GTr approaches where cells generally are cultured and expanded, but it will limit the use of retroviruses for *in vivo* GTr.

(b) Studies have demonstrated the gradual inactivation of expression from viral promoters (21, 22) even though the vector DNA remains present (22), thus repeated treatments may be necessary to maintain sufficient expression.

(c) Retroviruses always carry the risk of causing insertional mutagenesis of proto-oncogenes and tumour suppressor genes.

A more efficient system for *in vivo* GTr is that of the adenovirus (AV). Adenoviral vectors can efficiently transfer genes to both non-replicating and replicating cells (23, 24) making them suitable for *in vivo* GTr applications. One disadvantage with AV vectors is their limited duration of expression, typically in the range of weeks to months, because the genetic information transferred with AV remains episomal. A second disadvantage is the non-specific inflammation and anti-vector cellular immunity evoked by the AV vectors currently used (3).

Other properties of AV may also be useful for GTh purposes: for example the observation that AV disrupt endosome function was utilized in a system of receptor-mediated GTr to hepatocytes where the ligand was linked to the DNA via polylysine. Low transfection efficiencies were observed, but when the ligand–polylysine–DNA complex was linked to AV and subsequently underwent endocytosis, the AV protein disrupted the endosome before DNA damage occurred (25). In this system 100% transfection efficiency of the hepatocytes could be achieved.

Adeno-associated virus (AAV) will in the presence of a helper virus (e.g. AV) integrate into the host genome (26). The integration is not well understood but it seems to occur at only one site in the human cells (chromosome 19q13-qter) (27) which, in theory, limits the risk of insertional mutagenesis.

Herpes simplex virus and vaccinia virus are genetically more complex than retroviruses and adenoviruses, they have a wider host range, and can potentially accommodate more foreign DNA while retaining infectivity (16).

In summary, the major advantages with viral vectors are their high transduction efficiencies and the possibility of long-term expression (retrovirus, AAV). Drawbacks include risk of insertional mutagenesis (e.g. activation of proto-oncogenes or disruption of tumour suppressor genes), generation of replication-competent recombinant viruses, and the immune response evoked by viral proteins.

1.1.2 Chemically-mediated gene transfer

A variety of chemicals have been employed to facilitate uptake of DNA by cells. Among the commonly used are calcium phosphate (28), liposomes (29), and diethylaminoethyl–dextran (30). Plasmid DNAs form complexes with these agents upon mixing and uptake into cells is likely mediated by endocytosis or fusion with cell membrane. The techniques are easily performed and they have found great utility in the field of *in vitro* GTr for research purposes. The efficiency of these methods for obtaining stable integration of the gene is low, typically in the range between one in a thousand and a million (31). The plasmid DNA will persist in the nucleus and be expressed by many cells (up to 50%) for several days before disappearing (31). Even though mostly used for *in vitro* GTr, the feasibility of liposome-mediated *in vivo* GTr has been demonstrated in the hair follicles of skin (32), and in cancer and cystic fibrosis patients (3). One advantage of chemical techniques and other non-viral techniques over viral vectors is that there is no direct limitation to the size of the DNA which can be transferred. This is typically in the range of 5–30 kb for viruses.

1.1.3 Electroporation

This method for GTr uses a short high voltage electric field to induce reversible micropores in the plasma membrane thereby allowing uptake of DNA from the medium wherein the cells are suspended (31). This technique is effective with cultured cells in suspension, but is has not been applied widely to primary cell cultures or cells *in vivo*.

1.1.4 Injection of DNA

This technique can be performed *in vitro* and *in vivo*. Direct injection of plasmid DNA into cell nuclei of cultured cells is termed microinjection. It utilizes microcapillaries to inject the DNA and is commonly used to target cells in early stage embryos (for the production of transgenic animals) or cells that are particularly difficult to transfect.

Injection of naked DNA *in vivo* using a standard hypodermic syringe has been demonstrated to be a simple method to obtain expression in a variety of tissues including brain (33), heart (34), liver (35), muscle (36, 37), and skin (38, 39). The method may prove useful when long-term expression is not required as in vaccination (39).

1.1.5 Particle-mediated gene transfer

Particle-mediated GTr was originally intended for GTr to plant cells (40–42), but has more recently been shown to be applicable to a wide variety of mammalian cells *in vitro* and *in vivo* (43, 44). The method utilizes small gold particles (1–5 μm in diameter) that are coated with plasmid DNA. The particles are then accelerated and directed into the target cells by an electric arc discharge (evaporating a small water droplet) or compressed helium. With particle-mediated GTr, we have shown that transfer of an epidermal growth factor (EGF)-expressing plasmid, to partial thickness wounds leads to production of EGF over several days, and results in significantly accelerated re-epithelialization of the wounds (45). Another application of particle-mediated GTr is vaccination. Targeting intact skin with this method has been shown to be effective in vaccination against influenza virus infection (46).

1.2 Gene therapy with growth factors

The feasibility of human GTh has already been extensively demonstrated in several clinical cases and trials, including a few where cytokines have been utilized (3). This overview intends to focus on the GTh applications of growth factors and will be limited to what has been studied experimentally.

Cancer and cardiovascular disease represent two major causes of death in the industrialized world and much of the research in GTh is directed towards treatment of these disorders. One approach to treatment of cancers would be to direct the body's own immune response against the tumour cells. Some cancers secrete substances that defend them from being attacked by immune cells. One illustrative example of such a substance is transforming growth factor-β (TGFβ), known to be over-expressed by many common tumours, including colon, breast, and prostatic carcinoma (47). In an effort to promote the immune system to attack tumour cells an immunization can be performed. Fakhrai and co-workers (47) immunized gliosarcoma-affected rats with lethally irradiated cells (of the same tumour origin) which also had been transfected with an antisense TGFβ vector, thereby lowering their TGFβ production. Rats immunized in this fashion displayed 100% 12 week survival and no residual tumours as compared to 20% in control animals that had received non-transfected irradiated cells. Similar strategies and comparable results have also been demonstrated with antisense insulin growth factor 1 (IGF-1)-expressing cell lines in the treatment of teratocarcinoma (48) and glioblastoma (49, 50). The results of such GTh-enhanced tumour immunizations are encouraging, but further studies have to show the feasibility of this strategy in the treatment of autologous tumours, since the cancers in these animal models are derived from transplanted allogeneic or xenogeneic cells. Moreover this therapy would require the establishment of cell cultures from the patients' tumours, a procedure successful only in about 50% of the cases (8).

Other cancer GTh approaches include the use of tumour infiltrating lymphocytes that are genetically modified to secrete various cytokines (e.g. interleukin-2 and tumour necrosis factor-alpha) and then reinjected into the affected animal or patient. This promotes an immune response or tumour toxic effect by the cytokine secretion, but unfortunately it has been difficult to achieve sufficient cytokine production by such lymphocytes (8). Experiments have also been performed with the intention of modifying the actual tumour cells by GTr to decrease their growth potential. It was shown that tumour cells (C6 glioma) stably transfected with an antisense vascular endothelial growth factor (VEGF)-expressing plasmid, produced significantly smaller tumours with increased necrosis compared with tumours developed from untransfected tumour cells (51). However GTr methods that specifically target tumour cells would be necessary to develop such strategies. Another application is antisense VEGF GTh to treat proliferative retinopathy, a disease in which retinal neovascularization is believed to lead to blindness. In animal experiments it was shown that intraocular injections of an antisense oligonucleotide against VEGF could slow the progression of the disease (52), but the efficiency of GTr appeared to be low and not permanent, limiting its practical use.

Side-effects from radiotherapy and chemotherapy limit their use. Faster recovery from chemotherapeutic-induced thrombocytopenia and leucopenia has been demonstrated with gene-mediated delivery of fibroblast growth factor-4 (53) and macrophage colony stimulating factor (54), respectively. The application of this approach may allow the use of more aggressive cancer therapies.

Cardiovascular diseases are also targets for GTh. Among the growth factors which have been delivered are vascular endothelial growth factor (VEGF), transforming growth factor-β1 (TGFβ1), and fibroblast growth factor 5 (FGF-5). VEGF has been delivered by GTr both in animals and in patients with resulting angiogenesis (55, 56) and increased blood flow to an ischaemic leg (56). Even though the clinical case utilized a somewhat inefficient method of GTr, i.e. 2 mg plasmid DNA was delivered to the arterial wall by an angioplasty balloon, it does represent one of growing number of studies demonstrating the feasibility of human GTh. Another growth factor involved in angiogenesis is FGF-5. This growth factor was, in a pig ischaemia model, demonstrated to successfully ameliorate abnormalities in myocardial blood flow and function following *in vivo* GTr (57). Transfection was carried out by intracoronary injection of a FGF-5-expressing AV. Targeted transduction was achieved as demonstrated by the finding that more than 98% of the virus stayed in the myocardium. Furthermore no significant signs of inflammation could be detected in the transfected heart or in the liver. The increased functional capacity persisted for at least 12 weeks, but long-term follow-up studies would be desired before starting clinical trials.

GTr has also been used in neural diseases. Nerve growth factor (NGF, neutrophin) is a small (about 120 amino acids) protein which supports the survival of embryonic neurones (58). It has been used to treat various experimental disorders, including memory impairment (59), autoimmune neuritis (60), and ischaemia-induced neuronal damage (61). The evaluation of NGFs effect on memory was carried out by transplanting human NGF cDNA transfected fibroblasts (retroviral transfection) to the nucleus basalis magnocellularis of selected, memory-impaired, aged rats. The treatment was found to significantly improve learning and memory, as evaluated in a Morris water maze test (59), and may represent a new modality treatment of age-related memory loss. Another elegant study demonstrating the delivery of NGF by GTr is that of Kramer and co-workers (60). In a model of autoimmune neuritis, lymphocytes targeting the peripheral nervous system were injected into animals to induce experimental autoimmune neuritis. However, animals injected with lymphocytes that had been modified by retroviral infection to express NGF showed less severe neuritis by clinical and histological scoring, compared to animals injected with mock-infected or non-modified lymphocytes. Even though this system has little practical implication since transfected lymphocytes still induced mild neuritis (though much lower than unmodified lymphocytes) it could serve as a model for the targeted delivery of NGF.

TGFβ has immunosuppressive effects which, as mentioned before, may represent means by which cancer cells can escape immune defenses. GTh directed towards cancer cells with antisense TGFβ-expressing vectors is a possible application. Another would be local immunosuppression at transplantation sites. For example, if one were to transfect an organ with an immunosuppressive growth factor such as TGFβ, it may be possible to avoid the complications associated with systemically delivered immunosuppressants (infections, cancer, etc.). The feasibility of this approach was demonstrated in a cardiac allograft transplantation model: allografts that had been injected with a TGFβ-expressing plasmid survived 26 days, twice as long as non-treated grafts (62). Future studies need to address the question of whether longer survival can be achieved.

In general many interesting GTh approaches to the treatment of disease have been proven in principle, but they are often of limited clinical use because of low levels of gene expression, lack of sustained expression, and/or lack of methods to achieve targeted gene delivery.

2. Growth factor applications to enhance cutaneous wound healing

The skin is an attractive target for GTh for a number of reasons:

(a) Cells of the skin, e.g. fibroblasts and keratinocytes (63), can be easily obtained and cultured using standard procedures, thereby permitting use of autologous cells and *ex vivo* genetic modification and testing.

(b) Relatively simple systems are available for transplanting genetically-modified cells to the skin. For example, methods for transplanting keratinocytes as sheet grafts (64) or cell suspensions (65), and fibroblasts, in vehicle (66) or cell suspensions (67), to the skin are well established.

(c) If unwanted effects occur in the genetically-modified skin, they can be readily detected and removed.

(d) Several methods for direct *in vivo* GTr to skin exist (32, 38, 43, 68).

Our laboratory has utilized two different approaches to obtain expression of peptides in skin. One is the direct *in vivo* GTr technique using the gene gun (45) or a new technique developed in our laboratory, named microseeding (68). The other is an *ex vivo* technique employing transplantation of genetically-modified fibroblasts (67) or keratinocytes (65). The purpose has been to accelerate wound healing in skin by supplying growth factors to the wound. Because of the short healing phase of wounds, our goals have not been directed towards long-term expression since it has not been a necessity for wound healing. We have for example shown that transfection of partial thickness wounds with an hEGF-expressing plasmid construct using the gene gun, can accelerate the re-epthelialization process of partial thickness wounds by 20% (45). This was achieved despite a steady decrease of growth factor expression seen during the first four days following transfection.

3. Growth factor expression in skin

During the past several years, we have used both *ex vivo* and *in vivo* GTr methods to express growth factors in skin and wounds. The following will describe our use of the gene gun (*Figure 2*) technique for *in vivo* transfection of porcine partial thickness skin wounds.

The gene gun must prior to its use be loaded with the correct ammunition, i.e. cartridges containing DNA-coated gold beads. The preparation of such cartridges is outlined in *Protocol 1*.

Protocol 1. Gold bead/DNA preparation and gene delivery with the ACCELL® gene gun[a]

Equipment and reagents

- Water-bath sonicator and vortexer
- Syringe fitted with rubber adapter
- Tefzel® tubing: 1/8″ o.d. × 3/32″ i.d. (McMaster-Carr)
- Tube turner (Agracetus/Auragen, Inc.)
- Peristaltic pump
- Compressed N_2 gas cylinder
- Compressed helium gas cylinder
- ACCELL® gene delivery device (Agracetus/Auragen, Inc.)
- Polyvinylpyrrolidone (PVP, 360000) (Sigma)
- 0.95 μm or 1–3 μm gold beads (Agracetus/Auragen, Inc.)
- 0.1 M spermidine solution (molecular biology grade, free base) (Sigma)
- 2.5 M $CaCl_2$ solution

Protocol 1. *Continued*

Method

1. Weigh 5–50 mg of gold beads in a 1.5 ml microcentrifuge tube, then add 100 μl 0.1 M spermidine solution.
2. Sonicate gold and spermidine mixture for 3–5 sec using water-bath sonicator to break up gold clumps into a single bead-suspension.
3. Determine the volume of DNA required to achieve the desired DLR (DNA loading rate)[b] and add that volume to the spermidine and gold mixture.
4. Mix DNA, spermidine, and gold by vortexing, and add 200 μl of 2.5 M $CaCl_2$ solution dropwise to the mixture while gently vortexing (volume of $CaCl_2$ added is twice the volume of spermidine used in step 1).
5. Allow the mixture to precipitate at room temperature for 5–10 min, then briefly centrifuge (10–15 sec in a microcentrifuge), remove supernatant, and discard.
6. Wash the pellet three times with 500 μl 100% ethanol by using a short spin of 30 sec between each wash. Remove and discard supernatants with a pipette. Do not pour off supernatant.
7. Pipette the required volume of 100% ethanol to yield the desired BLR (bead loading rate)[c] into a labelled polypropylene culture tube.
8. After the final wash, resuspend pellet in 100% ethanol with or without PVP[d] by adding 500 μl of the ethanol (taken from the culture tube, step 7) to the gold pellet. Vortex, then sonicate briefly (2–3 sec). Transfer gold suspension to culture tube. Repeat the transfer of 500 μl ethanol between culture tube and microcentrifuge tube until all gold has been transferred to the culture tube. The beads are now ready for tube preparation.[e]
9. Before loading the tubes (Tefzel® tubing) with the DNA/gold beads, vortex and sonicate the beads briefly to achieve an even suspension.
10. Using a syringe fitted with a rubber adapter, draw the suspended beads into a Tefzel® tube of the approximate length of the tube turner. In high humidity conditions, run ethanol through the tube at 2–3 inches (approx. 5–8 cm) per second using a peristaltic pump before drawing up beads.
11. Slide loaded tube, with syringe attached, into the tube turner.
12. Allow beads to settle 3 min for 1–3 μm gold and 5 min for 0.95 μm gold. Detach the tube from the syringe and attach it to a peristaltic pump.
13. Draw off ethanol. (Use the peristaltic pump at a setting which draws ethanol from the tube at a rate of 2–3 inches per second.)

14. Detach peristaltic pump from tube.
15. Rotate tube at 20 r.p.m. in the tube turner.
16. Allow the gold to smear in the tube for 30 sec; then open N_2 valve to allow 350–400 ml/min flow rate to dry tube (3–5 min to dry completely).
17. Using a scalpel, cut off and discard uneven ends or pieces from the tube. Cut the remaining tube at 0.5″ (approx. 1.3 cm) intervals to create cartridges that can be loaded into the revolving cylinder of the gene gun. Store the cartridges, if not immediately used, at 4 °C with a desiccant pellet in Parafilm sealed and labelled vial. Cartridges can be stored for up to four months at 4 °C.
18. Plug the solenoid of the gene gun into a outlet and connect the gene gun to a supply of compressed helium and insert a tube (cartridge) loaded '12-shooter' cylinder into the barrel of the gun. Adjust pressure regulator to appropriate pressure.[f]
19. After putting on hearing protection and with an empty cylinder in place, point the device away from any by-standers, and depress the trigger two or three times to discharge the device (this step is required to fill the device's internal reservoir with the correct helium pressure).
20. The gene gun is ready to shoot. Aim the device at the target (let the spacer touch target area) and press trigger to shoot.[g]

[a] This a concise version of Agracetus/Auragen, Inc. ACCELL® gene gun manual. For safe handling and transfection of samples other than skin, please, read the original manual for more information.

[b] DNA loading rate (DLR). Typical DLRs range between 1–5 μg DNA/mg gold. A 28 mg preparation with 1 μg/mg DNA requires 28 μg of DNA. DNA stock concentration should be approx. 1 μg/μl.

[c] Bead loading rate (BLR). Typical bead loading rates (mass of gold to be delivered/target) range between 0.25 mg/cartridge and 0.5 mg/cartridge. A 1 ml suspension will fill a 7″ (approx. 17.8 cm) length of tubing and one cartridge is 0.5″ (approx. 1.3 cm) long. The minimum length of tubing required is 28″ (approx. 71.1 cm). For an 0.5 mg/target BLR, set bead preparation at 7 mg gold/ml ethanol. A 28″ length of tubing would require 28 mg of gold resuspended in a final DNA/gold mix of 4 ml ethanol. For best results, the researcher should optimize the BLR by experimenting with a range of BLRs.

[d] PVP serves as an adhesive during the cartridge preparation process. At higher discharge pressures, PVP can increase the total number of particles delivered and the depth of penetration when combined with 1–3 μm gold, but it has no effect on 0.95 μm gold. Therefore, PVP is not added to 0.95 μm gold bead preparations but may be used with 1–3 μm gold bead preparations when deeper bead penetration is desired (for shots 500–800 p.s.i., 0.3 mg/ml PVP).

[e] At this point, bead preparations can be stored for up to two months at –20 °C in sealed containers (e.g. centrifuge tubes with plug seal caps).

[f] We have had the highest levels of expression in pig skin when using 600–800 p.s.i.

[g] When transfecting partial thickness wounds (see *Protocol 2*): first, keep the wounds moist (with e.g. saline soaked gauze) until they are to be transfected, and secondly gently wipe off any blood or clots that have collected on the wound surface before shooting with the gene gun.

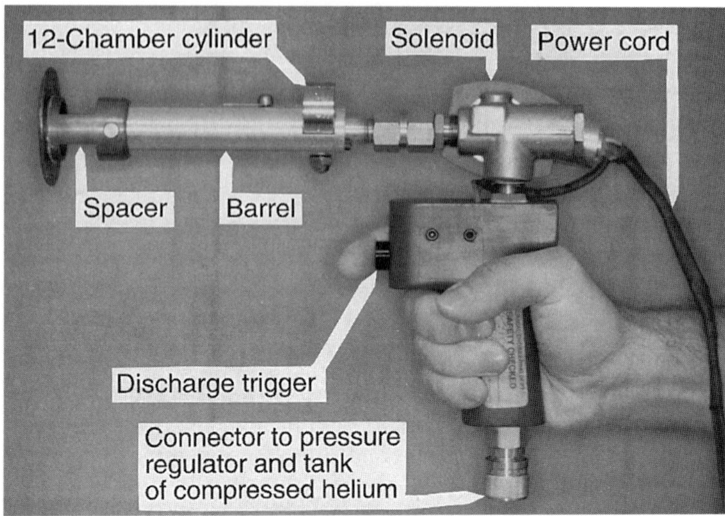

Figure 2. The gene gun utilizes particle-mediated GTr technique. The gene gun technology employs a helium pulse gene delivery device which can deliver gold particles coated with expression vectors intracellularly into the epidermis. The pulse gun depends on a high velocity stream of helium to accelerate the gold particles towards a target. The pressure release causes a rapid acceleration of the helium and the beads down the bore of the gun. Immediately beyond the acceleration channel, the barrel begins to widen as a cone. This causes the gas to expand while the beads maintain direction and high velocity. The penetration in skin is up to 200 μm, but can be modified by adjusting helium pressure and PVP (see *Protocol 1*).

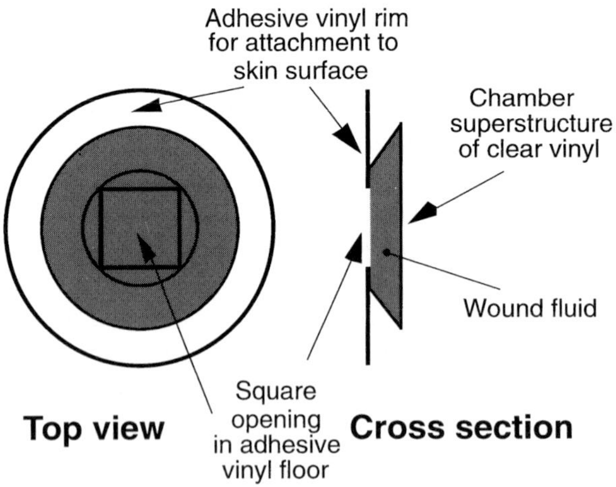

Figure 3. Schematic drawing of the chamber. It consists of an adhesive vinyl rim for attachment to the skin surface. There is a square opening for the wound in the adhesive floor and the wound can be observed through the clear vinyl superstructure. Reproduced with permission (70).

Partial thickness wounds in pig skin can be easily created after some practice. This is described in *Protocol 2*. To monitor the expression of hEGF liquid-tight chambers containing saline (0.9% NaCl) are applied to the wounds (see *Protocol 2*). The growth factors expressed in the wound will be secreted into the saline which is collected daily and replaced (45). The original intention of immersing the wound in saline was not that of monitoring the gene expression, but rather because it offers a very favourable environment to the wound, promoting re-epithelialization and minimizing tissue necrosis, compared with wounds left dry (69). The design of the wound chamber is shown in *Figure 3*.

Protocol 2. Creation of split-thickness skin wounds in pigs and wound chamber application[a]

Equipment and reagents

- Panepinto sling™ (Charles River Laboratories, Inc.)
- Young female Yorkshire pig (30–40 kg)
- Anaesthesia delivery/recovery system (DRAGER COMPACT), halothane (Ayerst Laboratories, Inc.), nitrous oxide, oxygen
- Puls oximeter (OHMEDA BIOX 3740)
- Hair clipper (GOLDEN A5, Oster®)
- Vacuum vented outside the OR (Mastercraft Industries, Inc.)
- Sterile drapes, drape clips, skin markers, gowns, gloves
- Padgett Electro Dermatome with 1.5 cm template and blade (Padgett Instruments)
- Surgical mask, nurses cap
- Skin preparation materials: razors (Schick), shaving cream (Barbasol®), scrub®160 and 201 (Becton Dickinson), 7.5% and 10% povidone-iodine soak (Clinipad®), 70% isopropanol, trichloroethane (ALDRICH Medical Company, Inc.), sterile gauze (PSS®), sterile bowls, sterile forceps
- Vinyl chambers with a 1.5 × 1.5 cm base opening (PAM, Inc.)
- Medical Adhesive (Hollister™)
- Sterile saline solution: 0.9% NaCl, 100 U/ml penicillin, 100 μg/ml streptomycin
- Sterile syringes and needles

Method

1. Place a fasted Yorkshire pig in a Panepinto hammock. Anaesthetize the animal using 1.0–2.5% halothane delivered in conjunction with a 30:50 mixture of oxygen and nitrous oxide via a facial mask. Monitor heart rate and blood oxygen saturation throughout the procedure.
2. Clip the skin hair from the rear flanks to the base of the neck. Apply water and shaving cream then carefully shave the remaining stubble. Wipe skin with paper towel then apply soap and water. Clean with scrub®160 then wipe again with paper towels.
3. Prepare the skin in a surgical fashion using 3 min applications of 7.5% povidone-iodine scrub solution, then 10% povidone-iodine soak solution, followed by 70% isopropanol. Lastly, apply trichloroethane for 2 min to defat the skin surface. Wear surgical clothes then drape the animal in standard surgical fashion with the entire dorsum exposed for surgery.
4. Outline the desired number of wounds[b] with a surgical marker. Create

Protocol 2. *Continued*

the wounds with a Padgett Electro Dermatome set at maximum depth (~ 0.9 mm).

5. Remove any blood around the wounds with saline and gauze.
6. Gently clean the skin area around the wounds with sets of trichloroethane soaked gauze.
7. Brush medical adhesive onto the skin where the chamber is to be placed. Firmly attach the chambers to the skin with the basal opening over the wound.
8. Use a syringe with a needle to fill the chambers with 1.2 ml of sterile saline solution. Seal the needle holes with a piece of vinyl tape.
9. Exchange chambers and fluid every 24 h[c] by repeating steps 6–8 with additional shaving every two days.

[a] This procedure must satisfy local and/or national legal requirements for the handling of animals.
[b] Up to 30 wounds, 1.5 × 1.5 cm, per animal is feasible.
[c] Depending on location and pig movement, most chambers will not adhere tightly for more than 24 h. Smooth wall cages will increase chamber survival.

The system employing liquid-tight chambers allows application to the wound surface of any liquid or suspension. In this way the feasibility of transplanting both fibroblasts (67) and keratinocytes (65) as single cell suspensions using the chamber has been demonstrated. Cell survival was verified by detecting growth factors produced by transfected cells and by detection of β-galactosidase activity in tissue sections. If necessary, one can also infuse cell culture medium together with the cells, thereby letting the chamber and wound serve as an *in vivo* incubator. Fibroblasts are easily cultured and transfected with lipofectin® which is why they are practical to use in *ex vivo* GTr approaches to deliver, for instance, growth factors to the wound. The process of culturing fibroblasts used in our laboratory is outlined in *Protocol 3*.

Protocol 3. Culture of porcine skin fibroblasts

Equipment[a] and reagents

- Equipment as for *Protocol 2*, steps 1–3
- Two sterile jeweler's forceps
- Screen cup and sieve screen (100 mesh, Sigma)
- Beckman CPR centrifuge
- PBS (Gibco BRL)
- Collagenase solution: enriched Waymouth's medium with 100 U/ml collagenase type 1A (Sigma)
- Hank's balanced salt solution (Sigma) with 100 U/ml penicillin (Gibco BRL) and 100 μg/ml streptomycin (Gibco BRL)

- Enriched Waymouth's medium:[b] Waymouth's medium MB 752/1 (Gibco BRL), supplemented with 15% fetal bovine serum (Sigma), 0.38 mg/ml L-arginine, 0.38 mg/ml sodium pyruvate (Sigma), 1.9 μg/ml putrescine (Sigma), 8 μg/ml insulin (Sigma), 8 μg/ml hydrocortisone (United States Biochemical Corporation), 10^{-10} M cholera toxin (Sigma), 100 U/ml penicillin, 100 μg/ml streptomycin, and 50 ng/ml amphotericin B (Sigma)
- Dispase solution: 5 mg/ml Dispase II (Boehringer Mannheim) in Hank's balanced salt solution, with 100 U/ml penicillin and 100 μg/ml streptomycin

Method

1. Follow the pig operative procedure described in *Protocol 2*, steps 1–3.
2. Harvest 0.012 inch (0.3 mm) split thickness skin sections from the dorsal neck region of the pigs with a Padgett Electro Dermatome.
3. Store the skin sections in containers with Hank's balanced salt solution. Samples can be stored on ice for up to 2 h.
4. Wash the sections twice in PBS. Transfer the samples into a dish and immerse them in the dispase solution[c] (~ 1–2 ml dispase/cm^2 skin). Incubate the dish for 2–3 h at 37 °C in a humidified environment of 5% CO_2.
5. Gently separate the dermis (white, shiny) from the overlying epidermis (looks brownish, not shiny) with two jeweler's forceps. If the epidermis is not easily removed from the underlying dermis one can incubate the section for another 1–2 h.
6. Wash the dermal portion in PBS and cut it into small pieces (1–2 mm^2) with scalpel or scissors. (The epidermal portion is discarded if it is not to be used for keratinocyte culture.)
7. Transfer mince corresponding to 7–8 cm^2 of skin dermis into a 100 mm diameter Petri dish containing 15 ml collagenase solution. Incubate the dish at 37 °C in a humidified environment of 5% CO_2.
8. After 4–12 h the dermis has been digested into a slurry. Gently pipette this slurry into a near single-cell suspension. Carefully monitor the digestion to avoid over-digestion of the tissue which will decrease plating efficiency and proliferative capabilities of the isolated fibroblasts.
9. Pour the cell suspension through a screen cup with attached sieve screen to remove undigested tissue and transfer it to 50 ml centrifuge tubes. Centrifuge the cells at 1200 r.p.m. for 5 min in a Beckman CPR centrifuge.
10. Carefully remove the supernatant above the cell pellet and gently resuspend the cells in 10 ml of enriched Waymouth's medium in the centrifuge tube.

11. Count cell numbers and plate in 20 ml enriched Waymouth's medium at a density of 25 000 cells/cm^2 on 100 mm tissue culture dishes.
12. After about three or four days the cultures should have reached confluence. The fibroblasts can be continously passaged at a 1:6 split ratio. Exchange the culture medium every two to three days.

[a] All equipment that will come in contact with cells or tissue samples must be sterile.
[b] This formulation was originally designed for culturing porcine keratinocytes. One can also culture the cells in DMEM (Sigma) with 10–15% fetal bovine serum, but we have experienced higher transfection efficiency (*Protocol 4*) and faster growth in cells cultured in enriched Waymouth's medium.
[c] This step utilizing dispase to separate epidermis from dermis, can be omitted if pure dermal grafts are obtained.

With this simple protocol for transfection of fibroblasts with an hEGF expression plasmid one can obtain relatively high levels of expression *in vivo* (around 10 ng/ml hEGF in wound fluid collected from wounds transplanted with one million transfected fibroblasts, unpublished data).

Protocol 4. Transfection of porcine fibroblasts with hEGF using lipofectin®

Reagents

- Confluent cultures of porcine fibroblasts (see *Protocol 3*)
- Waymouth's medium MB 752/1 (Sigma)
- Enriched Waymouth's medium (*Protocol 3*)
- hEGF plasmid DNA construct (e.g. as in ref. 45)
- Lipofectin® (Gibco BRL)

Method

1. Release confluent pig fibroblasts by trypsinization and plate 5×10^5 cells/60 mm tissue culture dish in 5 ml enriched Waymouth's medium.
2. Incubate the cells for 24 h at 37 °C in a humidified environment of 5% CO_2.
3. Suspend 2 μg of EGF plasmid DNA (for each plate to be transfected) in Waymouth's medium (no supplements, no antibiotics) to a final volume of 100 μl in capped 12 × 75 mm sterile plastic culture tubes.
4. Suspend 8 μl of lipofectin® in 92 μl Waymouth's medium in capped 12 × 75 mm sterile plastic culture tubes.
5. Allow the suspensions from steps 3 and 4 to sit at room temperature for 30 min, then gently mix them together by transferring the DNA solution to the lipofectin® solution using a 200 μl sterile pipette. Incubate at room temperature for 15 min.

6. Dilute the lipofectin®–DNA complex suspension to 3 ml with Waymouth's medium, and then add the suspension to a dish with fibroblasts (from which the culture medium has been removed).
7. Remove the transfection medium 6–24 h post-transfection and replace with 5 ml enriched Waymouth's medium.

The fibroblasts can be released two days post-transfection (counted from the addition of lipofectin®–DNA complex suspension), suspended in medium, and transferred to syringes at appropriate concentrations (10^4 to 10^6 fibroblasts/1.5 × 1.5 cm^2 wound have been transplanted in this fashion). The cells are then injected into a chamber that has been applied to a full thickness wound. It is important to keep the animal under anaesthesia for additional two hours after cell injection, to allow the cells to attach to the wound surface. *Protocol 5* describes the creation of the full thickness wounds onto which cells can be transplanted.

Protocol 5. Creation of full thickness wounds in pigs, and chamber application[a]

Equipment and reagents

- Surgical equipment and reagents (see *Protocol 2*)
- Animal marker and ink (Spaulding and Rogers)
- Concept cautery® (XOMED)
- No. 3 scalpel handle, No. 11 blades (Bard Parker), Adson forceps (STORZ)

Method

1. Follow *Protocol 2,* steps 1–3.
2. Outline the desired number of wounds (measuring 1.5 × 1.5 cm) with a surgical marker and tattoo the margins with the animal marker and ink.
3. Create wounds by excising the skin vertically, within the tattooed borders, down to the panniculus (~ 0.8 cm).
4. Achieve haemostasis by cauterization (bleeding will interfere with cell attachment).
5. Follow *Protocol 2,* steps 5–9.

[a] This procedure must satisfy local and/or national legal requirements for the handling of animals.

4. Detecting growth factor expression in target organs

Two relatively simple methods have been used to monitor gene expression and cell survival in the porcine wet healing model. Detection and quantification of hEGF in wound fluid is carried out by the ELISA described in *Protocol 6*.

Protocol 6. Detection of hEGF in wound fluid by ELISA

Equipment and reagents

- ELISA plate (Falcon 3072 microtest III tissue culture plate, Becton Dickinson)
- Microplate reader capable of measuring optical density at 450 nm
- hEGF monoclonal antibody (R & D)
- PBS (phosphate-buffered saline)
- Sodium azide (Sigma)
- Bovine serum albumin (BSA, Sigma)
- Wound fluid
- hEGF protein (R & D)
- Rabbit anti-EGF polyclonal antibody (SC275, Santa Cruz Biotechnology)
- Nonidet P-40 (NP-40, Sigma)
- Horse-radish peroxidase (HRP)-conjugated goat anti-rabbit polyclonal antibody (BioRad)
- TMB Peroxidase EIA Substrate Kit (BioRad)

Method

1. Add 125 ng of anti-hEGF monoclonal antibody in 100 μl PBS/0.02% sodium azide per well of a 96-well microtitre plate and incubate for 4–5 h at room temperature with the lid closed.
2. Remove unbound antibody by filling the wells with 300 μl of PBS containing 5% BSA and 0.02% Na azide. Incubate for 16–20 h at 4°C.[a]
3. Wash wells four times with PBS containing 1% BSA.
4. Add 200 μl of wound fluid[b] or EGF standards[c] in PBS containing 3% BSA, and incubate at 4°C overnight.
5. Wash wells four times with PBS containing 1% BSA.
6. Add 75 ng of anti-EGF polyclonal antibody to each well in the presence of PBS containing 3% BSA to a final volume of 100 μl per well, and incubate for 3 h at room temperature.
7. Wash the wells three times with PBS containing 1% BSA.
8. Add 100 μl of a 1:3000 dilution of HRP-conjugated goat anti-rabbit polyclonal antibody to each well, and incubate for 1.5 h at room temperature.
9. Wash the wells four times with PBS containing 0.1% NP-40 and perform peroxidase assays as described in the instructions provided by the suppliers of the TMB Peroxidase EIA Substrate Kit.

[a] Antibody coated plates can be stored after step 2 either in PBS with 0.02% Na azide at 4°C for one week, or indefinitely at –20°C after removing the blocking solution.
[b] Any cells and/or debris should first be removed by centrifugation. Dilute the sample with PBS containing 3% BSA if EGF concentration is expected do be higher than 1000 pg.
[c] The standard range used with this protocol has been 1–2000 pg (make serial dilutions).

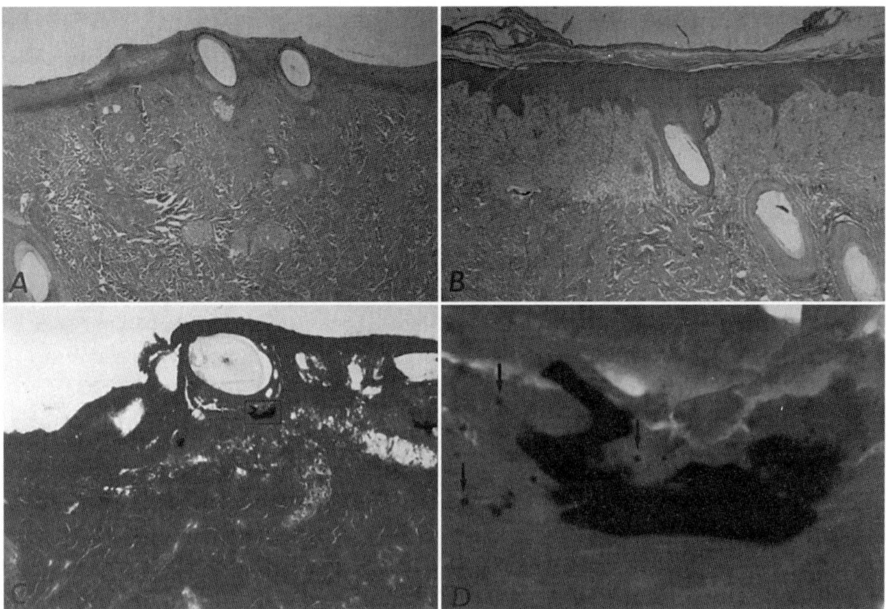

Figure 4. Particle-mediated gene transfer of hEGF and β-galactosidase to porcine partial thickness wounds. (A and B) Histologic sections of wounds after bombardment with the hEGF expression plasmid. Cross-sections of wounds (A) one day and (B) nine days after particle-mediated GTr are shown (× 38). Sections confirm regeneration of epidermis (B) as well as minimal tissue and inflammatory reaction. (Note that the gold beads are not seen at this magnification.) (C and D) *In situ* cytochemical staining of wounds after bombardment with pCMV β-galactosidase. (C) Cross-section of wound three days after particle-mediated GTr (× 75). (D) Enlarged rectangle from (C) shows transgenic cells (originally blue) and gold beads (arrows) (× 900). Reproduced with permission (45).

To detect transfected cells, to determine which cells were transfected, and the fate of transplanted, transfected, cells it is often useful to transfect with a marker gene, such as β-galactosidase. By the simple histochemical staining procedure described in *Protocol 7*, the cells that express the β-galactosidase gene can be marked in the biopsy sections taken from transfected wounds. *Figures 4C* and *4D* show cells that have been transfected with β-galactosidase DNA construct using the gene gun technique, staining blue.

Protocol 7. X-Gal staining of frozen tissue sections

Equipment and reagents

- Cryostat microtome
- Staining dishes (VWR)
- O.C.T. compound (Sakura Finetek)
- PBS (Gibco BRL)
- Magnesium chloride ($MgCl_2$, Sigma)
- Fix solution: methanol (EM Science), acetone (J. T. Baker, Inc.), 1:1 (v/v)
- Ethanol (Pharmco Products, Inc.)

Protocol 7. *Continued*

- 40 × X-Gal solution: 40 mg/ml X-Gal (Gibco BRL) in dimethyl formamide—store at −20°C protected from light
- X-Gal solution: add 0.50 ml 20 × KC solution and 0.25 ml 40 × X-Gal solution to 9.25 ml PBS containing 1 mM $MgCl_2$
- 20 × KC solution: add 0.83 g potassium ferricyanide and 1.05 g potassium ferrocyanide to 25 ml PBS—store at room temperature protected from light
- Xylene (Fisher)

Method

1. Freeze tissue[a] (not larger than 1 cm^2 × 0.4 cm) by immersing it in liquid nitrogen[b] for 60 sec. Transfer samples to pre-cooled (−70°C) capped vials (e.g. plastic scintillation vials) and store at −70°C until further use.
2. Place a drop of O.C.T on a pre-cooled (−20°C) microtome specimen holder, then quickly place the tissue in it and let it adjust to −20°C.
3. Cut 6–10 μm sections of the sample in a cryostat microtome set at −20°C. Collect the section by letting it adhere to and thaw on a warm (room temperature) glass slide.
4. Dry slide mounted sections in air for 10 min.
5. Wash the slides with the fresh tissue sections in PBS containing 1 mM $MgCl_2$ (ten dips).
6. Fix sections for 30 sec in fix solution at −20°C.
7. Wash the slides twice in two changes of PBS containing 1 mM $MgCl_2$ (ten dips each).
8. Use a pipette to pour approx. 0.6 ml of X-Gal solution[c] to each slide so that the sections are completely covered.
9. Place wet paper towels along the inner edges of Petri dishes (15 × 150 mm) to create humidified chambers and transfer the slides to the dishes.
10. Incubate the slides in the covered dishes at room temperature until satisfactory blue staining is obtained (up to 24 h).
11. Wash the slides in PBS containing 1 mM $MgCl_2$ (ten dips).
12. Counterstain (if desired) with haematoxylin and/or eosin as described by the manufacturer, but leave out any fixation step as the sections are already fixed.
13. Dehydrate the sections by immersing the slides in two changes of 95% ethanol (ten dips each), two changes of 100% ethanol (ten dips each), and finally in two changes of xylene (ten dips each).
14. Mount in resinous medium (e.g. Permount, Fisher) with cover glasses (Fisher).

[a] If the tissue samples are not immediately frozen after excision, store them in closed vials on ice until further processing.

[b] Alternatively the samples can be frozen in isopentane, pre-cooled to liquid nitrogen temperature. This will shorten the freezing time since isopentane will not bubble as the tissue is immersed.

[c] Prepare fresh X-Gal solution for each staining.

5. Determining the effects of growth factors on wound healing

The disruption of skin integrity by wounding exposes damaged vessels and dermal tissue with the release of blood and interstitial fluid. This wound fluid has a high protein content determined (after the cells have been removed by centrifugation) by a commercially available turbidimetric test (Stanbio Laboratory, Inc.). The consequent healing and reconstitution of the normal skin barrier, i.e. the re-epithelialization, was found to correlate with decreased total protein concentration in wound fluid collected daily (70). The wound fluid protein concentration for each day plotted on a logarithmical scale is shown as a function of time after wounding in *Figure 5*, since the wound fluid from the chamber is collected on a 24 hour basis and replaced with 1.2 ml of saline, the protein concentration represents the total protein loss over 24 hours. This graph has a characteristic declining phase paralleling the re-establishment of the epithelial barrier function (see *Figure 5*). A regression

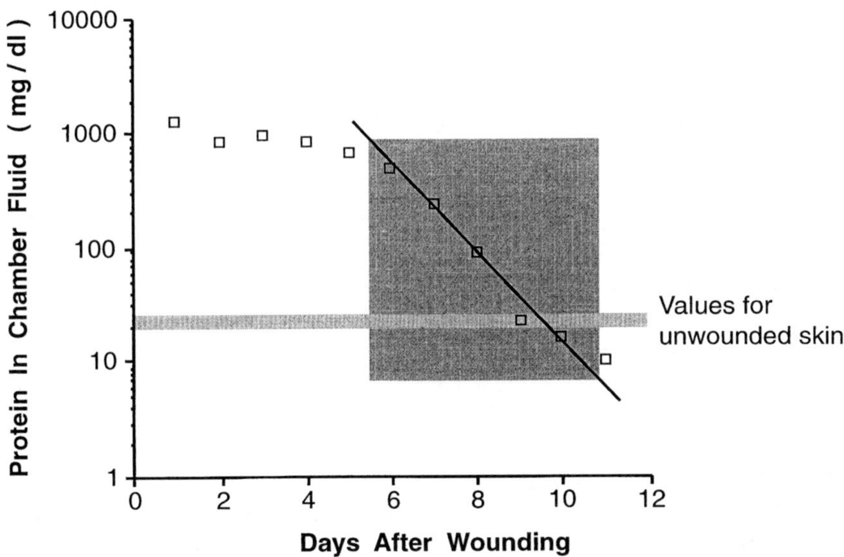

Figure 5. Total protein concentration in wound fluid as a non-invasive marker of re-epithelialization (i.e. return of epithelial barrier function). Since the decline in protein efflux seemed to display first order kinetics, the logarithm of the protein concentration in wound fluid (collected daily and replaced with saline) was plotted over time for each wound. A regression line was then calculated for the linear declining phase (darkly stippled box). The time point at which this regression line crossed the upper 95% confidence interval for values of unwounded skin (lightly stippled bar) was taken as the return of epithelial barrier function (here about 9.3 days after wounding). Reproduced with permission (70).

line was drawn through the declining phase of the curve; when this curve intercepts the mean values for unwounded skin, the wounds are considered healed and a healing time can be calculated from the time (x axis) coordinate of interception. The protein values of unwounded chamber enclosed skin are about 20 mg/ml (determined for 1.5 × 1.5 cm skin prepared according to *Protocol 2*, steps 1–3 and enclosed in 1.2 ml saline for 24 h). Wounds that have reached such levels display a three to four cell layer thick epidermis as verified by histology (70).

6. Summary

Gene transfer techniques have been discussed in general, together with their use in gene therapy. The methods have different characteristics that will affect selection of technique. Viral vectors are efficient in delivering the gene into the cell and the duration of expression can be prolonged, but safety issues remain to be investigated and solved. Non-viral gene transfer techniques are not directly limited by the size of the DNA to be transferred, but the frequency of integration is very low, limiting the duration of expression. The gene gun technique can be used for *in vivo* transfection or *in vitro* for cells that are particularly difficult to transfect by other means.

Gene therapy applications of growth factors have also been discussed. Treatment of cancer and cardiovascular disease are commonly targeted, both experimentally and in clinical trials. These experiments have in principal proven the feasibility of gene therapy, but several problems remain, including the chronological onset of cessation of transgene expression and the removal or down-regulation of the transferred gene after the desired period of gene therapy.

Gene delivery of growth factors to skin and skin wounds can be carried out using common gene transfer techniques. If defects occur in the genetically modified skin, they can be readily detected and, if necesarry, removed. We have demonstrated accelerated healing by transfecting skin wounds *in vivo* with an EGF gene. *In vivo* and *ex vivo* gene transfer techniques for skin and two simple assays used to detect the gene product, have been described in detail in this chapter.

References

1. Rosenberg, S.A., Aebersold, P., Cornetta, K., *et al.* (1990). *N. Engl. J. Med.*, **323**, 570.
2. *Human gene transfer protocols* from: Office of Recombinant DNA activities, National Institutes of Health, 6000 Executive Blvd, MSC 7010, Suite 302, Bethesda, MD 20892–7010.
3. Crystal, R.G. (1995). *Science*, **270**, 404.

4. Nicola, N.A. (1994). In *Guidebook to cytokines and their receptors* (ed. N.A. Nicola), p. 1. Oxford University Press Inc., New York.
5. Brown, G.L., Nanney, L.B., Griffen, J., *et al.* (1989). *N. Engl. J. Med.*, **321**, 76.
6. Tomao, S., Mozzicafreddo, A., Raffaele, M., Romiti, A., Papo, M.A., and Campisi, C. (1995). *Clin. Ter.*, **146**, 491.
7. Garbe, C. (1995). *Recent Results Cancer Res.*, **139**, 349.
8. Schmidt-Wolf, G.D. and Schmidt-Wolf, I.G.H. (1995). *Immunol. Today*, **16**, 173.
9. Holland, J.J. (1990). In *Virology* (ed. B.N. Fields, *et al.*), p. 151. Raven Press, New York.
10. Smith, K.T., Shepard, A.J., Boyd, J.E., and Lees, G.M. (1996). *Gene Ther.*, **3**, 190.
11. McLachlin, J.R., Cornetta, K., Eglitis, M.A., *et al.* (1990). *Prog. Nucleic Acid Res. Mol. Biol.*, **38**, 91.
12. Gilboa, E., Eglitis, M.A., Kantoff, P.W., *et al.* (1986). *Biotechniques*, **4**, 504.
13. Brody, S.L. and Crystal, R. (1994). *Ann. N. Y. Acad. Sci.*, **716**, 90.
14. Muzyczka, N. (1992). *Curr. Top. Microbiol. Immunol.*, **158**, 97.
15. McLaughlin, S.K., Collis, P., Hermonat, P.L., *et al.* (1988). *J. Virol.*, **62**, 1963.
16. Moss, B. and Flexner, C. (1987). *Annu. Rev. Immunol.*, **5**, 305.
17. Piccini, A. and Paoletti, E. (1988). *Adv. Virus Res.*, **34**, 43.
18. Breakefield, X.O. and DeLuca, N.A. (1991). *New Biol.*, **3**, 203.
19. Kreuger, G.G., Morgan, J.R., Jorgensen, C.M., *et al.* (1994). *J. Invest. Dermatol.*, **103**, 76S.
20. Miller, D.G., Adam, M.H., and Miller, A.D. (1990). *Mol. Cell. Biol.*, **10**, 4239.
21. Bowtell, D.L., Cory, S., Johnson, G.R., and Gonda, T.J. (1988). *J. Virol.*, **62**, 2464.
22. Palmer, T.D., Roseman, G.J., Osborne, W.R.A., and Miller, A.D. (1991). *Proc. Natl. Acad. Sci. USA*, **88**, 1330.
23. Rosenfeld, M.A., Siegfried, W., Yoshimurs, K., *et al.* (1991). *Science*, **252**, 431.
24. Bajocchi, G., Feldman, S.H., Crystal, R.G., and Mastrangeli, A. (1993). *Nat. Genet.*, **3**, 229.
25. Christiano, R.J., Smith, L.C., and Woo, S.L.C. (1993). *Proc. Natl. Acad. Sci. USA*, **90**, 2122.
26. Samulski, R., *et al.* (1993). *Curr. Opin. Gen. Dev.*, **3**, 74.
27. Kotin, R.M., Menninger, J.C., Ward, D.C., and Berns, K.I. (1991). *Genomics*, **10**, 831.
28. Graham, F.L. and van der Eb, A.J. (1973). *Virology*, **52**, 456.
29. Fraley, R., Subramani, S., Berg, P., and Papahadjopoulos, D. (1980). *J. Biol. Chem.*, **255**, 10431.
30. Sompayrac, L.M. and Danna, K.J. (1981). *Proc. Natl. Acad. Sci. USA*, **78**, 7575.
31. Watson, J.D., Gilman, M., Witkowski, J., and Zoller, M. (1992). *Recombinant DNA*, 2nd edn, p. 213. Scientific American Books, New York.
32. Li, L. and Hoffman, R.M. (1995). *Nat. Med.*, **1**, 705.
33. Ono, T., Fujino, Y., Tsuchiya, T., *et al.* (1990). *Neurosci. Lett.*, **117**, 259.
34. Lin, H., Parmacek, M., Morle, G., *et al.* (1990). *Circulation*, **82**, 2217.
35. Hickman, M.A., Malone, R.W., Lehmann-Bruinsma, K., *et al.* (1994). *Hum. Gene Ther.*, **5**, 1477.
36. Wolff, J.A., Malone, R.W., Williams, P., *et al.* (1990). *Science*, **247**, 1465.
37. Wolff, J.A., Williams, P., Acsadi, G., *et al.* (1991). *Biotechniques*, **11**, 474.

38. Hengge, U.R., Chan, E.F., Foster, R.A., Walker, P.S., and Vogel, J.C. (1995). *Nat. Genet.*, **10**, 161.
39. Katsumi, A., Emi, N., Abe, A., *et al.* (1994). *Hum. Gene Ther.*, **5**, 1335.
40. Klein, T.M., Wolf, E.D., Wu, R., *et al.* (1987). *Nature*, **327**, 70.
41. McCabe, D.E., Swain, W.F., Martinell, B.J., *et al.* (1988). *Biotechnology*, **6**, 923.
42. Christou, P. (1994). In *Particle bombardment technology for gene transfer* (ed. N.-S.Yang and P. Christou), p. 71. Oxford University Press, New York.
43. Yang, N.S., Burkholder, J., Roberts, B., *et al.* (1987). *Proc. Natl. Acad. Sci. USA*, **87**, 9568.
44. Williams, R.S., Johnston, S.A., Riedy, M., *et al.* (1991). *Proc. Natl. Acad. Sci. USA*, **88**, 2726.
45. Andree, C., Swain, W.F., Page, C., *et al.* (1994). *Proc. Natl. Acad. Sci. USA*, **91**, 12188.
46. Haynes, J.R., Fuller, D.H., and Eisenbraun, M. (1994). In *Particle bombardment technology for gene transfer* (ed. N.-S. Yang and P. Christou), p. 175. Oxford University Press, New York.
47. Fakhrai, H., Dorigo, O., Shawler, D.L., Lin, H., Mercola, D., Black, K.L., *et al.* (1996). *Proc. Natl. Acad. Sci. USA*, **93**, 2909.
48. Trojan, J., Johnson, T.R., Rudin, S.D., Blossey, B.K., Kelley, K.M., Shevelev, A., *et al.* (1994). *Proc. Natl. Acad. Sci. USA*, **91**, 6088.
49. Trojan, J., Blossey, B.K., Johnson, T.R., *et al.* (1992). *Proc. Natl. Acad. Sci. USA*, **89**, 4874.
50. Trojan, J., Johnson, T.R., Rudin, S.D., *et al.* (1993). *Science*, **259**, 94.
51. Saleh, M., Stacker, S.A., and Wilks, A.F. (1996). *Cancer Res.*, **56**, 393.
52. Robinson, G.S., Pierce, E.A., Rook, S.L., Foley, E., Webb, R., and Smith, L.E.H. (1996). *Proc. Natl. Acad. Sci. USA*, **93**, 4851.
53. Konishi, H., Ochiya, T., Sakamoto, H., Tsukamoto, M., Saito, I., Muto, T., *et al.* (1995). *J. Clin. Invest.*, **96**, 1125.
54. Rosenthal, R.M., Früh, R.M., Henschler, R., Veelken, H., Kulmberg, P., Mackensen, A., *et al.* (1994). *Blood*, **84**, 2960.
55. Mühlhauser, J., Merrill, M.J., Pili, R., Maeda, H., Bacic, M., Bewig, B., *et al.* (1995). *Circ. Res.*, **77**, 1077.
56. Isner, J.F., Pieczek, A., Schainfeld, R., Blair, A., Haley, L., Asahara, T., *et al.* (1996). *Lancet*, **348**, 370.
57. Giordano, F.J., Ping, P., McKirnan, M.D., Nozaki, S., Demaria, A.N., Dillmann, W.H., *et al* (1996). *Nat. Med.*, **2**, 534.
58. Barde, Y.-A. (1994). In *Guidebook to cytokines and their receptors* (ed. N.A. Nicola), p. 140. Oxford University Press Inc., New York.
59. Chen, K.S. and Gage, F.H. (1995). *J. Neurosci.*, **15**, 2819.
60. Kramer, R., Zhang, Y., Gehrmann, J., Gold, R., Thoenen, H., and Wekerle, H. (1995). *Nat. Med.*, **1**, 1162.
61. Pechan, P.A., Yoshida, T., Panahian, N., Moskowitz, M.A., and Breakefield, X.O. (1995). *NeuroReport*, **6**, 669.
62. Qin, L., Chavin, K.D., Ding, Y., *et al.* (1994). *Ann. Surg.*, **220**, 508.
63. Rheinwald, J.G. and Green, H. (1975). *Cell*, **6**, 331.
64. Oconnor, N.E., Mulliken, J.B., Banks-Schlegel, S., *et al.* (1981). *Lancet*, **i**, 75.
65. Vogt, P.M., Thompson, S., Andree, C., *et al.* (1994). *Proc. Natl. Acad. Sci. USA*, **91**, 9307.

66. Moullier, P., Bohl, D., Heard, J.-M., and Danos, O. (1993). *Nat. Genet.*, **4**, 154.
67. Svensjö, T., Winkler, T., Yao, F., and Eriksson, E. (1996). *Surg. Forum*, **47**, 715.
68. Eriksson, E., Yao, F., Svensjö, T., *et al.* Accepted for publication in *J. Surg. Res.* (1998).
69. Vogt, P.M., Andree, C., Breuing, K., *et al.* (1995). *Ann. Plast Surg.*, **34**, 493.
70. Brueing, K., Eriksson, E., Liu, P., and Miller, D.R. (1992). *J. Surg. Res.*, **52**, 50.

A1

Growth factors

The peptide factors which transmit signals between animal cells, including cytokines and chemokines, are all listed together here as **Growth factors**. We have classified them as large or small peptide factors and then further subdivided each classification into super-superfamilies, superfamilies, families, and individual factors. The reader should be aware that these divisions are rather arbitrary and in some cases rather tenuous, being based on computer predictions of homology, and of secondary and tertiary structures. Also factors which are not presently included in families may become so, if relatives are later identified. This list will inevitably become rapidly out of date and interested readers should survey the most recent literature for new factors or family associations. At time of going to press, a list of biology databases was available on the Internet at:

`http://www.science.gmu.edu/~michaels/Bioinformatics/database.html`

and a growth factor resource was available at:

`http://www.rndsystems.com`

The families and individual factors are listed alphabetically. References to the factors and, wherever possible, recent reviews of the ligands, their receptors, or both are included. Some of the commonly used non-peptide growth factors are also included.

1. Large peptide factors

A. Cystine knot super-superfamily (1).
   - (a) Neurotrophin family (2)
   - (b) Platelet-derived growth factor (PDGF) superfamily.
     - PDGF family (3)
     - Connective tissue growth factor (CTGF) family (4)
     - Vascular endothelial cell growth factor (VEGF) family (5, 6)
   - (c) Transforming growth factor-β (TGFβ) superfamily (7).
     - TGFβ family (8)

- Growth and differentiation factor (GDF) family
- Vgr-1 family
- Bone morphogenetic protein (BMP) family
- Osteogenin family
- Distantly related factors

B. Four helix bundle superfamily (9).
- Interleukin-6 (IL-6) family (10–12)
- Growth hormone (GH) family (9, 13–15)
- Interleukin-2 (IL-2) family (16, 17)

C. Chemokine superfamily (18).
- C–X–C family
- C–C family
- C family
- CX_3C family (19)

D. Other large peptide factor families.
- Eph/Eck ligands family (20, 21)
- Epidermal growth factor (EGF) family (22, 23)
- Fibroblast growth factor (FGF) family (24)
- Insulin-like growth factor (IGF) family (25)
- Interleukin-1 (IL-1) family (26–28)
- Interleukin-4 (IL-4) family (29, 30)
- Pleiotrophin (PTN) family (31, 32)
- Stem cell factor (SCF) family (33)
- Tumour necrosis factor (TNF) family (34, 35)

E. Individual large peptide factors.
- Angiogenin (ANG) (36)
- Cholera toxin (CT) (37)
- Growth arrest-specific gene 6 product (Gas 6) (38)
- Granulocyte/macrophage colony stimulating factor (GM-CSF) (27, 39)
- Growth potentiating factor (GPF) (40)
- Hepatocyte growth factor (scatter factor) (HGF) (SF) (41)
- Hepatocyte growth factor-like protein (HGF-LP) (42)
- Interferon alpha (IFNα) (27, 43)
- Interferon beta 1 (IFNβ1) (27)
- Interferon gamma (IFNγ) (27, 44)

- Interleukin-3 (IL-3) (27, 39)
- Interleukin-5 (IL-5) (27, 45)
- Interleukin-7 (IL-7) (27, 46)
- Interleukin-9 (IL-9) (27, 47)
- Interleukin-10 (IL-10) (27, 48)
- Interleukin-11 (IL-11) (27, 49)
- Interleukin-14 (IL-14) (50)
- Interleukin-17 (IL-17) (51)
- *Pasteurella multocida* toxin (PMT) (52), S100b (53)
- Platelet-derived endothelial cell growth factor (PD-ECGF) (54)

Note: PD-ECGF has been identified with thymidine phosphorylase which lacks a hydrophobic signal sequence and is not a classical secretory protein.

2. Small peptide factors

(a) Small peptide factor families.

- Bombesin family (55)
- Calcitonin family (56, 57)
- Endothelin (ET) family (58)
- Proopiomelanocortin family (59)

(b) Individual small peptide factors.

- Angiotensin II (60)
- Bradykinin (61)
- Cholecystokinin (62)
- Epidermal inhibitory pentapeptide (EIP) (63)
- Epithelin 1 (TGFe, granulin A) (64)
- Epithelin 2 (64)
- Galanin (65)
- Gastrin (62)
- Haemoregulatory pentapeptide (HP) (63)
- Mastoparan (66)
- Neurotensin (67)
- Oxytocin (61)
- Substance K (68)
- Substance P (67)
- Thrombin (69)
- Tuftsin (67)
- Vasopressin (61)

3. Commonly used non-peptide growth factors

(a) Antibiotics: e.g. A23187 (calcimycin).

(b) Corticosteroids: e.g. dexamethasone, hydrocortisone (70).

(c) Leukotrienes: e.g. LTB4.

(d) Phorbol esters: e.g. TPA, PDBu (61).

(e) Prostaglandins: e.g. PGE_1, $PGF_2\alpha$ (71).

References

1. Sun, P. D. and Davies, D. R. (1995). *Annu. Rev. Biophys. Biomol. Struct.*, **24**, 269.
2. Maness, L. M., Kastin, A. J., Weber, J. T., Banks, W. A., Beckman, B. S., and Zadina, J. E. (1994). *Neurosci. Biobehav. Rev.*, **18**, 143.
3. Meyer-Ingold, W. and Eichner, W. (1995). *Cell Biol. Int.*, **19**, 389.
4. Igarashi, A., Nashiro, K., Kiuchi, K., Sato, S., Ihn, H., Fujimoto, M., *et al.* (1996). *J. Invest. Dermatol.*, **106**, 729.
5. Kolch, W., Martiny-Baron, G., Kieser, A., and Marme, D. (1995). *Breast Cancer Res. Treat.*, **36**, 139.
6. Kukk, E., Lymboussaki, A., Taira, S., Kaipainen, A., Jeltsch, M., Joukov, V., *et al.* (1996). *Development*, **122**, 3829.
7. McPherron, A. C., Lawler, A. M., and Lee, S-J. (1997). *Nature*, **387**, 83.
8. Cox, D. A. (1995). *Cell Biol. Int.*, **19**, 357.
9. Bazan, J. F. (1990). *Immunol. Today*, **11**, 350.
10. Rose, T. M. and Bruce, A. G. (1991). *Proc. Natl. Acad. Sci. USA*, **88**, 8641.
11. Merberg, D. M., Wolf, S. F., and Clark, S. C. (1992). *Immunol. Today*, **13**, 77.
12. Bazan, J. F. (1991). *Neuron*, **7**, 197.
13. Bartley, T. D., Bogenberger, J., Hunt, P., Li, Y-S., Lu, H. S., Martin, F., *et al.* (1994). *Cell*, **77**, 1117.
14. Xu, B. C., Wang, X., Darus, C. J., and Kopchick, J. J. (1996). *J. Biol. Chem.*, **271**, 19768.
15. Edery, M., Levi-Meyrueis, C., Paly, J., Kelly, P. A., and Djiane, J. (1994). *Mol. Cell. Endocrinol.*, **102**, 39.
16. de Totero, D., Francia de Celle, P., Cignetti, A., and Foa, R. (1995). *Leukaemia*, **9**, 1425.
17. Cosman, D., Kumaki, S., Ahdieh, M., Eisenman, J., Grabstein, K. H., Paxton, R., *et al.* (1995). *Ciba Found. Symp.*, **195**, 221.
18. Premack, B. A. and Schall, T. J. (1996). *Nat. Med.*, **2**, 1174.
19. Bazan, J. F., Bacon, K. B., Hardiman, G., Wang, W., Soo, K., Rossi, D., *et al.* (1997). *Nature*, **385**, 640.
20. Tuzi, N. L. and Gullick, W. J. (1994). *Br. J. Cancer*, **69**, 417.
21. Pandey, A., Lindberg, R. A., and Dixit, V. M. (1995). *Curr. Biol.*, **5**, 986.
22. Kumar, V., Bustin, S. A., and McKay, I. A. (1995). *Cell Biol. Int.*, **19**, 373.
23. Burden, S. and Yarden, Y. (1997). *Neuron*, **18**, 847.
24. Slavin, J. (1995). *Cell Biol. Int.*, **19**, 431.
25. Schmid, C. (1995). *Cell Biol. Int.*, **19**, 445.

26. Conti, B., Jahng, J. W., Tinti, C., Son, J. H., and Joh, T. H. (1997). *J. Biol. Chem.*, **272**, 2035.
27. Balkwill, F. R. (ed.) (1995). *Cytokines: a practical approach*, 2nd edn. IRL Press, Oxford, UK.
28. Dinarello, C. A. (1996). *Blood*, **87**, 2095.
29. Caput, D., Laurent, P., Kaghad, M., Lelias, J. M., Lefort, S., Vita, N., *et al.* (1996). *J. Biol. Chem.*, **271**, 16921.
30. Keegan, A. D., Nelms, K., Wang, L. M., Pierce, J. H., and Paul, W. E. (1994). *Immunol. Today*, **15**, 423.
31. Li, Y. S., Milner, P. G., Chauhan, A. K., Watson, M. A., Hoffman, R. M., Kodner, C. M., *et al.* (1990). *Science*, **250**, 1690.
32. Nakagawara, A., Milbrandt, J., Muramatsu, T., Deuel, T. F., Zhao, H., Cnaan, A., *et al.* (1995). *Cancer Res.*, **55**, 1792.
33. Lyman, S. D., James, L., Vanden Bos, T., de Vries, P., Brasel, K., Gliniak, B., *et al.* (1993). *Cell*, **75**, 1157.
34. Hill, C. M. and Lunec, J. (1996). *Mol. Aspects Med.*, **17**, 455.
35. Smith, C. A., Farrah, T., and Goodwin, R. G. (1994). *Cell*, **76**, 959.
36. Fett, J. W., Strydom, D. J., Lobb, R. R., Alderman, E. M., Bethune, J. L., Riordan, J. F., *et al.* (1985). *Biochemistry*, **24**, 5480.
37. Davis, J. B. and Stroobant, P. (1990). *J. Cell Biol.*, **110**, 1353.
38. Nagata, K., Ohashi, K., Nakano, T., Arita, H., Zong, C., Hanafusa, H., *et al.* (1996). *J. Biol. Chem.*, **271**, 30022.
39. Sakamoto, K. M., Mignacca, R. C., and Gasson, J. C. (1994). *Receptors Channels*, **2**, 175.
40. Nakano, T., Higashino, K-I., Kikuchi, N., Kishino, J., Nomura, K., Fujita, H., *et al.* (1995). *J. Biol. Chem.*, **270**, 5702.
41. Naldini, L., Weidner, K. M., Vigna, E., Gaudino, G., Bardelli, A., Ponzetto, C., *et al.* (1991). *EMBO J.*, **10**, 2867.
42. Han, S., Stuart, L. A., and Friezener Degen, S. J. (1991). *Biochemistry*, **30**, 9768.
43. Yang, C. H., Shi, W., Basu, L., Murti, A., Constantinescu, S. N., Blatt, L., *et al.* (1996). *J. Biol. Chem.*, **271**, 8057.
44. Kotenko, S. V., Izotova, L. S., Pollack, B. P., Muthukumaran, G., Paukku, K., Silvennoinen, O., *et al.* (1996). *J. Biol. Chem.*, **271**, 17174.
45. Dickason, R. R., Huston, M. M., and Huston, D. P. (1996). *J. Immunol.*, **156**, 1030.
46. Ziegler, S. E., Morella, K. K., Anderson, D., Kumaki, N., Leonard, W. J., Cosman, D., *et al.* (1995). *Eur. J. Immunol.*, **25**, 399.
47. Demoulin, J. B., Uyttenhove, C., Van Roost, E., DeLestre, B., Donckers, D., Van Snick, J., *et al.* (1996). *Mol. Cell. Biol.*, **16**, 4710.
48. Liu, Y., Wei, S. H., Ho, A. S., de Waal Malefyt, R., and Moore, K. W. (1994). *J. Immunol.*, **152**, 1821.
49. Cherel, M., Sorel, M., Apiou, F., Lebeau, B., Dubois, S., Jacques, Y., *et al.* (1996). *Genomics*, **32**, 49.
50. Ford, R., Tamayo, A., Martin, B., Niu, K., Claypool, K., Cabanillas, F., *et al.* (1995). *Blood*, **86**, 283.
51. Fossiez, F., Djossou, O., Chomarat, P., Flores-Romo, L., Ait-Yahia, S., Maat, C., *et al.* (1996). *J. Exp. Med.*, **183**, 2593.
52. Rozengurt, E., Higgins, T., Chanter, N., Lax, A. J., and Staddon, J. M. (1990). *Proc. Natl. Acad. Sci. USA*, **87**, 123.

53. Selinfreund, R. H., Barger, S. W., Pledger, W. J., and Van Eldik, L. J. (1991). *Proc. Natl. Acad. Sci. USA*, **88**, 3554.
54. Furukawa, T., Yoshimura, A., Sumizawa, T., Haraguchi, M., Akiyama, S-I., Fukui, K., *et al.* (1992). *Nature*, **356**, 668.
55. Battey, J. and Wada, E. (1991). *Trends Neurosci.*, **14**, 524.
56. Muff, R., Born, W., and Fischer, J. A. (1995). *Eur. J. Endocrinol.*, **133**, 17.
57. Muff, R., Born, W., and Fischer, J. A. (1995). *Can. J. Physiol. Pharm.*, **73**, 963.
58. Sakurai, T., Yanagisawa, M., and Masaki, T. (1992). *Trends Pharmacol. Sci.*, **13**, 103.
59. Cossu, G., Cusella-De, A-M., Senni, M. I., De-Angelis, L., Vivarelli, E., Vella, S., *et al.* (1989). *Dev. Biol.*, **131**, 331.
60. Re, N. (1989). *J. Mol. Cell Cardiol.*, **21**, 63.
61. Corps, A. N. and Brown, K. D. (1993). In *Growth factors: a practical approach* (ed. I. McKay and I. Leigh), p. 35. IRL Press, Oxford.
62. Yamaguchi, K. (1990). *Hum. Cell.*, **3**, 23.
63. Robinson, P., Whitehead, P., and Hume, W. (1993). In *Growth factors: a practical approach* (ed. I. McKay and I. Leigh), p. 133. IRL Press, Oxford.
64. Shoyab, M., McDonald, V. L., Todaro, G. J., and Plowman, G. D. (1990). *Proc. Natl. Acad. Sci. USA*, **87**, 7912.
65. Sethi, T. and Rozengurt, E. (1991). *Cancer Res.*, **51**, 1674.
66. Gil, J., Higgins, T., and Rozengurt, E. (1991). *J. Cell Biol.*, **113**, 943.
67. Moore, R. N., Osmand, A. P., Dunn, J. A., Joshi, J. G., Koontz, J. W., and Rouse, B. T. (1989). *J. Immunol.*, **142**, 2689.
68. Kris, R. M., South, V., Saltzman, A., Felder, S., Ricca, G. A., Jaye, M., *et al.* (1991). *Cell Growth Differ.*, **2**, 15.
69. Mitsuhashi, T., Morris, R. C., and Ives, H. E. (1991). *J. Clin. Invest.*, **87**, 1889.
70. Green, B. and Leake, R. E. (ed.) (1987). *Steroid hormones: a practical approach*. IRL Press, Oxford.
71. New, R. R. C. (ed.) (1987). *Prostaglandins and related substances: a practical approach*. IRL Press, Oxford.

A2

Amino acid abbreviations

Amino acid	Three letter abbreviation	One letter code
Alanine	Ala	A
Arginine	Arg	R
Asparagine	Asn	N
Aspartic acid	Asp	D
Asn or Asp	Asx	B
Cysteine	Cys	C
Glutamine	Gln	Q
Glutamic acid	Glu	E
Gln or Glu	Glx	Z
Glycine	Gly	G
Histidine	His	H
Isoleucine	Ile	I
Leucine	Leu	L
Lysine	Lys	K
Methionine	Met	M
Phenylalanine	Phe	F
Proline	Pro	P
Serine	Ser	S
Threonine	Thr	T
Tryptophan	Trp	W
Tyrosine	Tyr	Y
Valine	Val	V

A3

Addresses of suppliers

Advanced Biotechnologies Ltd., Unit 7 Mole Business Park, 3 Randalls Road, Leatherhead, Surrey, KT22 7BA, UK.

Agracetus/Auragen, 8520 University Green, Middleton, Wisconsin 53562, USA.

Aldrich Inc., 1001 West Saint Paul Avenue, Milwaukee, WI 53233, USA.

American Type Culture Collection (ATCC), 12301 Parklawn Drive, Rockville, MD 20852–1776, USA.

Amicon Inc., 72 Cherry Hill Drive, Beverly, MA 01915, USA.

Anachem Ltd., Anachem House, 20 Charles Street, Luton, Bedfordshire LU2 0EB, UK.

Andreas Hettich, GmbH & Co., KG. Gartenstr. 100, D-78532, Tuttlingen, Germany.

Baxter Healthcare Corp., Scientfic Products Division, McGaw Park, IL 60085–6787, USA.

BioComp Instruments Inc., 650 Churchill Row, Fredericton, New Brunswick E3B 1P6, Canada.

C. A. Hendley (Essex) Ltd., Oakwood Hill Industrial Estate, Loughton, Essex, UK.

Charles River Laboratories, 251 Ballardvale Street, Wilmington, MA 01887, USA.

Clonetics, 9620 Chesapeake Drive, Suite 201, San Diego, CA 92123, USA.

Corning-Costar Corporation, 1 Alewife Center, Cambridge, MA 02140, USA.

Coulter

Coulter Corporation, PO Box 169015, Hialeah, FL 33011–1690, USA.

Coulter Electronics Ltd., Northwell Drive, Luton, Bedfordshire LU3 3RH, UK.

Dakopatts Ltd., 16 Manor Court Yard, Hughenden Avenue, High Wycombe, Buckinghamshire HP13 5RE, UK.

Du Pont NEN Research Products, 549 Albany Street, Boston, MA 02118, USA.

Dynal, Dynal A.S., Head Office, PO Box 158, SkÌyen, N-0212 Oslo, Norway.

EM Science, 480 S Democrat Road, Gibbstown, NJ 08027, USA.

Flowgen Instruments Ltd., Lynne Lane, Shenstone, Lichfield, Staffordshire WS14 0EE, UK.

Forma Scientific, Box 649, Marietta, Ohio 45750, USA.

Fuji Medical Systems USA, 333 Ludlow Street, Stamford, CT 06450, USA.
Greiner Labortechnik Ltd., Station Road, Dursley, Gloucestershire GL11 5NS, UK.
Harvard Apparatus Canada, 610 Vanden Abeele, Saint-Laurent, Quebec H4S 1R9, Canada.
Hoefer Scientific Instruments, 654 Minnesota Street, PO Box 77387, San Francisco, CA 94107, USA.
ICN
ICN Biomedicals Ltd., Eagle House, Peregrine Business Park, Gomm Road, High Wycombe, Buckinghamshire HP13 7DL, UK.
ICN Pharmaceuticals Inc., 3300 Hyland Avenue, Costa Mesa, CA 92626, USA.
J. T. Baker Inc., 222 Red School Lane, Phillipsburg, NJ 08865, USA.
Kapak Corp., 5305 Parkdale Drive, Minneapolis, MN 55416, USA.
Kojair Tech OY, Teollisuustie 3, SF-35700 Vilpulla, Finland.
Labsystems, Labsystems Oy, Pulttitie 8, 00880 Helsinki, Finland.
Leica Instruments GmbH, Postfach 1120, Heidelberger Str. 17–19, Germany.
Molecular Dynamics, 880 East Arques Avenue, Sunnyvale, CA 94086, USA.
Nalge Corp., 75 Panorama Creek Drive, PO Box 20965, Rochester, NY 14602–0365, USA.
Neuro Probe Inc., PO Box 400, Cabin John, MD 20818, USA.
New Brunswick Scientific Co. Inc., PO Box 4005, 44 Talmadge Road, Edison, NJ 08818–4005, USA.
Nucleopore Corp., 1 Alewife Center, Cambridge, MA 02140, USA.
Nunc, P280, DK-4000 Roskilde, Denmark.
Ohmeda
Ohmeda, 450 Raritan Center Parkway, Raritan Center, Edison, NJ 08837, USA.
Ohmeda House, 71 Great North Road, Hatfield, Hertfordshire AL9 5EN, UK.
Oncogene Science
Oncogene Science Inc., 80 Rogers Street, Cambridge, MA 02142, USA.
Cambridge Bioscience, 25 Signet Court, Stourbridge Common Business Centre, Swanns Road, Cambridge CB5 8LA, UK.
OSTER Professional Products, Division of Sunbeam Corporation, Milwaukee, WI 53217, USA.
Packard Instrument Company, 800 Research Parkway, Meriden, CT 06450, USA.
Padget Instruments Inc., 1730 Walnut Street, Kansas City, MO 64108–1384, USA.
PAM Inc., 915 Carter Creek Pike, Coumbia, TN 38401, USA.
Pharmco Products Inc., DSP-CT-18 Brookfield, CT, USA.
Pierce Chemical Co., 3747 N Meridan Road, PO Box 117, Rockford, IL 61105, USA.

Polysciences Inc., 400 Valley Road, Warrington, PA 18976, USA.
R & D Systems, 614 McKinley Place NE, Minneapolis, MN 55413, USA.
Santa Cruz Biotechnology Inc., 2161 Delaware Avenue, Santa Cruz, CA 95060, USA.
Sakura Finetek Co. Ltd., Tokyo, 103 Japan, Torrance, CA 90504, USA.
Sartorius, Sartorius GmbH, Postfach 3243, Weender Landstrasse 94–108, 3400, Goettingen, Germany.
Seescan plc. – now Cambridge Advanced Electronics plc., Cambridge, UK.
Spaulding and Rogers Inc., PO Box 439, Rt 85, New Scotland Road, Vaarheesville, NY 12186, USA.
Stanbio Laboratory Inc., 2930 East Houston Street, San Antonio, TX 78202, USA.
STORZ, 3365 Tree Court Industrial Boulevard, St Louis, MO 63122, USA.
The Separations Group, 17434 Mojave Street, Hesperia, CA 92345, USA.
Upchurch Scientific, 619 West Oak Street, Oak Harbor, WA 98277, USA.
Upstate Biotechnology Inc., 199 Saranac Avenue, Lake Placid, NY 12946, USA.
VWR, PO Box 232, Boston, MA 02101, USA.
Waters Chromatography, 34 Maple Street, Milford, Massachussetts, MA 01757, USA.
Whatman, Whatman International Ltd., Whatman House, St Leonard's Road, 20/20 Maidstone, Kent ME16 OLS, UK.
Wyerth-Ayerst Inc., Philadelphia, PA 19101, USA.
XOMED Inc., 6743 Southpoint Drive North, Jacksonville, FL 32216, USA.

Index